EVOLUTION AND HUMAN ORIGINS

EVOLUTION AND HUMAN ORIGINS SECOND EDITION

An Introduction to Physical Anthropology

B. J. WILLIAMS
UNIVERSITY OF CALIFORNIA, LOS ANGELES

HARPER & ROW, PUBLISHERS
New York, Hagerstown, Philadelphia, San Francisco, London

Sponsoring Editor: Richard Heffron
Project Editor: Joyce Marshall
Designer: Helen Iranyi
Production Manager: Stefania J. Taflinska
Compositor: Ruttle, Shaw & Wetherill, Inc.
Printer and Binder: Halliday Lithograph Corporation
Art Studio: Danmark & Michaels, Inc.

EVOLUTION AND HUMAN ORIGINS
An Introduction to Physical Anthropology, Second Edition

Library of Congress Cataloging in Publication Data

Williams, Bobby Joe, Date
Evolution and human origins.
Includes index.
1. Human evolution. 2. Somatology. I. Title.
GN281.W54 1979 573 78-21947
ISBN 0-06-047121-2

Contents

Preface to the Second Edition

Humans are a part of the animal world, but they are also unique in the animal world. They level forests, cultivate the soil, and build cities and highways. Their activities often determine which of the other animal species shall flourish and which shall die out.

We call the life of a people their culture. The variety of cultures and the effect of culture on the environment and on the people themselves is a major concern of anthropology. Another great concern of anthropology are the age-old questions, What it is in "human nature" that differs so from other animals? What is it that makes culture possible as well as the variety of cultures that are evident today and in the past?

The question regarding human nature is a particular concern of the physical or biological anthropologist. The approach of the biological anthropologist is to investigate the differences between humans and nonhuman primates, the origin of these differences, and the process that produced these differences. The latter we call the evolutionary process.

The hope of such studies is to produce a more profound understanding of ourselves, our species, and our relation to the world around us. It is the author's hope that the following pages will help the reader along this path.

The reader should note that there have been changes made in this new, second edition which merit mention here.

In this edition a number of chapters have

been retained with only small changes from their presentation in the first edition. These are the chapters which users of the first edition found highly useful in their present form. Most of the changes in these chapters represent a simplification of vocabulary by eliminating technical terms when the ideas they convey can be presented as well without technical terminology.

A major revision begins with Chapter 3, Evolution in Populations. The usual presentation of evolution at the population level requires quite a bit of mathematical elaboration to describe the quantitative relationships between evolutionary variables. In this edition a quantitative presentation has been, for the most part, eliminated. A minimum of simple quantitative relationships are presented. The principles of population genetics are presented extensively but with a nonquantitative verbal elaboration which should be welcomed by students with no background in science.

The first edition included an extensive section on the evolution of life from prebiotic conditions. This material was rather difficult for the nonscience student and, as it did not contribute directly to subsequent chapters, has been eliminated entirely in this edition. The elimination of this material has permitted the addition of new material of more immediate significance to anthropology.

The major addition of totally new material in this edition is Chapter 9, Primate Behavior and Evolution. Nonhuman primate behavior is intrinsically interesting. For this reason it is easy to present a section on primate behavior in a textbook on evolution without regard to the connection between this material and larger evolutionary issues. In this chapter some care has been taken to relate findings from the study of nonhuman behavior to issues in the understanding of human evolution. For this reason Chapter 9 also includes material on present-day human hunting groups for the insights they reveal.

The discussion of behavior and genetics was more extensively presented in the first edition of this book than in comparable textbooks. As this material was well received and relevant to a number of social issues, it has been retained and extended a bit in this edition. It is extended to a discussion of "racial" differences in IQ and the so-called Jensen hypothesis on IQ differences between Blacks and Whites in the United States.

Finally there has been a revision in the order in which material is presented. This, too, is in response to comments by users of the first edition. Material on the evolution of present-day populations which was formerly presented in two parts, prior to and subsequent to the material on the fossil record, is now grouped together as Part Two of this edition. An extensive glossary is included in this edition, although an attempt has been made also to define each term that may be new to the reader, when the term first occurs in the book.

B. J. WILLIAMS

Preface to the First Edition

This is a book about human biology, especially human biology in its relation to human culture. It is written for the serious student or reader but does not presuppose specialized knowledge of biology. In particular it is intended as a text for an introductory undergraduate college course in human evolution for nonscience majors or as a supplementary text for more specialized introductory science courses. It presents, at an introductory level approaches and major findings from many fields that add up to a balanced view of human evolution.

The first half of the book presents a generally accepted theory of organic evolution and an explication of various evolutionary processes that are critical in light of the general theory. The processes of organic evolution are genetic processes. Therefore, considerable attention is paid to the genetics of man in the first seven chapters. Some of the genetic systems described here will be used in later chapters to document the occurrence of ongoing-evolution in general, and natural selection in modern populations in particular.

Chapters 8 to 14 deal with the story of human evolution, the chronology of events that led up to man and to modern human populations. It is hoped that the understanding of evolutionary processes gained from earlier chapters will be of value in the interpretation of past and continuing change. Only such an understanding can keep an introductory presentation of man's fossil ancestors from becoming a recitation of "funny fossils I have known." Such an understanding can also help resolve some current common controversies on biology and behavior.

The overall objectives of this book are to familiarize the reader with evolutionary theory, with some of the methods used in the study of human evolution, with human ancestry to the extent it is now known, and, finally, to point out some of the unresolved problems in understanding man's past and present condition. These are broad objectives. Correspondingly the topics presented are broad ranging. It is my hope that I have been able to present the findings of many specialties in rather simplified form without, in the process of simplifying, introducing significant error. Secondly, I hope the book leads the reader through this diversity of topics without subjecting him to the jarring shifts of pace or seeming discontinuities that are, unfortunately, a product of our compartmentalized academic method but are in no way inherent in the phenomenon we study, human evolution.

B. J. WILLIAMS

EVOLUTION AND HUMAN ORIGINS

PART ONE

GENETICS AND THE MECHANISMS OF EVOLUTION

Evolution is a process—a process of change. It is change, from generation to generation in the inherited characteristics of a population. Human populations are no exception to this. The process of evolutionary change is continuous in our species.

Before looking at specific changes that have occurred in the human species it will be useful to know something about the basic mechanisms which make evolutionary change possible and, as you will see, inevitable. These mechanisms we call genetic mechanisms. The first 3 chapters, Part One of this book, briefly outline these mechanisms and give some background on how science came to know of them. With this knowledge you will be better able to understand and appreciate the small as well as the dramatic evolutionary changes that have occurred in our species.

1

The Study of Human Evolution

Why are there such great differences in skin color between different peoples? Why do males and females differ in size? Do they also differ behaviorally due to inherited characteristics? Are there genetic differences between ethnic groups or "races" which affect behavior? These are questions which have long been asked and are being asked by people today. Such questions are asked by all of us, in all walks of life. They are asked "naturally" because it seems to be a human characteristic to wonder about differences. But the questions, as well as the answers, have important social and political implications which affect the lives of all of us.

When students have learned something about evolution, they ask other questions. How did skin color differences originate? Is there a biological basis for language, for the human way of life, and when did this arise? What was the common ancestor of man and the apes like? Do we still have behavioral tendencies that we can attribute to this common heritage? These are fascinating questions. Much of this book deals with ways of getting answers.

The study of human evolution is as wide ranging as it is fascinating. One student of human evolution may spend 30 to 40 years excavating for the remains of early man. Another researcher may confine himself for long days and nights in the laboratory studying the structure of a single protein. Yet another may be concerned with the factors influencing the choice of marriage partners within specific groups. The surroundings in which these researchers work may be just as varied as their investigations. While one may be working in a laboratory, another may be excavating next to a superhighway or in some of the most remote and inaccessible parts of the world. These persons, as well as others engaged in very different kinds of activities, all contribute to a broad understanding of man's past and possibly his future.

PHYSICAL ANTHROPOLOGY

The field of physical anthropology, or biological anthropology, is concerned with the extent and origin of biological diversity in the human species. As such, physical anthropology is the study of human biology with a strong evolutionary orientation. Many physical anthropologists devote their energies to the study of human biology without directly addressing the question of origins—the question of how differences arise or how change occurs. But the student will find that an evolutionary framework provides an orientation which is essential to understanding the remarkable diversity of our species. An evolutionary framework provides what has been called a "view of life" which can be gained in no other way. Because of this, all the chapters which follow are devoted to, and organized around, the central concern of physical anthropology, human evolution.

The studies that are most relevant to our understanding of human evolution include growth, nutrition, physiological adaptation, comparative and functional anatomy, primate behavior, human paleontology, human genetics, population genetics, and population ecology. This looks like a formidable list, and it is. No one person can be an expert in all of these areas. Persons who wish to perform at a professional level in the study of human evolution must pick a restricted area of specialization within which they will try to advance the frontiers of knowledge. But if their work is not to be intellectually narrow, and perhaps irrelevant, they must have an appreciation of the major findings of many other specialists.

In basic aspects the study of human evolution does not differ from that of any other animal species. Evolutionary change in living organisms is a change in the genetic structure of a population. Many genetic processes at work in man are at work in all living organisms. The ways in which the genetic material of these organisms can change are, in general terms, common to all species. Factors regulating the number of individuals in an area are, in a general way, the same for all species. Therefore many of the statements in the first half of this book are applicable to many species other than *Homo sapiens.*

Our concern here is *Homo sapiens,* and if we are to understand its particular course of development, we must also be concerned with the ways in which it differs from other animals. One could say that, as far as his evolutionary history is concerned, man differs from other mammals only in certain ecological considerations. However this would be a great understatement. Population ecology does indeed deal with the factors influencing the numbers and kinds of organisms in a species, but in man these factors include not only the physical environment, but the whole range of activities, beliefs, technologies, and institutions normally designated as human culture.

All mammals alter their physical environment, which in turn affects the development of the species. All mammals learn behaviorial responses to some extent. These responses, in turn, influence the species' possibilities for evolutionary change by their effects on population density and on mating patterns. Man, however, differs quantitatively from other mammals in his abilities to learn to such an extent that we recognize this as a new level of behavioral abilities. Man, through the system of sounds we call language, is able to transmit his experience to others and to subsequent generations. Information, including, we should note, error, is accumulated in this way. But over long periods of time, the errors tend to be weeded out in a process analogous to natural selection. This store of information allows individuals to simulate with mental processes the external world and to predict outcomes of hypothetical situations. This permits foresight, or planning for events that have not yet occurred. Information stored in the form of social customs permits specialization of function among individuals and, consequently, further accumulation of knowledge.

In short, man has made a biological adaptation which permits further and widespread cultural adaptation and which does not necessitate genetic changes. Such cultural adaptations have broad consequences for genetic changes. They can affect the selection of mates, family size, rates of disease transmission, availability of food and other resources and, in sum, become the major consideration among the ecological factors affecting human evolution. As will be demonstrated later, this situation has existed as far back as we have been able to identify the ancestral line unique to modern man. A consequence of this situation is that there is no such thing as a "state of nature" for man without benefit of technology and other cultural arrangements. Although tool and language use are a result of biological adaptation, man is no longer able to live without them.

Theories of human nature based on a hypothetical state of nature without culture inevitably prove false. Moreover it takes very little documentation to assert that, for a long time, there has been an interaction or *feedback* between the processes of biological evolution and cultural evolution. But these broad generalizations are not what is most fascinating about the study of human evolution. The fascination lies in the detailed, and sometimes complicated, findings behind such generalizations.

It is fortunate that many persons are fascinated with the details of man's biological nature as well as the details of how man got to where he is. Knowledge is always used. We have many examples from the past of knowledge being used to maintain an inequitable social system or the ideology supporting such a system. We can be sure that there will be examples of this in the future. The only proven way to moderate such possibilities lies in the widest possible dissemination of knowledge. The responsible members of each generation must remain in the forefront of intellectual activity if knowledge is not to be misused in fallacious arguments having real consequences for groups of people.

The first fossils to be recognized as such by Renaissance Europeans were used as often to argue against evolutionary theories as to argue for them. In the last pages of this book, we will return briefly to how findings on human biology are used to support one ideology or another. For a person to assess the validity of the arguments involved in each of these cases, and many like them, requires more than platitudes or the broader generalizations about human evolution. It requires a fairly detailed knowledge of process and history.

EVOLUTIONARY THOUGHT: THE PERSPECTIVES OF HISTORY

Science concerns itself with the assessment of evidence used to support statements or hypotheses. There are fairly explicit ground rules for the evaluation of such scientific hypotheses. One approach, which everyone agrees has no place in science, is the simple appeal to authority. Thus statements that have been long repeated should have no more weight than novel statements unless accompanied by evidence in the form of confirmable observations. This means that science is in one sense not time dependent: the sanctity of time has nothing to do with the assessment of evidence.

If all the above is true, why then should we be concerned with the history of a science? Simple curiosity should be adequate motivation, but we have other good reasons for presenting a short look at history in this chapter. We do not seek a simple catalog of hypotheses on human evolution, both those which are currently accepted and those which have been rejected. Such a listing would be very dry reading. Of equal or greater importance to any thriving field of inquiry is the process by which the inquiry is advanced. This process involves not only the techniques presently available to carry out investigations but also the theories and approaches that influence the kinds of questions being asked. These theories and approaches are in some cases explicit and well recognized; in others they are only implicit in a heritage of thought.

A part of the rationale for scientific research is the assumption that understanding of natural phenomena can, in some ways, give us a more conscious influence over those phenomena. If this is so, then making explicit the scientific process itself, including how we got to where we are now, should be important to the further progress of our inquiries. In short, we desire some historical perspective on what we are doing.

Where then do we start in gaining a historical perspective? A conventional solution to this question is to start with the Greek philosophers. This is not because the Greeks had the first words to say on all important questions but because theirs are the first written words to which we can refer, thanks to the Arab scholars who preserved their works. Fortunately we do not need to follow this custom. Our starting point is dictated by a simple question: Why was the concept of evolution so revolutionary to Western science? Why was evolutionary biology born in heated debate and turmoil that went far beyond the academic cloisters?

The question almost answers itself. The concept of evolution must have been in strong contrast to conventional concepts or beliefs, and these conventional beliefs must have gone far beyond casual academic discourse. More precisely, the belief that biological change within a species could proceed, perhaps even in a single direction, without known limit was diametrically opposed to what has been called the "conventional understandings" enshrined in the medieval political, religious, and scientific views.

The conventional or orthodox Western view that was common until well into the nineteenth century was what has been termed a belief in the "fixity of type" or "fixity of species." The phrase *fixity of type* is a *post hoc* way of summarizing a set of beliefs. These included the belief that nonarbitrary types (in the plant and animal kingdoms as well as in other natural phenomena) existed, and that, although these types or species showed some variation, it was quite limited in range. And since variation was closely bounded, there was no possibility of a species becoming truly differentiated into more than one species. Indeed, variation constituted only an obstacle to be gotten around in identifying the "true type."

Why should this have been important to the Western world? The Western world was the Christian world, and it has been argued that the fixity of species concept was a necessary accompaniment to the literal interpretation of the Christian book of Genesis. The question arises, why should Europeans of the latter Middle Ages have insisted on a literal interpretations of Genesis? Were they congenitally less capable of the range of interpretations given biblical writings today? Obviously they were not. The "Hellenizing" of Hebraic teachings in the first century of the Christian movement and, particularly, the contribution of allegorical interpretation of the Old Testament, associated with Philo of Alexandria, lent to Hebraic traditions the flexibility necessary to allow for the introduction of Greek concepts of evolution in a partly naturalistic interpretation of creation.

There is good evidence that the early Christian thinkers continued Greek traditions of speculative thought that included theories of evolution. Many were biological as well as intellectual descendants of the earlier Greek philosophers. St. Augustine (A.D. 353–430), at least, specifically included man in a naturalistic development.

Throughout the Middle Ages we can find philosophers and theologians continuing evolutionary ideas (Osborn, 1922). During the same time, however, an orthodoxy was forming that would eventually declare evolutionary thought to be heresy. While closely defined orthodoxy may have been important in integrating the empire of the Roman Christian church of Medieval Europe, the rise of another imperial proselytizing religion outside of Europe was even more important.

The spread of Islam and Arab leadership in the Middle East effectively cut off Europe from the rest of the Old World. The efforts of the Christian crusaders to establish a series of forts guarding trade routes across Asia Minor could hardly be counted as a success. The last of these major efforts occurred in the thirteenth century.

Greek science, including evolutionary thought, had found a more congenial home in the Moslem world than in Europe. The flowering of Arab intellectual life occurred as Europe went through the so-called Dark Ages. As a result, Greek thought, especially the teachings of Aristotle, began to be reintroduced to Europe from Arab sources. This was partially due to contacts established during the Crusades themselves, but even more importantly it was due to the influence of newly established universities in Italy and Spain. The ideas of Arab scholars were the greatest single influence on Thomas Aquinas, who did much to integrate Aristotelian ideas into Christian theology.

However, the identification of Greek naturalism with Moorish thought was to be its final blow so far as Europe was concerned. It is generally true that nations and all groupings of men have symbols of group identification. When such groups are threatened externally they tend to tighten up internally, and their members must tread a narrower path in demonstrating symbolically that they are "good" group members. Thus, in thirteenth-century Europe, speculation about evolution began to be associated with that which was bad and destructive. Free variation in this area of thought began to decline. The last great naturalist theologian to present evolutionary interpretations of the Old Testament was the Italian Giordano Bruno (1548–1600), who was burned at the stake for continuing this ideological heresy long after it had become intolerable to established authority.

This is a little different from the naive view that the conflict between Christianity and evolutionary ideas existed unchanged through time. The reaction to scientific thought in general, and evolutionary ideas in particular, began to develop strongly only in the latter part of the Middle Ages. It developed as a result of political and economic conditions, the eventual result of which was a rigidly enforced system of belief. But forces that were to reverse the ideological climate and permit science and evolutionary thought to blossom in Europe had been set in motion.

Europe broke out of the Moslem encirclement, partly by military victories in Europe, but largely by taking to the sea to explore new trade routes. Throughout the seventeenth century, exploration and trade by sea took Europeans to all parts of the known world and also added vast previously unknown portions to the map of the world. In these voyages Europeans encountered entirely new kinds of animals and plants. They encountered peoples who differed both physically and culturally from themselves.

All of this diversity had to be ordered in some way, and named accordingly, to permit people to discuss and to deal with this new world. This was the birth of what can best be called natural history, which included zoology, botany, and the earth sciences. Young men from wealthy families now had an adventurous but respectable outlet for their energies; they became explorers, naturalists, and, finally, taxonomists.

Taxonomy has been defined as the art, or science, of ordering complex data. Some authorities today consider taxonomy to be the core of a much larger field of inquiry called systematics. As Simpson has defined it, "Systematics is the scientific study of the kinds and diversity of organisms and of all relationships among them" (Simpson, 1961 : 7). This defines systematics to be a rather broad concern; in fact, it is broader than evolutionary biology itself, as all of the social sciences must be included in systematics. For the present we will concentrate on taxonomy and the birth of scientific taxonomy in the seventeenth and eighteenth centuries.

As mentioned, the rise of natural history including taxonomy was intimately connected with the expansion of Western economic and political interests. The man considered the father of modern taxonomy was Carl von Linné, born in 1707 in Sweden and educated as a physician in Holland. He is better known by his Latinized name, Carolus Linnaeus. While still a student, Linnaeus wrote a number of treatises on botany and mineralogy which brought him some note as a scholar and lecturer. The first edition of what was to be recognized as a landmark work, *Systema naturae,* appeared in 1735.

How does one show various degrees of similarity between organisms through a system of names? Linnaeus did this by using binomials. In other words, he gave each organism two names. The first denoted the *genus,* which indicated broader morphological relationships than did the second, which named the *species.* A familiar example is that of the *Canidae,* the grouping of animals including dogs, wolves, foxes, and jackals (Figure 1.1). The domestic dog is designated *Canis familiaris.* The second, the species name, is never capitalized. The wolf, which is also a doglike animal, is designated *Canis lupus.*

In this fashion we indicate in the species name that wolves differ from domestic dogs. However, at a broader level they are similar, as indicated by the genus name. The concept of *levels* of classification has been quite important in the development of taxonomy. It has led to viewing classifications as hierarchies of order. To illustrate this we have spoken of two levels in this hierarchy.

FIGURE 1.1

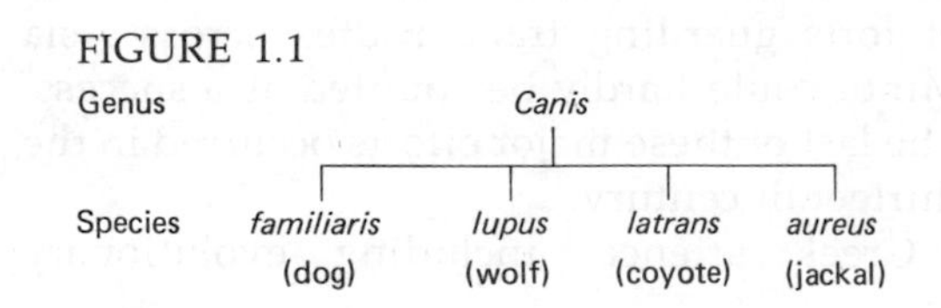

On the other hand, taxonomists consider the red fox and the desert dwelling kit fox, although doglike, to be somewhat different from the animals shown in Figure 1.1. Therefore they are put in a different genus termed *Vulpes*. *V. fulva* is the red fox, *V. macrotis* is the kit fox. Their similarity to *Canis* is shown by grouping them together at a higher level, the *family*, as shown in Figure 1.2. Thus Figure 1.1 shows two taxonomic levels; Figure 1.2 shows three.

By way of introducing the term, it should also be pointed out that Figure 1.1 shows five taxa. Figure 1.2 shows nine taxa. In other words, each name in the figure represents one taxon. A taxon is a group of organisms recognized as belonging together in a classificatory unit. Therefore the genus *Canis* is a taxon, and the species *C. latrans* is a taxon.

The tenth edition of *Systema naturae* (1758) is now used as the basic work to which all later emendations or additions must refer. This edition approximated the present use of seven basic levels of classification. They are

Kingdom
Phylum
Class
Order
Family
Genus
Species

As classifications become more detailed, intervening levels must often be interpolated between those shown above. For instance, if

FIGURE 1.2

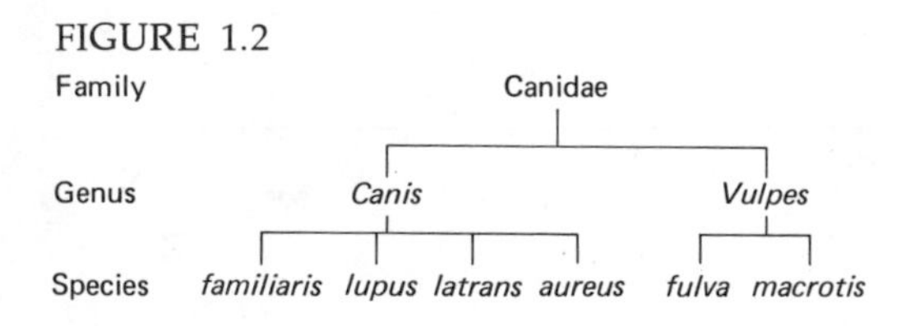

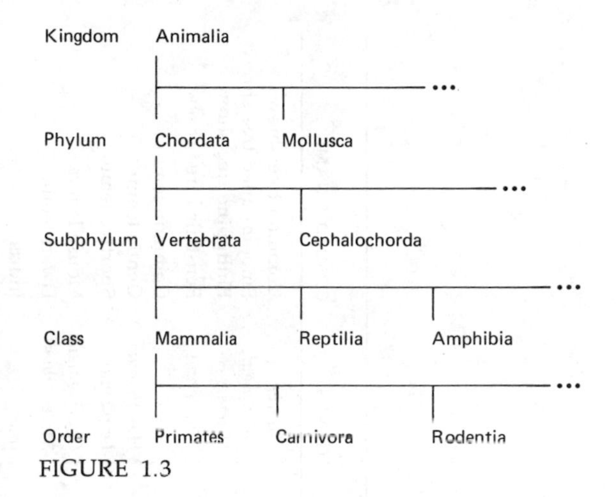

FIGURE 1.3

we need to make finer distinctions in showing similarity between several orders within a class, we can use

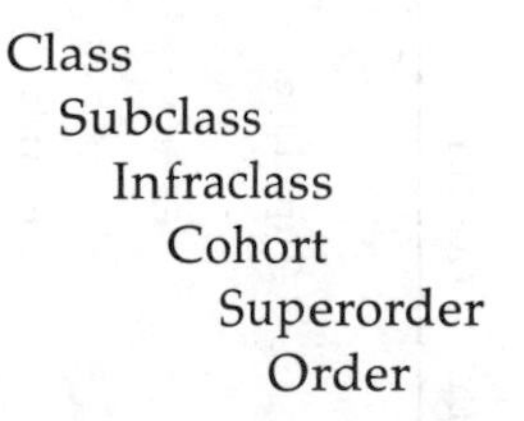

By the use of such intermediate levels we have up to 34 levels, including subspecies, in a classificatory hierarchy.

Since our interest is man, how is he classified? Linnaeus recognized man's similarities to monkeys and apes and put them all in the order *Primates*. (The word rhymes with *Euphrates*, thereby distinguishing this order from Anglican church officials.) This order is placed as shown in Figure 1.3.

A full taxonomy of the primate order is given in Table 1.1. Hardly anyone today can contemplate such a classification without also thinking of the *phylogeny*, or descent relationships, suggested. In other words, such a classification of man today implies a great deal about his ancestry. This has not always

TABLE 1.1
Taxonomy of the Primates

ORDER	SUBORDER	INFRAORDER	SUPERFAMILY	FAMILY	SUBFAMILY	GENUS	COMMON NAME
	Prosimii	Lemuriformes	Tupaioidea	Tupaiidae	Tupaiinae	*Tupaia*	Common tree shrew
						Dendrogale	Smooth-tailed tree shrew
						Urogale	Philippine tree shrew
					Ptilocercinae	*Ptilocercus*	Pen-tailed tree shrew
			Lemuroidea	Lemuridae	Lemurinae	*Lemur*	Common lemur
						Hapalemur	Gentle lemur
						Lepilemur	Sportive lemur
					Cheirogaleinae	*Cheirogaleus*	Mouse lemur
						Microcebus	Dwarf lemur
				Indridae		*Indri*	Indris
						Lichanotus	Avahi
						Propithecus	Sifaka
				Daubentoniidae		*Daubentonia*	Aye-aye
		Lorisiformes	Lorisoidea	Lorisidae		*Loris*	Slender loris
						Nycticebus	Slow loris
						Arctocebus	Angwantibo
						Perodicticus	Potto
				Galagidae		*Galago*	Bush baby
		Tarsiiformes	Tarsioidea	Tarsiidae		*Tarsius*	Tarsier
				Callithricidae		*Callithrix*	Plumed and pygmy marmoset
						Leontocebus	Tamarin
					Callimiconinae	*Callimico*	Goeldi's marmoset
					Aotinae	*Aotes*	Douroucouli
						Callicebus	Titi

ORDER	SUBORDER	INFRAORDER	SUPERFAMILY	FAMILY	SUBFAMILY	GENUS	COMMON NAME
Primates	Anthropoidea	Platyrrhini	Ceboidea	Cebidae	Pithecinae	*Pithecia*	Saki
						Chiropotes	Saki
						Cacajao	Uakari
					Alouattinae	*Alouatta*	Howler
					Cebinae	*Cebus*	Capuchin
						Saimiri	Squirrel monkey
					Atelinae	*Ateles*	Spider monkey
						Brachyteles	Woolly spider monkey
						Lagothrix	Woolly monkey
		Catarrhini	Cercopithecoidea	Cercopithecidae	Cercopithecinae	*Macaca*	Macaque
						Cynopithecus	Black ape
						Cercocebus	Mangabey
						Papio	Baboon Drill
						Theropithecus	Gelada
						Cercopithecus	Guenon
						Erythrocebus	Patas monkey
					Colobinae	*Presbytis*	Common langur
						Pygathrix	Douc langur
						Rhinopithecus	Snub-nosed langur
						Simias	Pagi Island langur
						Nasalis	Proboscis monkey
						Colobus	Gueraza
			Hominoidea	Hylobatidae		*Hylobates*	Gibbon
						Symphalangus	Siamang
				Pongidae		*Pongo*	Orangutan
						Pan	Chimpanzee
						Gorilla	Gorilla
				Hominidae		*Homo*	Man

been so, and was not so for Linnaeus. For him a classification reflected morphological similarity but said nothing about common descent. As the period in which he lived in Europe was the so-called Age of Enlightenment, evolutionary theories could be, and were being, advanced again. But Linnaeus remained a steadfast nonevolutionist, although in later years his own work forced him to modify his views.

As classification proceeded, more and more taxa were found that had to be fitted between earlier seemingly distinct taxa. As this went on, the distinctions between the closest taxa became more and more minute. As in some cases the variation between groups of organisms was taking on the aspect of continuous variation, Linnaeus made what amounts to a minimal concession. He concluded that certain species had resulted from hybridization between other species or genera. This is a minimal concession in that it still leaves intact the idea that most species were created separately, were quite distinct, and required no further explanation as to origin. This bit of intellectual skillfulness must be understood as a part of what will be discussed at great length later as the *archetypal mentality*. As will be shown in Chapter 11, the archtypal mentality, especially in the form of "race history," has plagued anthropology up to the present day.

Linnaeus can be considered a contributor to evolutionary thought, however, simply because the evolutionary implications of his taxonomy were, in the long run, undeniable. The only other system of binomial nomenclature, well known for its spread at that time among the masses in Europe, was the system of using a family name, or surname, in addition to the given name. This binomial system was designed to trace descent. Whether or not this had any direct influence on the interpretation of animal and plant taxonomy we do not know. We do know that evolutionary inferences began to be made on the basis of Linnaean taxonomy. Indeed, many evolutionary studies even today use such classifications as a point of departure.

At the same time that Linnaeus was writing, and immediately thereafter, a number of men were advancing imaginative, and sometimes fairly systematic, evolutionary theories. Calling them evolutionary can be justified specifically because they contained the premise of the transmutability of species and, more generally, because they used genetic explanations. Genetic in this sense means that explanations of present forms were in terms of change from previously different forms due to specific conditions causing change. This is the idea contained in the phrase "descent with modification." A list of the most notable of these thinkers would have to include Georges Buffon (1707–1788), a contemporary of Linnaeus, Erasmus Darwin (1731–1802), grandfather of Charles Darwin, and Jean Baptiste de Lamarck (1744–1829), whose ideas on the causes of change had great appeal to the socially mobile classes throughout the Western world.

It is with the name of Lamarck that we associate the phrase "the inheritance of acquired characteristics." Actually the major points of theory in the works of Lamarck and of Erasmus Darwin were remarkably alike, but since subsequent history has associated Lamarck's name with these ideas we will discuss him. Lamarck's outstanding theoretical work was *Philosophie Zoologique,* published in 1809. He proposed a theory in which change, although set in motion by a Creator, was unlimited and undirected except for the effects of natural forces in the environment.

What caused these changes? In the case of animals, a change in the environment caused a change in their habits. New exertions by an animal or the greater employment of an organ caused that organ to grow. This increase was transmitted to descendants. Thus direct inheritance of morphological change, better known as the *inheritance of acquired characteristics*, was a fundamental part of Lamarck's theory. In the case of plants, which have no nervous system, he could hardly speak of new habits. Therefore he assumed plants to be more directly influenced by such factors as temperature, light, humidity, and so on. Changes induced by variation in these factors were again supposed to be transmitted to the next generation. For animals the lack of use of a muscle or organ led, according to this theory, to a reduction that was, similarly, transmitted to subsequent generations.

Lamarck's ideas stimulated a great deal of discussion but almost no acceptance during his lifetime. Surprisingly it was only after the work of the great Charles Darwin had made evolutionary explanation generally acceptable in the world of learning that Lamarck's idea came into some vogue. There are two or three reasons for much of the latter-day acceptance of Lamarckian explanation. One good reason was that Charles Darwin, himself unable to satisfactorily explain the origin of new variation, leaned more and more toward the Lamarckian explanation. Secondly, a Lamarckian approach is intuitively believable, as the inheritance of acquired characteristics does in fact hold with regard to cultural change, that is, the social heredity of man. And this change, being relatively rapid, is easily observable by the layman, whereas biological change in species is not easily observable.

Finally the struggle for social equality in the Western world made Lamarckian explanation quite attractive. In Europe the class-restricted inheritance of wealth and political power had come to be given an ideological justification that was, in part, biological. People were born to their station in life. The economic and social changes that swept the West in the nineteenth century swept away this rationale, or at least moved it out to the colonies. The rising middle class rejected the old theories, biological and social, that would deny them the possibility of rising as high on the social scale as the aristocracy. This is why much of educational theory in the United States today still assumes a kind of infinite perfectibility of all. The USSR, which had to go through a much more violent revolution than did the United States, eventually got rid of all modern genetic theory and research (until the mid-1950s) in favor of Lamarckianism because they felt that genetics, as we know it, perpetuated a theory of social inequality.

But Lamarck and the other evolutionists of the time represented no great advance over the Greek thinkers who antedated them by almost 2500 years. Theirs were speculative theories which made good stories but contained little in the way of confirmation. Finally they all suffered from the fact that they had to assume that evolution was a very slow process. Since the span of recorded history showed no sensible changes in animal species, except for some domestic ones, and even mummified specimens from Egypt differed not at all from modern forms, the assumption of a slow process was demanded. But biblical scholars had determined, based on genealogical studies in the Old Testament, that the world was a little under 6000 years of age.

Obviously we must now look to developments in geology and paleontology and the answers they are beginning to provide as to the earth's age. Petrified forms which were similar to animal remains had been known and discussed by Greek thinkers, had been shrewdly commented upon by Leonardo da Vinci (1452–1519), and were a subject of debate in the general intellectual awakening of natural science from the seventeenth through the nineteenth centuries. By the end of the seventeenth century, most naturalists had rejected the earlier suggestion that these things grew in the earth like gallstones. Not only were fossils recognized as mineralized organic remains but their occurrence in exposed earth strata often showed a clear progression or order from lower to higher strata.

Georges Buffon was one of the first to reject the idea that marine fossils occurring in mountain ranges could be explained by deposition in the biblical flood. He assumed that mountain ranges were thrust up as a result of natural forces buckling and distorting the earth's crust. He assumed that sedimentary strata were built up slowly and, on the basis of this assumption, estimated the earth's age at about 75,000 years.

The notion that the earth's features observable today are primarily the result of gradual and continuous change was developed in some detail by the Scottish physician and geologist James Hutton (1726–1797) whose first draft of *Theory of the Earth* appeared in 1788. Hutton viewed the land masses of the earth as being continually and simultaneously destroyed and rebuilt. The forces involved in these two processes were, he claimed, the same in the past as they are today: vulcanism, wind and water erosion, consolidation of sedimentary deposits into rock under the influence of heat and pressure, and so on. This concept came to be known as *uniformitarianism* in contrast to *catastrophism,* which explained changes in the fossil fauna of different strata as a result of the sudden destruction of the earlier forms in brief, unusual, and catacylsmic events.

An articulate spokesman for catastrophism was Georges Cuvier (1763–1832). Cuvier is recognized as one of the founders of paleontology but, as paradoxically as in the case of Linnaeus, Cuvier was a leading opponent of evolutionary thought. He spent many happy hours ridiculing the thoughts of his less well-off contemporary, Lamarck.

It should be noted at this point that the rapid extinction of a single species, or of several species in a single geological event, is indeed catastrophic for the species in question, but this is not what is meant by catastrophism. The extinction of the passenger pigeon in the United States was, on a geological time scale, abrupt and catastrophic. Recognizing this occurrence, however, does not make one a catastrophist. Catastrophism legitimately denotes only the approach that explains all change in the fossil record as a result of abrupt, cataclysmic change.

Why should the opponents of evolution have been forced to this kind of approach at this time? Because fossil forms did not just differ from living forms, they also differed between strata. That fossil species were characteristic enough to be used in indexing or labeling different strata was suggested by Abbé Sauvages de la Croix in 1749. The concept of the *index fossil* was elaborated by William Smith in England. Such index fossils permitted investigators to relate geographically separated strata over large regions and to build up a large series of paleontologically distinct *horizons.*

This indisputable wealth of change between geological horizons required either an evolutionary explanation or repeated broad-scale catastrophies with subsequent new creations or immigration of new species from unknown regions. Hutton's reasonably sound work was buried under an avalanche of the latter types of explanation. Cuvier proposed 26 such catastrophies with subsequent repopulation each time to explain the fossil record he helped reveal. Some of the explanations of the time were far more fanciful than would have been put forward at an earlier period and can only be understood as notions advanced by persons desperate for some explanation of change other than that inevitably approaching. But this interpretation is available to us only as hindsight. At the time that Cuvier wrote, a conservative reaction to the French Revolution had set in, and he was being hailed as the great man who had once and for all reconciled the paleontological record with the biblical accounts.

In geology the death blow to catastrophism was the appearance of the works of another Scottish geologist, Charles Lyell (1797–1875). His three-volume *Principles of Geology* was published between 1830 and 1833. The evidence he gathered in these volumes finally convinced reasonable scholars that the geological forces active in the past are the same as those that act in the present. Although Lyell was quite convincing in showing how the earth's crust had evolved, he refused to extend evolutionary principles to the fossil sequence he observed. He believed in the fixity of species. He was quite clear in his belief that man represents a special and very late creation in the animal kingdom. As for other fauna, he vacillated between explaining the fossil record as being a result of migrational changes in the species of a region and their extinction versus invoking additional special creations. The theory of animal evolution with which he was familiar was that of Lamarck. On weighing Lamarck's arguments and evidence, Lyell felt that he had to reject biological evolution entirely.

In 1831 Charles R. Darwin (1809–1882), then 22 years old, departed from England as the naturalist on the *H.M.S. Beagle* and took along the first volume of Lyell's *Principles.* The second volume reached him in Montevideo in 1832. This voyage of the *Beagle* lasted five years. Darwin seemed particularly aware of phenomena, no matter how small, that had evolutionary significance, and his observations during this time were to have a deep influence on his thoughts on evolution. Although he little recognized it, Darwin must have been influenced by the writings of his grandfather, Erasmus Darwin, and the discussions of these ideas in the Darwin household. Lyell's second volume was devoted to a critical analysis of Lamarck's theories, to zoogeography, the study of the geographical distribution of animals, and to Lyell's explanation of the fossil record.

As they sailed south along the east coast of South America, Darwin noted that certain animal varieties slowly replaced others with change in latitude. Later he noted that the fossil fauna of a region was often physically closely related to the modern fauna of the same region. This Darwin called the "law of succession of types." He hinted that the differences in fauna and flora of the east and west coasts of South America had come about subsequent to the rise of the Andean mountain chain. Still Darwin was not prepared for what he was to find in the Galapagos Islands west of Ecuador. These are small, relatively barren islands situated only a short distance apart. The Galapagos turtles provided an unexpected challenge for Darwin. Since evolutionary change was not considered a possibility in the zoogeography of that day, the "explanation" of why the particular turtles were found on the islands was simply an historical problem that can be stated as follows: What were the migrations or dispersals that brought these turtles to the islands? To answer this question, Darwin collected and described them in order to identify the center of dispersion from which the Galapagos Islands had been populated. Only toward the end of his stay did he discover that local residents could tell him from which island each turtle came. Darwin had not expected identifiable and consistent variation between islands that were so close together and climatically undifferentiated. He found similar small variation in finches and other fauna. How could dispersion from a single center have resulted in differences between islands? Darwin hinted then that he might be considering evolutionary explanations, but did not come forward with his full theory until 1859.

In 1859 the first edition of Darwin's *Origin of Species* appeared. In the intervening years Darwin had become one of the most respected naturalists of his day and had been quietly collecting data on what was called "the species problem." He investigated the origin of variation in domestic breeds. He became a pigeon fancier, joining two London pigeon clubs in the pursuit of science. He convinced himself that although the domestic pigeon varieties were as diverse as those that would be classified as separate species in the wild, they were all descended from the rock pigeon.

Darwin became aware of the concept that has been termed "the struggle for existence" from Lyell's third volume. This phrase came from Candolle (1778–1841) and from Malthus (1766–1834). For Candolle, a Swiss botanist, it meant that more plants sprouted in a given area than could survive there. For Malthus, an economist, it meant that in animal populations there would always be competition for food such that not all would survive. The basic thesis of Malthus was that animal populations tend to increase their numbers as a geometric progression, while their food supply increases as an arithmetic progression. This would imply that, regardless of their starting point, the animal populations will eventually outrun their food supply. Malthus' assumptions were faulty, but it is well confirmed that food supply is an important limiting factor in the growth of many animal species.

Darwin had ample documentation of the effects of selection on domestic animal and plant populations through the work of animal and plant breeders. He had ample evidence of the occurrence of variation within wild species. If certain varieties within a species were better equipped to survive and to reproduce in the competition suggested by Malthus, this would result in that variant becoming more common. And with successive varieties doing the same thing, the process would be analogous to the artificial selection practiced by breeders. Since the process Darwin now envisaged was carried out without the help of man he termed it "natural selection" and this became the basis of the origin of species.

How long Darwin would have continued to gather his evidence, had not an unexpected event speeded up affairs, we do not know. The unexpected event was the receipt by Darwin of a manuscript from Alfred Russel Wallace (1823–1913) which briefly detailed a theory of evolution almost identical to that which Darwin had developed. Wallace was a naturalist working in Southeast Asia. Like Darwin he had been impressed in the course of fieldwork with the succession of varieties correlated with geographic change. He had read at least some of Darwin's published works and was in correspondence with him. Both men had read from a similar body of theories and criticisms, and both acknowledged the stimulus of reading Malthus' *Essay on Population*. The ideas were, in a sense, in the air and waiting for someone to put them together. Darwin and Wallace did so almost simultaneously.

Both men were sharp observers who were, as it is said, at the right places at the right time. As an observer of man Wallace was far more prominent than Darwin. Why then does Darwin receive most of the credit for the theory of natural selection? Primarily because it was Darwin who amassed the evidence that convinced people. It was he who did for biology what Lyell had done for geology. As mentioned previously, all of the elements of the theory had been present in the writings of others. In this connection it is interesting to note that natural selection was first suggested as early as 1813 by an expatriate American, William Wells, who suggested that the distribution of skin color differences in man could be explained by a tendency of darker skinned varieties of man to increase in the tropics due to an association between dark skin and disease resistance in these regions!

Darwin received Wallace's manuscript in 1858 and, on the advice of friends, immediately presented Wallace's paper and one of his own before the Linnean Society of London. In 1859 Darwin's book appeared, entitled *The Origin of Species by Means of Natural Selection, or the Preservation of Favoured Races in the Struggle for Life.* As important as this work has been for scientific thought it must not be assumed that Darwin presented a complete theory of evolution as it is known today. He demonstrated the preeminent role of natural selection operating on variation within a species as a cause of evolutionary change. One of the weak points of his theory was that Darwin could not adequately account for the initial occurrence of variation, for what animal and plant breeders call *sports.* Therefore he did not have a satisfactory theory of inheritance.

The importance of natural selection depended upon the assumed occurrence of rare, fortuitous sports, or mutants, which became the raw material for selection. Darwin boldly made this assumption in 1859 but slowly retreated from it. When he dealt with the evidence of man's evolution in *The Descent of Man* (1871), his data were better than those of Lamarck but his theory was no longer so different because he now deemphasized natural selection and placed consequently greater emphasis on Lamarckian processes.

This retrogression was due to a criticism that Darwin was never able to handle. Darwin shared the common sense theory of inheritance of his time which assumed that offspring were, in hereditary features, a blend of parental features. Obviously this did not hold in the determination of sex but this seems to have bothered no one. Biologists tended to avoid thinking in quantitative terms or posing quantitative problems; blending inheritance thus remained a secure assumption. Fleeming Jenkin, an English engineer, did see a quantitative problem in *The Origin of Species.*

We have illustrated Jenkin's argument in Figure 1.4. Let the bars represent the height of individuals. The taller bar, A, represents a mutant. Assume that the remaining individuals of a population of 10,000 are all of

FIGURE 1.4 Decrease in offspring height (bar length) due to blending of parental heights.

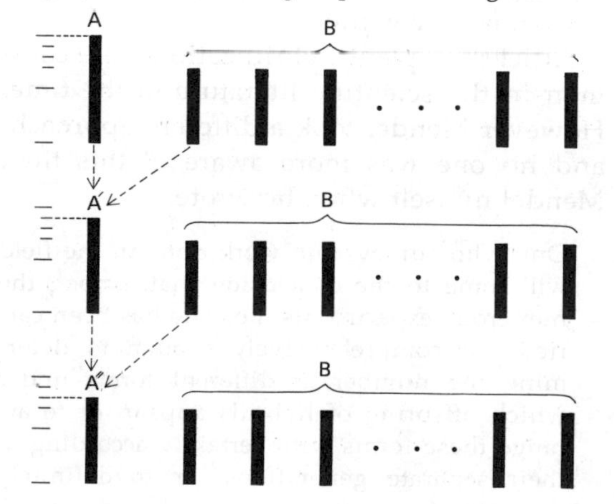

height B. A then mates with a *B* individual to produce offspring whose height is *A*′, intermediate between *A* and *B*. These *A*′ will in the next generation probably mate with *B* individuals, further reducing the height of their offspring *A*″.

Unless selection is very rapid, more rapid than is known to be the case, the effect of the favorable mutant is soon swamped. If, on the other hand, mutants in this direction are numerous, then natural selection is of much less importance to begin with. Darwin tried in subsequent years to deal with this criticism but was never able to do so successfully. The argument was undeniable. As long as the difference between a mutant and normal forms was infinitely subdivisible, mutant effects would soon disappear into the larger population.

In 1865, a year before Jenkin's criticism appeared, a paper that was read in Brünn, Czechoslovakia reported results of certain experiments and their interpretation. This report forms the basis of genetic theory today. These results would have taken care of the criticism by Jenkin and others had the importance of the paper been recognized. It was the work of Gregor Mendel (1822–1884), an Augustinian monk who carried out breeding experiments on common garden peas in the monastery garden.

Articles on plant hybridization were common in the scientific literature of the time. However Mendel took a different approach, and no one was more aware of this than Mendel himself when he wrote

> One who surveys the work done in the field will come to the conclusion that, among the numerous experiments, not one has been carried out comprehensively enough to determine the number of different forms under which offspring of hybrids appear, or to arrange these forms with certainty according to their separate generations, or to definitely ascertain their statistical relations (Bennett, 1965:8).

In other, almost offhand statements, Mendel revealed his belief in organic evolution and hinted that his work on heredity held important implications for evolutionary theory. Unlike Linnaeus and Cuvier, Mendel cannot be classified as one making an unintended contribution.

To carry out his program of research, Mendel chose discrete, usually two-valued, traits for which he could obtain true-breeding strains and whose frequency he could count in the hybrid offspring of various crosses. Such traits as seed shape (round or wrinkled), height of the plant (tall or dwarf), and cotyledon color (yellow or green) were included in the seven traits observed. In the F_1 (the first "filial" or first offspring) generation each hybrid showed only one form, which was termed the *dominant* trait (Figure 1.5).

When the F_1 hybrids were crossed with other F_1 hybrids, the resultant F_2 included both dominant round forms and *recessive* wrinkled forms in the approximate ratio of three to one. Similar results were found for other traits. Mendel noted that plants having a recessive character, when crossed, "remain constant" in subsequent generations. Of the F_2 dominants two-thirds, when crossed, again produced the three-to-one ratio; one-third produced only dominants.

Mendel's explanation of these results

FIGURE 1.5

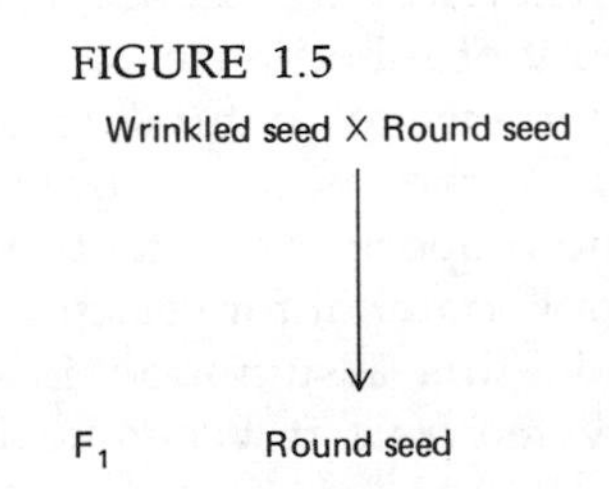

postulated factors in the pollen and ova of the pea plants which combined in random fashion to form the new generation. He let A represent the dominant trait, a the recessive trait. If we let these symbols represent the factors (genes) which are being transmitted, we can show the production of the F_1 generations as in Figure 1.6. Self-fertilizing plants of the F_1 generation yields the F_2 generation (Figure 1.7). The observed three-to-one ratio would be a result of the AA and Aa individuals both showing the dominant trait. The one-third of the dominants reproducing to give only dominants are the AA plants.

Letting A and a represent factors for round and wrinkled seeds and B and b factors for yellow and green seeds, respectively, Mendel found that the F_2 of plants hybrid for both characteristics showed the ratios

$$AA + 2Aa + aa$$

and

$$BB + 2Bb + bb$$

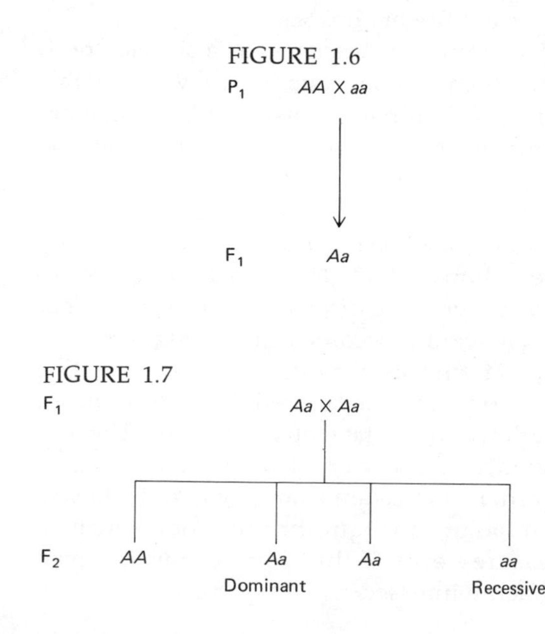

FIGURE 1.6

FIGURE 1.7

He further observed that the combination of seed shape and color occurred in frequencies suggesting that the factors determining them were assorting independently of one another. In other words, if one-half of all the seeds were Aa and if one-half of all the seeds were Bb then one half of the Aa were also Bb. The same explanation was found to hold true for the frequency of all other combinations.

Two major generalizations from this work have been called the *law of segregation* and the *law of independent assortment*. The law of segregation simply says that the hereditary factors, to which Johannsen gave the name *gene* in 1909, do not blend as a result of being united in one cell. They are, in a sense, atoms of heredity that segregate out in subsequent generations. The law of independent assortment says that pairs of genes affecting a trait will assort independently of other pairs of genes. (Another way of stating Mendel's second law is given in Chapter 2.)

Mendel represents the classic case of being intellectually too far ahead of his contemporaries. Perhaps he was also handicapped by the fact that he presented his ideas in a simple and straightforward manner. Adornment and bombast might have drawn more attention to his views, but such pretention was alien to Mendel. Furthermore, in an act of civil disobedience, Mendel refused to pay a new tax levied on the monastery in 1874 and became involved in a struggle that lasted until his death in 1884. This struggle, added to his duties as a prelate, left Mendel little time or energy to pursue his plant experiments.

A little-noted fact mentioned by Mendel's biographer, Hugo Iltis, suggests another contributing factor in the failure of biologists to perceive the importance of Mendel's work. He delivered his report in two parts at successive sessions. The first report, describing his experiments, elicited considerable interest and comment. At the subsequent meeting, where he presented his numerical analysis, the attention of the audience seemed to wander, and at the end there was not a single question asked (Iltis, 1932). Science probably paid a price that was due in part to a passive resistance to numbers. This is a phenomenon not unfamiliar to the teacher of evolutionary theory today when he opens the topic of population genetics.

By 1900 science was more receptive, and Mendelian principles were rediscovered almost simultaneously by three or four independent investigators. A search of the literature resulted in the rediscovery of Mendel's work, and the study of inheritance patterns in discrete traits has since been called Mendelian genetics.

Mendel's was a particulate theory of inheritance and was compatible with selection as an important factor in evolutionary change. But the harmonious union of Darwinian and Mendelian principles that, in retrospect, we would expect, was found in the works of only a few biologists at the beginning of this century. There are reasons for this both inside and outside of science. Social forces outside science include the depression of the 1870s and the subsequent rise in power of labor unions and Marxist social philosophy in the 1880s. A generalized conservative reaction to these developments carried with it a reaction to much of the evolutionary perspective that had become fashionable after Darwin. In the social sciences, physical sciences, and in much of biology attention reverted to "pure" empiricism and to study of functional relations between isolated variables.

Within science, evolutionary theory suffered from an excess of single-factor explanations. Some investigators were "mutationists" and considered mutations to be major events creating, in one step, species differences. Others considered variation to be almost indefinitely available at any time or place, thus making selection the only important factor in determining what variants would become common. Occasionally such tendencies can still be identified in the writings of some investigators. Speculation, in some cases far-fetched, on adaptive features in animals ran wild.

The resolution of the conflict between various single-factor views awaited the work of other men who, like Mendel, converted the questions to a form that could be given quantitative answers. These men were the founders of population genetics. They include J. B. S. Haldane, R. A. Fisher, and Sewall Wright. The basis of population genetics was established in 1908, shortly after the rediscovery of Mendel's work. This was a period of pedigree collecting in an effort to find simple genetic traits in man. One such trait, occurring in unusual frequency in an isolated new England village, was polydactyly, a condition characterized by excess fingers or toes.

This trait was described as a simple dominant. Some geneticists asked why, if this trait was dominant, it had not become more common in human populations than the supposedly recessive five-digit form. The answer to this query was supplied simultaneously by G. H. Hardy and W. Weinberg. They showed that dominance or recessiveness has nothing directly to do with whether or not a trait becomes more frequent in a population (see Chapter 3). To prove this assertion they established the condition for equilibrium, a state of no change. The importance of this cannot be overemphasized. In order to measure change you must have a zero point, an equilibrium from which to measure—even if this equilibrium is never, in fact, witnessed.

Furthermore, if you can state the necessary and sufficient set of conditions for an equilibrium, you have at the same time given a complete listing of factors (processes) that can cause change. In other words, any condition necessary to equilibrium which is not met constitutes a "cause" for change. The papers by Hardy and Weinberg, with later modifications by Haldane, Fisher, and Wright, provide such a set of conditions for genetic change. Evolution is a change in inherited features. If all inherited features are controlled by genes, then the conditions for genetic change are the conditions for evolutionary change. These are conditions that can be quantified, and thus their resultant interactions can be predicted.

The utility of such an approach will be illustrated by analogy. Suppose a man is rowing across a river (Figure 1.8). What can we predict? As long as he continues to row in direction *A*, he will reach the opposite shore. This is a qualitative statement. It is equivalent to saying that if selection favors a certain form, the population will change in that direction. Or if mutation occurs repeatedly in one direction, change will occur. But going back to the boat analogy, most people would agree that it is more useful to know not just that the boat will land on the opposite shore but *where* on the opposite shore it will land. We can determine this if we can quantify the processes by **A** and **B** representing the direction and speed of the boat being propelled and the stream flowing, respectively. The resultant vector, **R,** indicates the path actually taken by the boat.

FIGURE 1.8 Boat analogy showing that change R is the outcome of two interacting processes, represented by A and B.

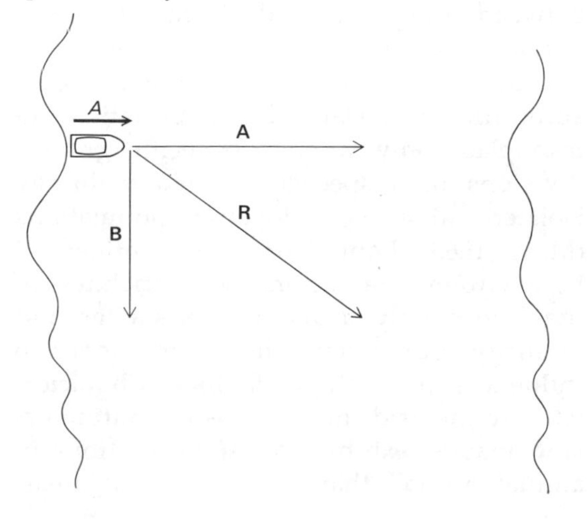

Similarly population models, as will be shown in Chapter 3, provide us with ways of quantifying processes of evolution and their resultant measured in terms of gene frequency change. One consequence of this is that evolution often is defined today as a change in gene frequency. And since it makes no sense to speak of gene frequencies in an individual it is obvious that it is the population that evolves and not individuals. The concept of population will be explored further in subsequent chapters.

Population genetics also contributed to the demise of the concept of progress as a part of the definition of evolution. The concept of progress became a part of the definition of evolution in the optimistic philosophies of the eighteenth century. Its value connotations aside, progress can be made an objective concept if progressive change is defined as a form of unidirectional change, specifically what the mathematician might call a monotonically increasing function. This is a function that represents change in a single direction only. For example, assume that curve A in Figure 1.9 represents change of head length with time. Since the head may grow at different rates at different times the curve is not a straight line, but unless the head can shrink the curve can never change direction. In other words, the function A

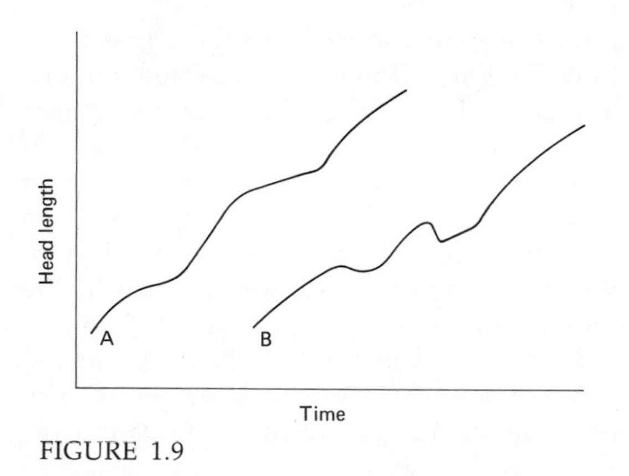

FIGURE 1.9

does decelerate at times, but does not reverse direction. Curve B, which is not monotonically increasing, does change direction.

The concept of progress as part of the definition of evolution would require change to occur as pictured in curve A. This, however, is not necessary. If one can explain change of the type represented in curve B, one can certainly explain change of the type represented in curve A. Therefore progress, in the only sense in which it could be made objective, becomes rather unimportant.

By 1930–1932 the basic population genetics models had been developed. These would permit the quantitative investigation of evolution in living organisms free of the polemics characteristic of the beginning of the century. Of course, it took a while for such approaches to penetrate the various fields concerned with evolution, including human evolution. Many refinements are still being made in an effort to make the analytic approaches more and more appropriate to the reality of living populations. Finally, population genetics and the study of population dynamics, generally, added impetus to the study of ecology.

Ecology has been defined as the study of the relationships between an organism and its environment. Population ecology, specifically, deals with environmental factors affecting the number and distribution of individuals. Why should developments in population studies affect ecology? Population genetics and related developments in theories of population dynamics lead to the view of a species, or an interbreeding population, as a homeostatic system, a system that tends to move in most processes toward an equilibrium state. Evolutionary change viewed from within the population, so to speak, is essentially conservative. It is change in the direction that would limit further change. In such systems the force or the push leading to long-term or continuing change must come from the outside. The environment can be defined as a set of outside influences that can act on a homeostatic system. This means that studies of animal ecology are a necessary and important part of the study of animal evolution.

It is impossible, in advance, to specify all of the features in the physical setting of a species that will constitute the effective environment. It is possible to categorize these features roughly by their effect on the species. Geographic barriers to migration and mating are important to genetic isolation between populations and are, therefore, important in speciation, the evolutionary process by which species are formed. Some authorities have claimed that speciation can take place only when subspecific groups, divisions of a species, are geographically isolated, that is, allopatric populations (Mayr, 1963). Population size is influenced by environmental factors such as richness of the food supply, availability of shelter, and intensity of predation. The environment also influences population stability. Characteristically the arid inland areas of continental land masses exhibit greater fluctuations in annual rainfall than do the coastal areas.

Fluctuations in water supply and food resources may be reflected in fluctuations in population size. The richness and diversity of a region, a greater number of nutritional levels, a more complicated food chain characteristic of tropical as opposed to arctic or subarctic regions, will provide an increased number of possibilities for specialization and speciation.

It is the environment that provides selective factors such as epidemic disease, parasites, predation, and the climatic and nutritional levels that test the efficiency of biological adaptations both between and within species. Finally, for man an important set of environmental factors resides in the existence of human culture. Later we will see culture itself as a product of biological evolution. But, retroflexively, culture affects biological evolution. Technology has vastly altered the exploitive ability of *Homo sapiens*. The employment of various technologies has altered the physical environment, sometimes drastically. Nongenetically determined social institutions influence population stability and size, influence geographic patterns of mating and, as reflected in individual attitudes, influence choice of mate and even fertility.

As evolutionary theory now stands there are still significant gaps. For instance, we assume that all quantitative (continuous) variation has the same basis in delimitable structural genes as qualitative (discontinuous) variation. But we have no adequate methodology for obtaining a "Mendelian" analysis of quantitative variation and cannot, therefore, turn our assumption into a testable hypothesis. As evolutionary theory now stands it includes much more than was included by Darwin. It also discards some major features of Darwin's theory. To differentiate the two and to designate the present-day approach Julian Huxley suggested the phrase, the "synthetic theory" of evolution (Huxley, 1942). The term "synthetic" is used as a way of recognizing that evolutionary theory today represents a synthesis of many fields and many approaches, principally those that have been outlined in this chapter.

REFERENCES CITED

Bennett, J. H. (ed.) 1965 *Experiments in Plant Hybridization*. Edinburgh: Oliver and Boyd. This is Mendel's original paper in English translation.

Huxley, J. S. 1942 *Evolution, the Modern Synthesis*. London: Allen and Unwin.

Iltis, H. 1932 *Life of Mendel*. Eden and Cedar Paul (trans.). London: Allen and Unwin.

Mayr, E. 1963 *Animal Species and Evolution*. Cambridge, Mass.: Harvard University Press.

Osborn, H. F. 1922 *From the Greeks to Darwin*. New York: Scribner's.

Simpson, G. G. 1961 *Principles of Animal Taxonomy*. New York: Columbia University Press.

FURTHER READING

Darwin, Charles R. 1951 *Origin of Species*, reprint of first edition (1859). New York: Philosophical Library.

Eiseley, Loren 1961 *Darwin's Century*. Garden City, N. Y.: Doubleday.

Greene, John C. 1959 *The Death of Adam*. Ames, Iowa: Iowa State University Press.

Irvine, William 1959 *Apes, Angels, and Victorians*. New York: World Publishing.

2

The Genetic Basis of Evolution

In the preceding chapter we saw that the most important part of Mendel's theory of inheritance was embodied in two laws. The first is the *law of segregation,* which states that traits do not blend but remain discrete and reappear in original form in subsequent generations regardless of having been combined with other traits in an ancestral individual. The second is the *law of independent assortment,* which states that traits are passed on to descendants in combinations that are independent of the combinations in which they arrived from the previous (parental) generation.

To understand the physical basis of the laws described by Mendel we must look briefly at the mechanics of cell division. We will be concerned with two kinds of cell division. One kind of cell division, which in higher forms of life is unique to the germinal cells, is called *meiosis.* The ordinary kind of cell division, which is characteristic of growth due to cell duplication, is called *mitosis.* To describe the differences between these two processes of division let us look at the structure of the cell.

With the exception of viruses, all living organisms are made up of cells. The *Protozoa* are unicellular or have cells forming a single tissue. The *Metazoa* have cells that are differentiated into more than one kind of tissue. Generalized cell structure can be depicted

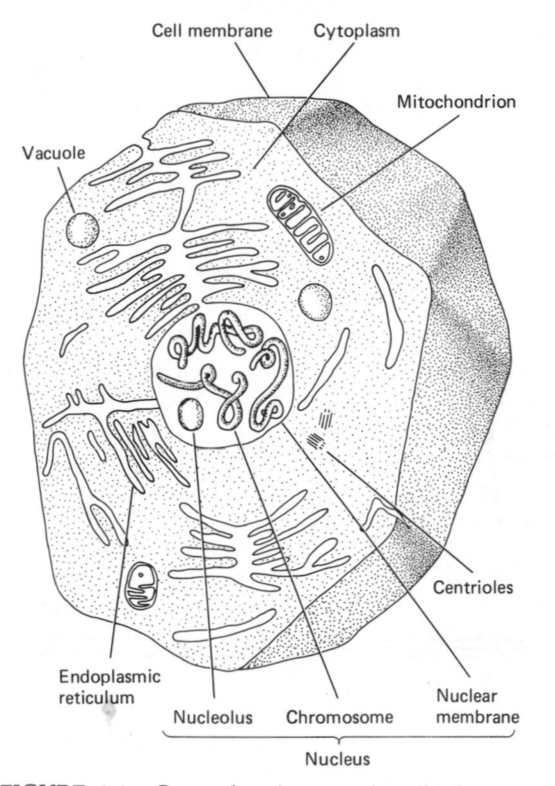

FIGURE 2.1 Generalized animal cell showing the nucleus and some structures in the cytoplasm.

as in Figure 2.1. In a typical *somatic* (body) cell in man, the nucleus (a spherical body within the cell containing the chromosomes and regulating the functions of the cell) is much smaller relative to total cell size than shown in the figure. The nucleus is in the cytoplasm, which consists of water, inorganic salts, carbohydrates, fats, and proteins, and contains a number of organelles (small organs within the cell) concerned with cellular metabolism. The cytoplasm is surrounded by a semipermeable cell membrane permitting the differential diffusion or transport of cytoplasmic components.

The nucleus contains *chromatin*, a material so called because certain dyes stain it deeply, permitting its observation under the microscope. At certain stages of cell division, the chromatin assumes the shape of separate stringlike bodies called *chromosomes*. As will be shown later, chromatin is a nucleoprotein that makes up the material of heredity. In man it occurs only in the cell nucleus and in small self-duplicating organelles called mitochondria. So far as is now known only the nuclear material is involved in man's genetic development.

Among the primates, as in most higher organisms, the cell nuclei of each species contain a specific number of chromosomes that is characteristic of the species. Since these chromosomes occur in pairs, the cells are called *diploid* cells, that is, twofold (after the Greek *di-* meaning "two," *-ploos* meaning "-fold"). The diploid number of various primate species ranges from 44 to 80. In man the number of chromosomes is 46, and this number is referred to as the diploid number. The 46 chromosomes occur in 23 pairs. Since this is one-half the total number of chromosomes, it is known as the *haploid* number (after the Greek *haploos* meaning "single").

Of the 23 pairs 22 are termed *autosomes*, and one pair the sex chromosomes. With the exception of the pair determining sex the two members of each pair are morphologically similar and are, therefore, called *homologous* chromosomes. The two sex chromosomes are denoted the *X* and *Y* chromosomes. Strictly speaking, the *X* and *Y* chromosomes are not homologous but do pair up at certain times, as do the homologous autosomal pairs.

Genes, which might be thought of as words in a hereditary communication, are segments of the strands of a nucleic acid molecule, deoxyribonucleic acid (DNA). They are the smallest segments of this molecule that can act as functional units encoding the structure of proteins produced in the cell.

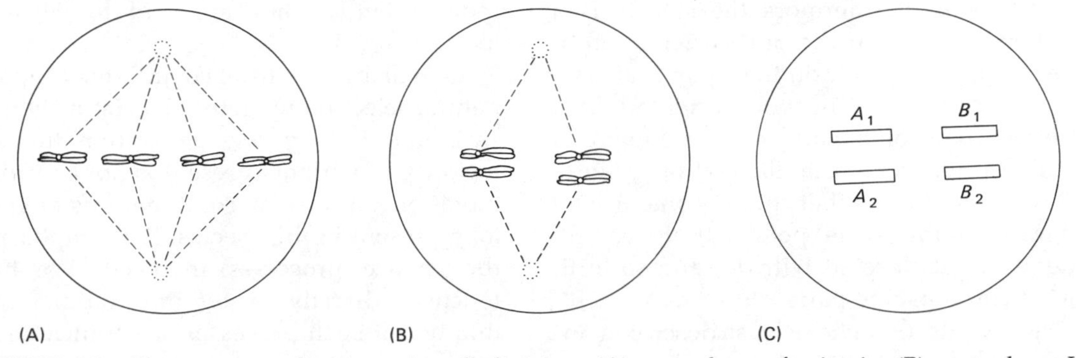

FIGURE 2.2 Chromosome alignment in cell division: (A) metaphase of mitosis, (B) metaphase I of meiosis, (C) same with two homologous pairs labeled. (Only four of the chromosomes are shown.)

Chromosomes are basically DNA plus a protein coat. Genes, as segments of the DNA molecule, occur in linear array down the length of the chromosomes. Obviously then the behavior of chromosomes at cell division, observable under the microscope, can tell us much about the behavior of genes.

Two kinds of cell division occur in human cells. These two kinds—or processes—of cell division are termed *mitosis* and *meiosis*. Mitosis is the cell division that occurs in the process of growth through cell proliferation. All cells, therefore, undergo mitotic division. The daughter cells produced in the mitotic process have chromosomes that are exact copies of the chromosomes of the parent cell. This means they are genetically identical.

Meiosis occurs only in the germinal cells, that is, those cells that are destined to become spermatozoa (if in the male) or ova (if in the female). The meiotic process produces daughter cells that have only one of each chromosome from a homologous pair. Having only half the normal number of chromosomes, they are called *haploid* cells. The mature haploid germ cell is called a *gamete*. Male and female gametes unite at fertilization to produce a diploid *zygote*.

The details of cell division need not concern us here. For descriptive purposes cell division is divided into a number of stages. One such stage which provides a crucial difference between mitosis and meiosis is called *metaphase*. In the mitotic metaphase [Figure 2.2(A)] all 46 chromosomes are aligned on a plane, the *equatorial plate*, between two poles of the cell. Each of the 46 chromosomes has duplicated itself before this. These duplicates split apart when contractile protein fibers begin to pull them toward opposite poles of the cell. They then form the nuclei of two genetically identical cells.

In the meiotic metaphase I [Figure 2.2(B)] the 46 chromosomes are aligned in homologous pairs on the equatorial plate. The contractile protein fibers attach to one of each pair in such a way that homologous, but genetically nonidentical, chromosomes move toward opposite poles. This first division in meiosis is followed by a second division that is much like mitosis.

The important point is that the metaphase I alignment of chromosome pairs produces genetically nonidentical daughter cells. This represents a reshuffling of the genetic deck, so to speak. Furthermore the shuffle is a random one in that the polar orientation of one chromosome pair on the equatorial plate is independent of the polar orientation of other chromosome pairs on the equatorial plate. In terms of the illustration [Figure 2.2 (C)], this means that chromosome A_1 will migrate to the same pole with B_2 as frequently as it does so with B_1, and so forth for all chromosome pairs.

The events of meiosis just described insure the reassortment of genes located on different chromosomes. It does not provide for the reassortment of genes located in different regions of the same chromosome. But this mechanism alone provides a surprising amount of genetic novelty in each generation, It means that each individual can produce potentially 2^{23} genetically distinct gametes. This is no small number, approximately 8.3 million different gametes.

But this is only a small part of the potential for genetic diversity through reassortment of genes. Genes on the same chromosome may assort independently as a result of *crossing over,* the exchange of chromosomal segments between homologous chromosomes. More than one theory attempts to explain crossovers. Without going into these we can say that crossovers occur prior to metaphase when homologous pairs are joined side-by-side. This is termed *synapsis.* The frequency of crossover is roughly proportional to the distance between genetic *loci* (the position or location of a gene) on the chromosome and is, in fact, used in "mapping" genes on chromosomes. A crossover frequency of 10 percent means that in meiosis genes at two loci are exchanged, over the long run, 10 percent of the time. A crossover frequency of 50 percent means that genes are crossing over with such frequency as to be independently assorting. If genes at two loci are on the same chromosome and are not assorting independently, they are said to be *linked* loci.

As will be emphasized in later chapters, natural selection operates on existing genetic variation. In any one generation the vast majority of variants present is not a result of mutation but of new combinations of genes long present in the species. This emphasizes the chance processes involved. We have touched, directly or by implication, on a number of such processes. They include the random assortment of chromosomes and the occurrence of crossovers between homologous chromosomes. Additional random events occur in the persistence of one out of four oöcytes (egg cells prior to maturation) to become the ovum (the mature egg cell), the choice of mate, and in the fact that one out of many spermatozoa fertilizes the ovum.

Genetic variation depends on the presence of functionally different forms of genes that occupy the same locus on homologous chromosomes. These alternative forms are termed allelomorphs or simply *alleles.* To talk of variation we use the terms *genotype* and *phenotype.* The genotype, which is inferred, is the set of genes present at one or more loci. The phenotype, which is the phenomenon observed, is the physiological or morphological manifestation of a set of genes. In describing the genotype of an individual, we say that he is *homozygous* if the two alleles at a given locus are identical, *heterozygous* if they differ. When one of the heterozygous genotypes is phenotypically indistinguishable from one of the homozygous genotypes we say, following Mendel, that the trait is dominant. When speaking inexactly we often speak of the gene as dominant although, strictly speaking, dominance is a characteristic of the phenotype.

An example of a dominant trait in man is "taster" ability. To some persons phenylthiourea, or phenylthiocarbimide (PTC), has a distinct bitter taste; to others it has no taste at all. This then is a two-valued trait much like those investigated by Mendel in peas. The two phenotypes are taster and nontaster. Taste ability is dominant to nontaste, which can be represented as

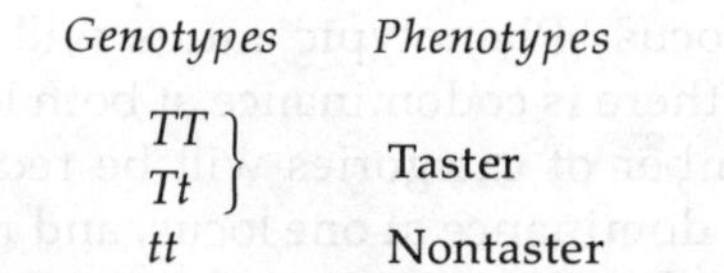

Representing the alleles involved by *T* and *t*, we can see that the homozygotes can produce only one kind of gamete each and that the heterozygote can produce both kinds of gametes. A mating between the two homozygotes can be shown as in Figure 2.3. The union of gametes in this mating can produce only heterozygotes. If, on the other hand, the mating is as shown in Figure 2.4, then two kinds of individuals can be produced genotypically.

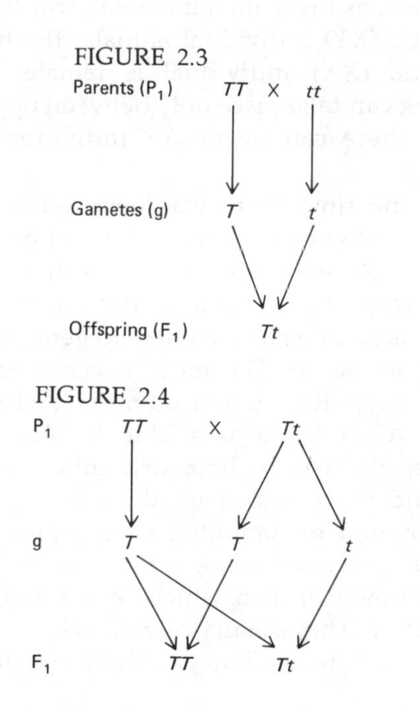

FIGURE 2.3

FIGURE 2.4

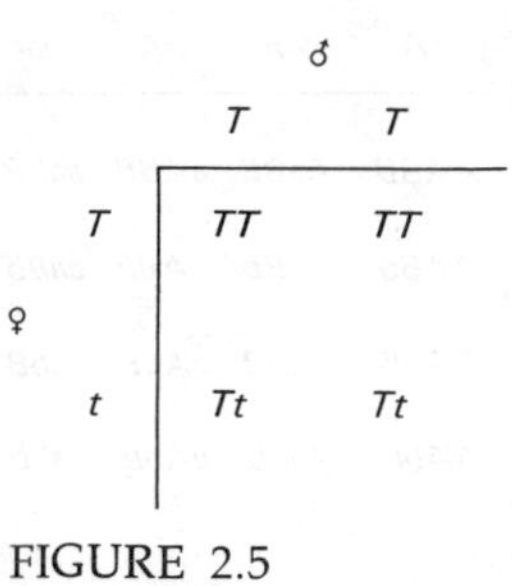

FIGURE 2.5

The ratios in which these genotypes are expected can be shown in a 2×2 matrix in which the female (♀) gametes are shown on one margin, the male (♂) gametes on the other. This is a way of representing the random union of gametes. Both alleles of the male are *T*. If the female produces *T* and *t* gametes in equal frequency, they can be represented as two rows and their intersection with columns representing the male gametes gives genotypes in the frequency expected from the mating. In this case we have the familiar 1 : 1 ratio for genotypes, but phenotypically this is a mating of taster by taster that produces only tasters. The matrix in Figure 2.6 represents the heterozygote by heterozygote mating.

Where the traits are codominant, that is,

FIGURE 2.6

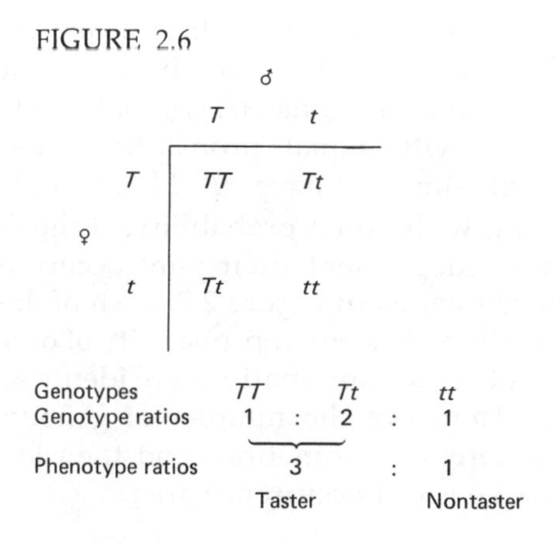

	AA	Aa	aA	aa
BB	AABB	AaBB	aABB	aaBB
Bb	AABb	AaBb	aABb	aaBb
bB	AAbB	AabB	aAbB	aabB
bb	AAbb	Aabb	aAbb	aabb

FIGURE 2.7

where the heterozygote can be distinguished from both homozygotes, there is a one-to-one relation between genotype and phenotype. Consequently, genotypic and phenotypic segregation ratios will be the same. Where one trait, such as taste ability, is dominant one must carefully note whether genotypic or phenotypic ratios are referred to, as they will differ.

The segregation ratios illustrated in the preceding figures refer to alleles at a single locus. In a population there can be more than two alleles, but for any given individual there can be only two alleles present at one time, one on each chromosome of the homologous pair. For a given mating, however, there are alleles at other loci assorting independently of the first locus. The latter situation is illustrated in Figure 2.7.

Let *A* and *a* be alleles at the **a** locus and *B* and *b* alleles at the **b** locus. The **a** locus will have the combinations *AA, Aa, aA,* and *aa* occurring with equal probability. The **b** locus will similarly have *BB, Bb, bB,* and *bb* occurring with equal probability. If the two loci are independent their joint occurrence can be shown as in Figure 2.7. Each of these combinations has equal probability of occurrence but some combinations are identical to others. Therefore the number of genotypically different combinations and their long-run frequency of occurrence are

$$1 : AABB \quad 2 : AaBb \quad 1 : aaBB$$
$$2 : AABb \quad 4 : AaBb \quad 2 : aa\ Bb$$
$$1 : AAbb \quad 2 : Aabb \quad 1 : aabb$$

These are the nine entries of the intersection of a 1 : 2 : 1 distribution for the **a** locus and independently a 1 : 2 : 1 distribution for the **b** locus. Phenotypic ratios will be the same if there is codominance at both loci, but the number of categories will be reduced if there is dominance at one locus, and reduced even further if there is dominance at both loci. The reader can easily work out the phenotypic ratios for the latter cases where dominance is involved.

The segregation ratios just discussed refer to the autosomal chromosomes. The situation is somewhat different for the sex chromosomes. The sex chromosomes of *Homo* are called the *X* and *Y* chromosomes. The *Y* chromosome is much smaller than the *X* chromosome but they pair as do the homologous chromosomes in cell division. In *Homo,* as in all other mammals, the heterogametic (*XY*) individual is male, the homogametic (*XX*) individual is female. Since mating can take place only between opposite sexes there can be no *YY* individual produced.

At one time there was believed to be a number of genes transmitted either on the *X* or the *Y* chromosome suggesting that indeed there was a homologous portion of the pair. Sex-linked inheritance refers to genes on the *X* chromosome. The term *holandric* inheritance is applied to genes on the *Y* chromosome. Curt Stern was able to show that almost all cases of holandric inheritance of specific traits rested on dubious pedigree information or had alternative explanations (Stern, 1957). At present there is only one trait known in man which seems definitely holandric. This is hairy pinnae, which can be defined as the tendency of fairly long hair to grow from the outer rim of the external ear (pinna). The criteria of such inheritance are obvious from the pattern of transmission of the *Y* chromosome. Since the presence of the *Y* chromosome determines that a zygote will develop into a male, the chromosome can only pass from father to son. And if the gene is fully penetrant, that is, those having the gene always show the trait, the trait will never skip a generation in the male line.

Traits of sex-linked inheritance are quite commonly known, although allelic forms at any one locus are usually rare because they are deleterious. Examples of sex-linked inheritance can be seen in red-green color blindness, hemophilia, and at least one blood group antigen system, *Xg*, which represents a common polymorphism (multiple form trait) in a sex-linked system.

In females sex-linked inheritance is much like autosomal inheritance in that the female can be heterozygous or homozygous. She receives sex-linked genes from both the father and mother. But the male, who has to receive the *Y* chromosome from his father, must always receive sex-linked genes from his mother and can transmit them only to inheritance in Figure 2.8. A female of the F_1 generation could be color blind only if the father was color blind and the mother had at least one gene for color blindness. The color-blind male is phenotypically like the homozygous recessive female. Having only one allele, he is spoken of as being *hemizygous.*

FIGURE 2.8

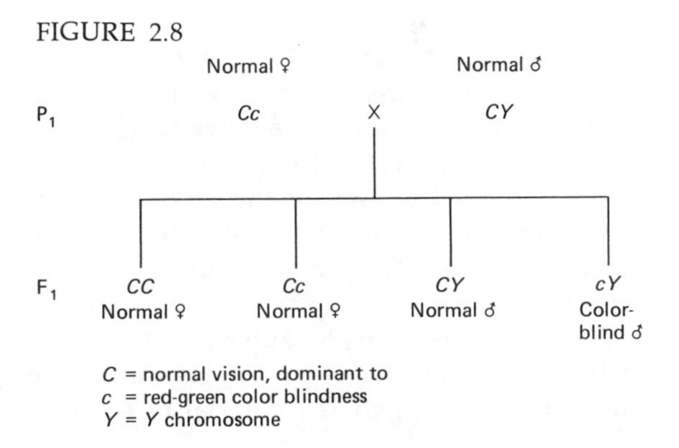

It is obvious that we must have alleles of a gene to carry out a Mendelian analysis of inheritance, but even when such alleles are present such an analysis is not always easy, especially in man where test matings are not possible. A major reason for this is that in only a few genetic systems is there an invariant relation between a single-locus genotype and a phenotypic trait. Many traits are affected by genes at more than one locus, the so-called *polygenic* traits. All phenotypic traits are a product of the interaction of genotype and environment. Those traits for which we have a Mendelian analysis are those in which environmental variation does not obscure the relation between genotype and phenotype. But undoubtedly the largest number of genetic systems in man are those in which environmental variation does have a pronounced effect.

Much biological variation in man, as in other animals, is seen as a continuum within the population. There are no distinct divisions. Examples are weight, stature, head length and breadth, skin color, and many other anthropometric and physiological variables. Since, through such means as twin studies (see Chapter 13), these variables are known to have a hereditary component, but do not show distinct classes that can be associated with particular genotypes, we assume them to be polygenic in character. The reasoning for this can be illustrated as follows. Assume that some anthropometric (e.g., nose length) is determined by two alleles at a single locus. Three classes of individuals are then possible. These might be represented as in Figure 2.9. In such a case an environmental contribution of ± 5 mm will not obscure the separate classes. But if the

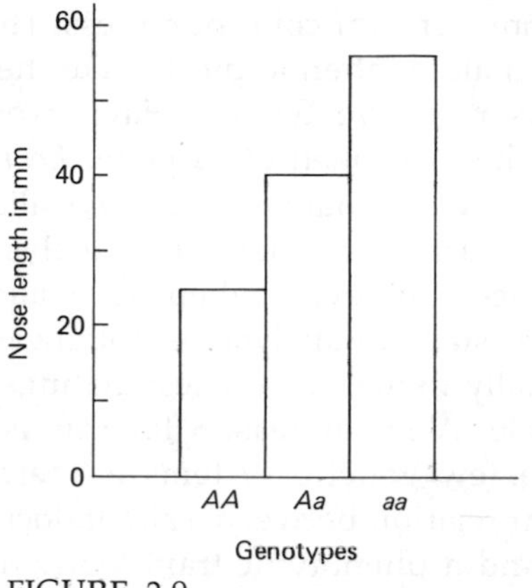

FIGURE 2.9

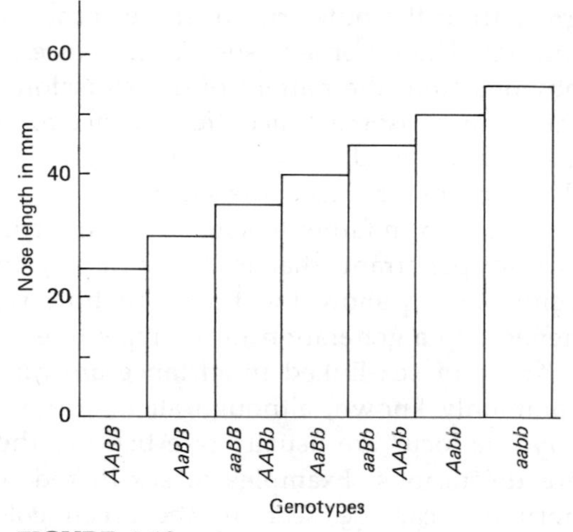

FIGURE 2.10

TABLE 2.1

GENOTYPE		*AA*	*Aa*	*aa*
	value	5	10	15
BB	20	25	30	35
Bb	30	35	40	45
bb	40	45	50	55

trait has the same range of values but is determined by two alleles at each of two loci, each having additive effects on the trait, then there can be seven classes determined by genotype alone that can be shown as in Table 2.1. Graphing as in the single-locus case produces Figure 2.10. In this case an environmental component of ±5 mm can completely obscure the genotypic classes. Continuous variation does not, therefore, yield to Mendelian analysis through pedigree studies and many of the powerful models of population genetics, dealing with evolution in terms of gene frequency change, are not directly applicable. The area of investigation known as quantitative genetics attempts to partition such variation into components due to additive gene effects, dominance effects, environmental effects, and so on. At present only the smallest beginning has been made in applying the approaches of quantitative genetics to man. The assumption remains, however, that the basic genetic mechanisms are the same for continuous as for discontinuous variation. Therefore our theories and approaches to the study of human evolution must at least be consistent with what we know to be true from Mendelian traits.

GENE STRUCTURE AND GENE ACTION

In order to understand the laws of genetic inheritance discussed in the preceding paragraphs, we have to look at the "mechanics" of the system. Some of the mechanics of chromosome transfer from generation to generation have been reviewed. This accounts for the regularity in the transfer of genes which are parts of the chromosomes. We must now look at the structure of the gene and how it acts to affect the biology of individuals.

The simplest way to approach this is to list a small number of principles that relate the structure of genetic material to some of the points developed in previous paragraphs. These principles will be expanded and supported by a further set of statements about genetic structure. This will provide a model, now well confirmed, of the mode of action of at least those genes showing Mendelian segregation.

Three general principles that can be stated are

1. Allelic differences are based on qualitative differences in the molecular structure of DNA (deoxyribonucleic acid).
2. The primary gene product that originates outside the nucleus of the cell is a polypeptide chain, a chain of amino acids, the basic building blocks of proteins.
3. Genes do not act continuously.

The implication of the first principle is that *selection* is basic to all genetic change. This selection is not synonymous with natural selection but includes mutation. That genetic differences are qualitative differences means that genetic change occurs not as a continuum, but by the substitution of one constituent for another in a finite set. The substitution of one constituent for another is the act of selection. When one constituent is substituted for another in the individual molecule we term this mutation. When a change occurs on a population basis as a result of the selection of particular molecules we term this natural selection.

The second principle is best illustrated by the structure of the hemoglobin molecule discussed in Chapter 4. Hemoglobin is the red pigment in the red blood cell responsible for oxygen transport in the blood stream. In most hemoglobin variants a difference between two alleles is expressed as a difference in one of the chains of amino acids (polypeptide chains) that unite to form the larger molecule of hemoglobin. This does not mean, however, that the phenotypes we observe in man, even those observed at the molecular level, necessarily represent primary gene products. Allelic variants in a protein may reflect a genetic difference in an enzyme that has acted upon a protein precursor in a long sequence of such changes. An enzyme is a protein which catalyzes (facilitates, without itself being used up) a change in a biological molecule. Our assumption is that the enzyme, which may not be observed directly, itself represents a genetic polymorphism.

The third principle, that genes do not act continuously, is a logical necessity to explain the differentiation of cells and tissues. In a given individual all cells having nuclei have the same genes as they are descendants of the original zygote through a series of mitotic divisions. For the original cell to differentiate into osteocytes (bone cells), neurons (nerve cells), epithelial cells (covering membranes), and so forth, providing they do not differ in their genes, they must differ in the rate of operation of these genes. In short, some genes have to be turned off some of the time in order to get differentiated tissues. Since differentiation of tissues and the development of an individual are regular and reasonably predictable there must be feedback mechanisms in which gene action is controlled by changes in the cytoplasmic environment of the nucleus and, in metazoans, by changes in other cells.

The three general principles just given can be elaborated by a set of statements that deal more directly with gene action. These are

1. The functional unit of inheritance, the gene, is a precisely delimited segment of the DNA molecule.
2. Genes are linear, as they are sequential segments of the DNA molecule.
3. There is a one-to-one relationship between changes in a gene and changes in the polypeptide chain whose formation it codes. Therefore, it follows:

4. There is colinearity between gene and polypeptide.
5. Molecules of RNA (ribonucleic acid) transmit the genetic message from the DNA of the nucleus to the site of protein synthesis in the cytoplasm of the cell.

This leads to a scheme of gene action that has come to be called the "central dogma" of molecular genetics and is shown simplified in Figure 2.11. This schematic representation shows the steps involved from gene to protein. The dashed line represents a feedback loop in which proteins can influence the rate of gene action.

There are numerous organisms to which this does not apply strictly. There are cells, the prokaryotes as opposed to the eukaryotes, that do not have nuclei. There are some bacteria and many viruses which have only RNA. But all higher organisms have DNA as the genetic material within a nucleus and this obviously is the model that applies to man.

Specification of a protein by a gene can be thought of as a two-part process. The first process is *transcription,* in which the structure of DNA specifies the structure of messenger RNA (mRNA). The second process is *translation,* in which the structure of mRNA specifies the primary structure (the sequence of amino acids) of the protein. In other words, we can consider Figure 2.11 to be divisible into the distinct steps

$$\text{DNA} \longrightarrow \text{mRNA} \quad \text{(transcription)}$$
$$\text{mRNA} \longrightarrow \text{protein} \quad \text{(translation)}$$

The DNA molecule, in the Watson-Crick model that is widely accepted today, is a long double-stranded helix. Each chain of this helix has the same kinds of units repeated many times down its length. The backbone of the helix is an alternating sequence of sugar (deoxyribose) and phosphate groups. To each sugar is attached a nitrogenous base to complete the chain. Figure 2.12 pictures

FIGURE 2.11

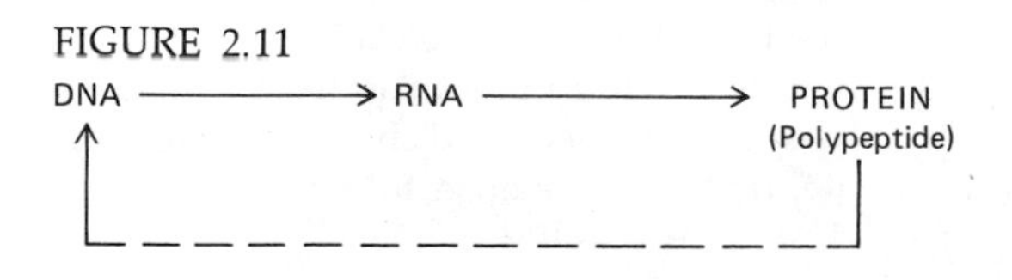

FIGURE 2.12 Segment of the DNA helix schematic.

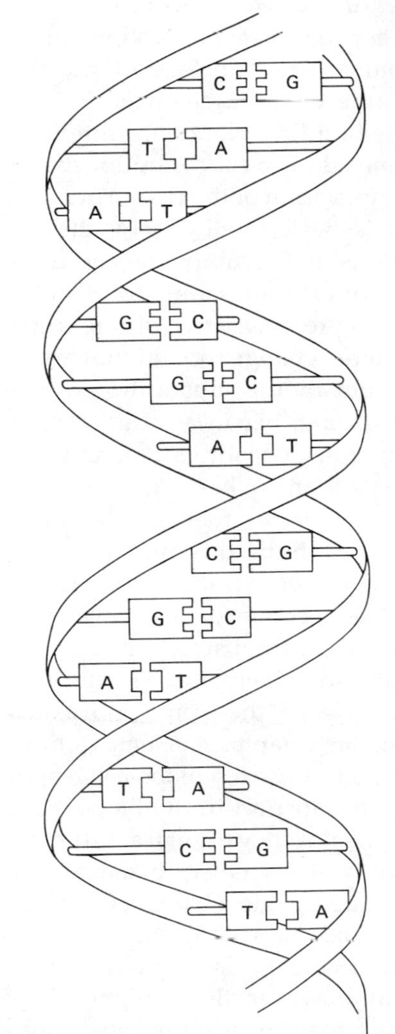

schematically the two complementary chains joined by hydrogen bonds down the axis about which coiling occurs.

The structure of the backbone of each chain of the helix is invariant; therefore no genetic (allelic) information can reside there. There are four nitrogen bases that occur in DNA. These are the two purines, adenine (A) and guanine (G), and two pyrimidines, thymine (T) and cytosine (C). A nitrogen base compound joined to a sugar is called a "nucleoside." A unit composed of base, sugar, and phosphate is called a "nucleotide." Each half of the DNA molecule is then a sequence of nucleotides. The four possible nucleotides can occur in any sequence down the length of the chain and variation in this sequence is the physical basis of genetic variation.

We can be even more explicit about this structure. Because of size differences, purines (A and G) being larger than pyrimidines (T and C), the helix can only be formed, normally, by the joining of a large and a small base, a purine with a pyrimidine. Further these units are partitioned in another way. One purine, adenine, and one pyrimidine, thymine, join by double hydrogen bonds. The other purine, guanine, and the other pyrimidine, cytosine, join by triple hydrogen bonds. This means that the two strands of the helix are invariant with respect to one another, as they will normally reflect one another (Figure 2.13). In other words, when one side of the nucleotide pair is determined the other is fixed. Genetic variation resides not in the pairing but in the sequence of nucleotide pairs.

Mispairings of nucleotides do occur. When they occur, the DNA of daughter cells can reflect this substitute nucleotide. If this substitution results in a functional change in a protein produced by the cell, this will be perceived by investigators as a mutation.

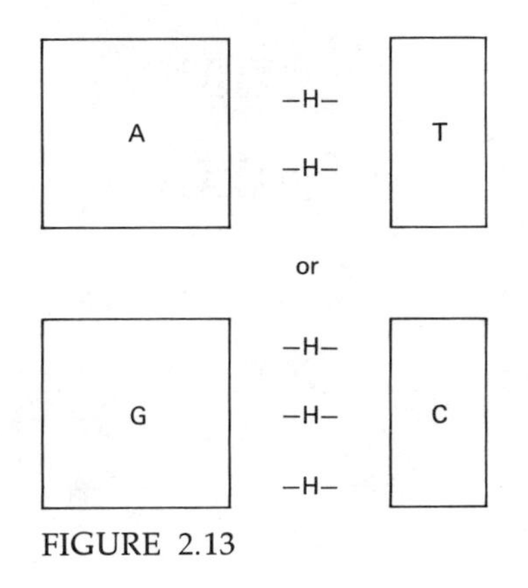

FIGURE 2.13

For translation of the genetic message to occur, the double helix of the DNA molecule must be "unzipped" so that one chain constitutes a template for the production of messenger RNA. A similar process occurs in cell division when DNA serves as a template for the production of DNA. A schematic picture of a segment of the DNA helix is drawn in Figure 2.12. The production of daughter strands of DNA, one from each half of the opened helix, is shown in Figure 2.14.

Enzymes that promote the production of DNA or RNA on the DNA template are known, appropriately enough, as DNA polymerase and RNA polymerase, respectively. A polymer is simply a substance formed by repetition of structural subunits. RNA polymerase is active when the cell is carrying on its normal metabolic functions. RNA differs from DNA in having ribose rather than deoxyribose as the sugar and having uracil rather than thymine. Therefore messenger RNA is complementary to DNA and maintains the genetic code.

The mRNA molecule is a small single-stranded molecule transcribed from a section of the DNA molecule. It can pass through the nuclear membrane into the cytoplasm. In the cytoplasm, and specifically on groups of organelles known as ribosomes,

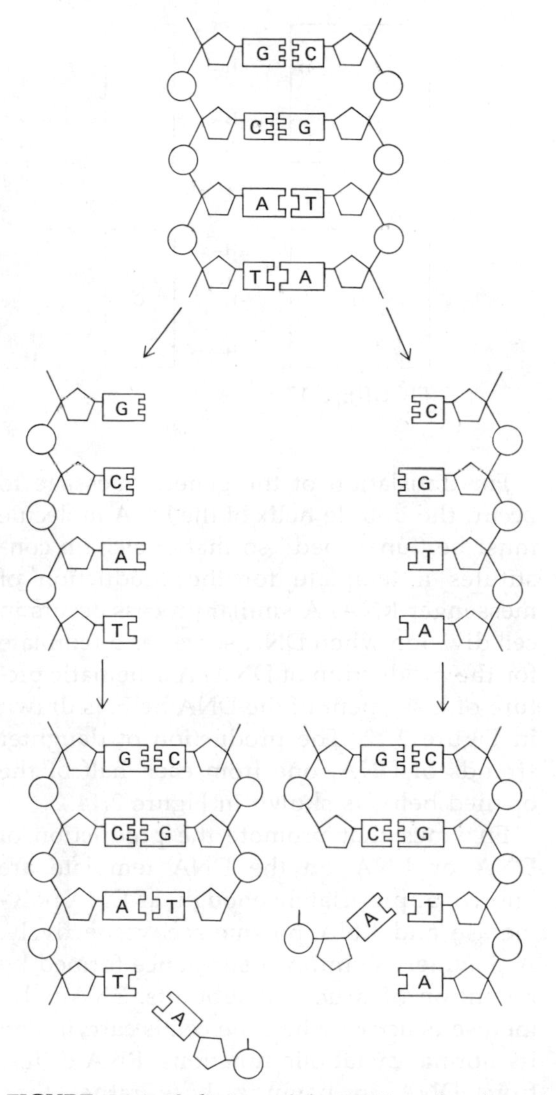

FIGURE 2.14 Splitting of the DNA molecule and synthesis of daughter strands of DNA: pentagons represent pentose sugars; circles represent phosphate groups.

mRNA constitutes in its turn a template for the ordering of amino acids into the polypeptide chains of proteins. This is the translation process. There are 20 amino acids that are commonly found in proteins. In man, as in other animals, some amino acids cannot be synthesized but are the products of the breakdown of food. But now we must ask, how do four possibilities (the nucleotides) specify which of 20 possibilities (the amino acids) will be placed in each position in a linear sequence?

Once the structure of the DNA molecule became known, it was quickly suggested that what is called a "three-letter, nonoverlapping code" was sufficient as a genetic code. A three-letter code means that nucleotides, taken in groups of threes, can provide at least 20 combinations. This can be visualized as drawn in Figure 2.15 where the three-letter codes are represented as three slots. Each slot can be filled with any of four nucleotides. This gives $4 \times 4 \times 4 = 64$ different combinations. This is more than enough to specify 20 amino acids, but a two-letter code giving $4 \times 4 = 16$ would be insufficient. That the code is nonoverlapping means that these *codons,* units of three nucleotides, are read as units having a specific beginning and end. This in turn implies that "punctuation" exists in the genetic message. Furthermore, since 64 possibilities lead to only 20 separate results, some different codons must specify the same amino acid. The genetic code is termed, for this reason, a *degenerate* code.

This purely deductive approach has been borne out completely by experiment. The full catalog of RNA codons and their amino acids was completed in 1968. This is given as Figure 2.16. So far as is known this code is universal for all living organisms.

FIGURE 2.15 Three of the 64 possibilities of four letters (nucleotides) taken three at a time.

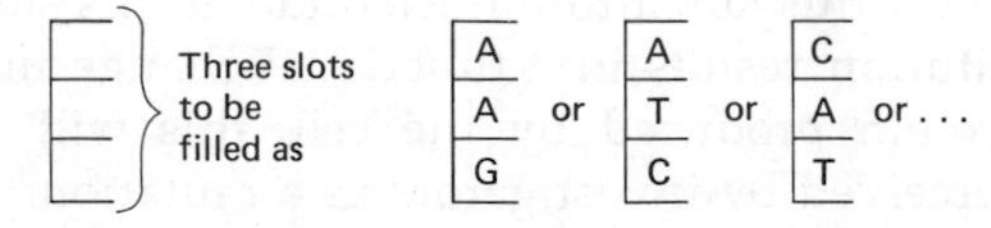

Codons	Amino acid	Codons	Amino acid
UUU, UUC	Phenylalanine	UAU, UAC	Tyrosine
UUA, UUG, CUU, CUC, CUA, CUG	Leucine	CAU, CAC	Histidine
AUU, AUC, AUA	Isoleucine	CAA, CAG	Glutamine
AUG*	Methionine	AAU, AAC	Asparagine
GUU, GUC, GUA, GUG*	Valine	AAA, AAG	Lysine
UCU, UCC, UCA, UCG	Serine	GAU, GAC	Aspartic acid
CCU, CCC, CCA, CCG	Proline	GAA, GAG	Glutamic acid
ACU, ACC, ACA, ACG	Threonine	UGU, UGC	Cysteine
GCU, GCC, GCA, GCG	Alanine	UGG	Tryptophan
		CGU, CGC, CGA, CGG	Arginine
		AGU, AGC	Serine
		AGA, AGG	Arginine
		GGU, GGC, GGA, GGG	Glycine
		UAA**, UAG**, UGA**	Stop

FIGURE 2.16 Codons and the amino acids they specify. Triplets known to function as "punctuation" are indicated by * for polypeptide initiation and ** for termination. (A polypeptide is a string of 10 to 100 amino acids.)

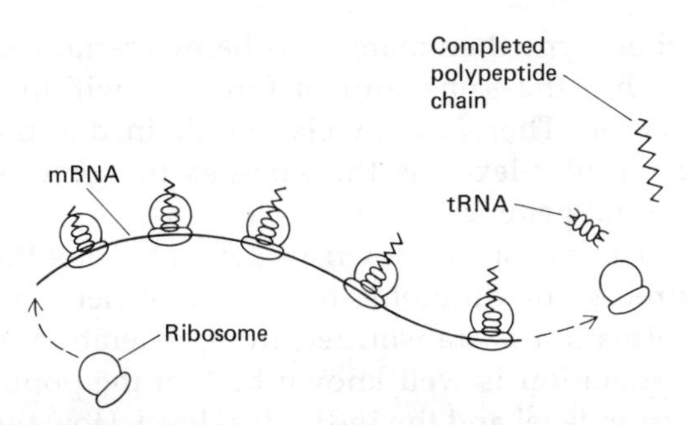

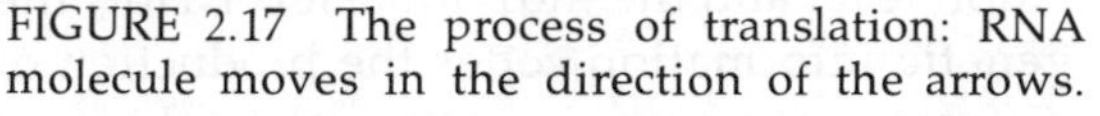

FIGURE 2.17 The process of translation: RNA molecule moves in the direction of the arrows.

Amino acids within the cell do not link up directly with mRNA. Another variety of RNA, a much smaller molecule than mRNA that is called transfer RNA (tRNA) or soluble RNA (sRNA), exists in a variety of forms in the cytoplasm. These tRNA molecules are also produced by DNA. Each kind of tRNA will attach to a specific amino acid at one end and to a specific codon of mRNA at the other end. Therefore the sequence of codons of the mRNA strand determines the sequence in which tRNA will align the amino acids. This occurs, with the mediation of enzymes, only at the surface of ribosomes that are, therefore, the "factory" sites for the production of proteins. Figure 2.17 depicts this process of translation.

Several ribosomes move down the length of the mRNA, each coiling off a polypeptide chain. Many of the proteins produced in this manner are enzymes that go on to promote the many metabolic steps characteristic of the continuous production and breakdown of substances within the cell.

In microorganisms where millions of offspring may be produced in a short period of time, it is possible to map structure within the gene. In such cases the limits of the unit commonly called the gene can be quite closely determined. This operationally defined gene has been termed the *cistron*. When the commonest alleles at each of two locations on the DNA molecule are on the same chromosome they are in the *cis* position. When they are on the opposite (homologous) chromosome they are in the *trans* position. An example of this is given as Figure 4.1 in Chapter 4.

If the *cis* configuration produces the same phenotype as the *trans* configuration, the two loci are in functionally independent regions of the DNA molecule and are considered different *cistrons*. If *cis* versus *trans* configurations produce a difference in phenotype then material is being rearranged within the same unit of function, within a cistron. Therefore the cistron, defined at the molecular level, is the same as the gene of formal genetics.

The major unknown in gene action is the process of differentiation. How genetic information is transmitted from generation to generation is well known both at the population level and the individual level. How the genetic information codes the production of different proteins is now relatively well known as described in this chapter. The feedback relationships between the environment, particularly the intracellular environment, and the genes is not well known. Such feedback regulation of genes must exist to develop specialized cells and for tissue differentiation to occur.

We do know that genes affect the rate of action of other genes, as does the environment. This has led to a distinction between *structural* genes and *controller* or *regulator* genes. The distinction is a bit artificial since both genes code protein, but the protein of regulator genes has as its primary function the repression of other genes. The model of genetic control receiving the most attention today is that of the *operon*. One or a number of adjacent genes controlled by an operator is called an operon. The operator, a segment of the DNA, acts as a gate permitting or preventing transcription from all of the structural genes within that operon. Diffusable repressor substances that form complexes with specific operator regions prevent the mRNA transcription. These repressor substances are produced by other genes called *regulator* genes.

It is the regulator gene that is, in turn, influenced by the intracellular environment. Specifically the regulator genes are influences by the concentrations of certain metabolites. A metabolite that is an *inducer* is so named because it inactivates the repressor gene, permitting the synthesis of mRNA. A *corepressor* activates the repressor gene therefore shutting down transcription. This system is depicted in Figure 2.18. The inducers may be ingested substance or hormones or proteins produced by other genes.

A number of experiments with microorganisms have confirmed this model of the operon, but how general this form of genetic control may be we do not know. Nor do we know how many other controller devices may be present in microorganisms or in man. It may well be that much more of the genetic material in higher organisms is devoted to regulation than in the microorga-

FIGURE 2.18 Operon schematic containing two structural genes, *A* and *B*. Arrows indicate the direction of influence.

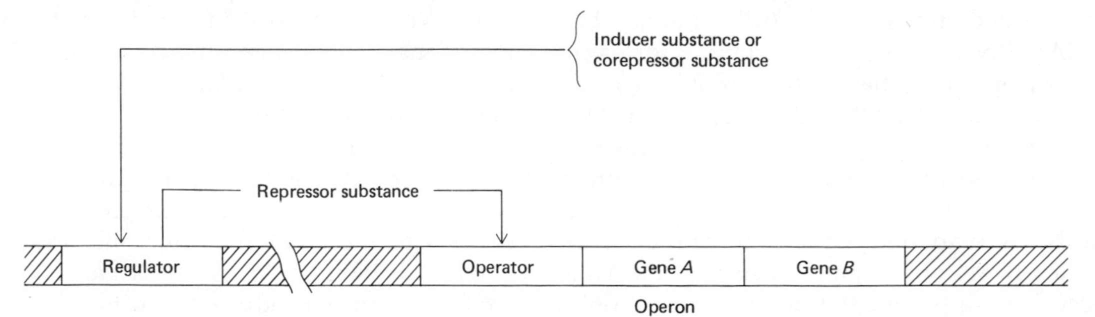

nisms in which these systems are most readily investigated. The one model presented here, that of the operon, is sufficient to show how the environment may regulate gene action by turning on genes (by means of inducer substances) or turning off genes (by means of corepressor substances).

The potential amount of phenotypic variation, of a Mendelian sort, is dependent on the number of structural genes in the species. One of the earliest estimates of the number of genes in man was 10,000 to 15,000. Some later estimates of the number of genes in man are 10^6 to 10^7. These later estimates were obtained by assuming the average gene to be approximately 1500 nucleotides in length and comparing this to the amount of DNA in the human chromosomes. But not all DNA may be involved in mRNA production, and of that which is, a large portion may have regulatory functions.

Another important implication of this view of gene action is that simple deductive hypotheses relating, in one-to-one fashion, a single allele and a single environmental variable can only rarely be expected to be fully valid. Most phenotypes, whether defined at a morphological or a molecular level, will be affected by a number of genes. This may be at the controller gene stage where one repressor substance may influence more than one operon. It may be at the stage of interacting gene products where a given metabolic substrate is changed in a sequence of enzyme controlled steps that are related, in turn, through networks of such steps. This is not said in order to discourage. What it does mean is that a realistic goal in the study of natural selection is the explanation of some part of the observed variation in frequency of a gene. Complete explanation will involve, in most cases, multiple variables and will not be soon in coming.

The view of gene action presented in this chapter, interestingly enough, has some implications for the descriptive rules used, during the past century, to interpret fossils. These will be mentioned in later chapters.

REFERENCES CITED

Benzer, S. 1956 The elementary units of heredity. *In* W. D. McElroy and Bentley Glass (eds.), *The Chemical Basis of Heredity*. pp. 70–93. Baltimore: Johns Hopkins Press.

Stern, C. 1957 The problem of complete Y-linkage in man. *American Journal of Human Genetics* 9:147–166.

FURTHER READINGS

Srb, A., R. Owen, and R. S. Edgar 1965 *General Genetics*, 2nd ed. San Francisco: Freeman.

3

Evolution in Populations

Evolution is a change in the inherited characteristics of a species or a population. Since genes are the material of inheritance, we may speak of evolution as a change in the genetic composition of a population. The genetic composition of a population can be described totally if we know the relative frequency of each allele at all loci. Of course, we can never have such complete data, but the idea provides the basis of a commonly used definition of evolution: Evolution is a change in gene frequency in a population.

This simple definition of evolution is very important. It means that a theory of evolution, at the genetic level, can be handled completely by a model of gene frequency change. That is, if we can specify all of the factors that can contribute to change in gene frequency in a population, we have all of the "causes" of evolutionary change. Further, remembering the boat analogy of Chapter 1, if we can quantify the effects of each of these factors we have a very powerful explanatory model; one permitting tests of specific hypotheses about change in a given population at a given time.

Such a model is really quite simple. There are only five factors which can contribute to gene frequency change within any population. These are

1. Selection
2. Mutation
3. Random gene drift

4. Gene flow
5. Nonrandom mating

These will be described in turn and illustrated. But before defining and discussing each of these factors let us look more critically at this undefined unit we have been talking about, the population.

THE BREEDING POPULATION

What is meant by "population" in this context? Obviously it is the unit which evolves. Individuals are not thought of as evolving during their lifetime. Gene frequency change occurs at the level of the population, specifically the *breeding population.* The breeding population is a group of individuals who intermarry or intermate. This is used as the basic unit of our evolutionary model because it constitutes the "genetic deck" which is reshuffled and dealt out anew each generation. In other words, it is the collection of genes from which new individuals are made up in each subsequent generation. The new generation is a sample of genes from the breeding population, sorted into new combinations.

In many animal species it is difficult to know the size or extent of the breeding population. Physical anthropologists are lucky in this respect in that the mating patterns in human groups are easily studied. We can often specify the size and location, within limits, of the intermarrying group, the breeding population. On an island or in an isolated area, the breeding population may be only a village or a collection of a few villages. And all the fertile, adult members of the village may be members of the same breeding population. In large metropolitan areas the structure of breeding populations is more complex. A number of such populations will live in the same geographic area but limit their intermarriage along religious, educational, skin color, and other ethnic differences. Breeding population sizes in man may range from 50 individuals to tens of thousands of individuals.

In the genetic model of evolution developed in this chapter, we will start with the simplest kind of breeding population, the random mating population. A random mating population is sometimes designated a *panmictic* population, and these terms will be used as synonyms.

Random mating within human breeding populations is not as extreme an assumption as might be assumed at the first encounter with this concept. A number of people today know their ABO blood type. But this almost never influences their choice of mate. Therefore within their breeding population they marry at random with respect to ABO blood type. The same is true for the dozen or so other blood group systems, and for thousands of loci within the human genotype. An exception to this may occur in those genetic systems affecting the external appearance of a person. In this case choice of a mate can be influenced by such features and result in a form of nonrandom mating called *assortative mating,* a concept we will return to later.

The importance of random mating, for our model, is as follows: Random mating means that the gametes (therefore the genes) contributed by the female members of the breeding population unite at random with the gametes (therefore the genes) contributed by the males of that breeding population. We are speaking, of course, of the males and females of a single generation. This means that the zygotes formed—the individuals of the next generation—are chance combinations of the genes of the parental generation.

What is so important about this? It is

simply that, knowing gene frequencies in the parental generation, we can predict genotype frequencies in the offspring generation. From this point we can build fairly elaborate quantitative models of gene frequency change (evolution) on the basis of the five factors of change listed previously.

FROM GENE TO GENOTYPE

An illustration is now in order. Let us assume a breeding population of 50 married couples. In this hypothetical population we look at a single hypothetical locus, the G locus. There are two alleles at the G locus which we will designate *G* and *g*. The proportion of alleles (the gene frequency) of *G* is 0.60. Since there are only two alleles the frequency of *g* must be 0.40. In other words, the frequency (or proportion) of all the alleles at any locus must add up to 1.00. By convention we usually indicate gene frequency by a bar over the symbol for allele, that is, $\overline{G} = 0.60$, $\overline{g} = 0.40$

Random mating can be depicted by means of a 2 × 2 matrix (Figure 3.1). The genes contributed by the females are shown in the two columns. The genes contributed by the males are shown in the two rows. Each intersection of a row and a column indicates the zygote produced by the union of those gametes.

FIGURE 3.1

	♀ G (0.6)	♀ g (0.4)
♂ G (0.6)	GG (0.36)	Gg (0.24)
♂ g (0.4)	Gg (0.24)	gg (0.16)

The numbers at the head of each row and column are the gene frequencies, but they are also the probability of that gene being contributed in any randomly selected mating. Where column and row intersect the product of the two probabilities is shown. This is the probability of producing the genotype shown.

In other words, given the *gene* frequencies

$$\overline{G} = 0.60$$
$$\overline{g} = 0.40$$

in the parental generation, we expect the *genotype* frequencies

$$\overline{GG} = 0.36$$
$$\overline{Gg} = 0.48$$
$$\overline{gg} = 0.16$$

in the offspring generation. The frequency of *Gg*, the heterozygote, is obtained by summing the two ways of forming a heterozygote.

Please note that the genotype frequencies sum to 100 percent, like the gene frequencies, as they must since we are talking about relative frequency.

The reader will note that there are similarities between the matrix of uniting gametes just shown and those shown in Chapter 2. There is an important difference. The matrices of Chapter 2 depicted the gametes of one couple. In such a case the two alleles contributed by one individual are always assumed to be in equal proportions. This leads to the simple 1 : 2 : 1 ratio, or the 3 : 1 ratio, or other simple ratios in the production of zygotes. But we are now talking about populations and random mating. In this case the proportion or relative frequency of the zygotes produced is determined only by the gene frequencies in the parents. This will seldom be in the simple ratios of Chapter 2 and, if so, only by chance.

Going back to the numbers in our illustration, please note that all that we have done is to square a binomial. In other words

$$(0.6 + 0.4)^2 = (0.36 + 0.48 + 0.16)$$

The numbers on the left of the equation represent gene frequencies in the parental population. The numbers on the right represent genotype frequencies in the offspring generation. To show this in yet another way, we write

$$\overline{GG} = \overline{G}^2$$
$$\overline{Gg} = 2(\overline{G})(\overline{g})$$
$$\overline{gg} = \overline{g}^2$$

where the (generalized) numbers to the left represent the genotype frequencies of the offspring generation; those to the right, the gene frequencies of the parent generation.

GROM GENOTYPE TO GENE

Now let us go the other way. Assume that the 100 parents of the preceding illustration produced 100 offspring in the expected proportions. What is the gene frequency in this offspring generation?

If the hypothetical G locus represents a system in which we can identify the heterozygous *Gg* individuals separately from the homozygous *GG* and *gg* individuals, we can get gene frequencies very easily. We simply count them and calculate the proportions. In our illustration

36 *GG* individuals contain 2(36) = 72 *G* alleles.
48 *Gg* individuals contain 48 *G* alleles and 48 *g* alleles.
16 *gg* individuals contain 2(16) = 32 *g* alleles.

So there are a total of 120 *G* alleles and 80 *g* alleles in the offspring generation for a total of 200 alleles, as there must be at an autosomal locus among 100 individuals. The gene frequencies are

$$\overline{G} = 120/200 = 0.60$$
$$\overline{g} = 80/200 = 0.40$$

We conclude that there has been no change in gene frequency from that of the parental population.

THE HARDY-WEINBERG LAW

It is an important fact that the preceding conclusion is not changed by the presence of dominance in the genetic system. All that changes is our way of calculating gene frequencies. To see what happens let us now assume that *G* is dominant to *g*. The dominant phenotype we will simply represent as *G*__ to indicate that we know that at least one allele is *G*, but we do not know about the other. The numbers of individuals in the offspring generation would be

*G*__ = 84
gg = 16

since *GG* and *Gg* are indistinguishable.

If we make the assumption that the proportions of zygotes expected from random mating are, in fact, present in this sample, we can still calculate gene frequency. To do this we note that the one distinguishable homozygous category, *gg*, has the frequency $gg = 0.16$. And, given our assumption $\overline{gg}$ should equal $(\overline{g})^2$. In other words

$$\sqrt{\overline{gg}} = \sqrt{0.16} = \overline{g} = 0.40$$

We then get the frequency of the other allele as the remainder, $\overline{G} = 1.00 - 0.40 = 0.60$.

What we have just developed is a rather wordy version of the Hardy-Weinberg law. In 1908 the mathematician Hardy and the geneticist Weinberg independently published articles showing that dominance alone will not change gene frequencies. They showed that, as long as certain conditions are met, gene frequencies will be maintained in a stable equilibrium from generation to generation with the simple relationship shown above between gene frequency and genotype frequency. The genotype frequencies are said to be in Hardy-Weinberg proportions when

genes

$$G, g \leftrightarrow GG, Gg, gg$$

genotypes

gene frequencies

$$a, b \leftrightarrow a^2, 2ab, b^2$$

genotype frequencies

holds true.

FACTORS OF CHANGE

Hardy-Weinberg proportions will occur, and no change in gene frequency will occur, if the following conditions are met by a population:

1. All genotypes survive equally well and are equally fertile.
2. Genes or chromosomes do not spontaneously change structure.
3. The breeding population is very large.
4. There is no migration or mating with different populations.
5. Mating is random.

When any one of these conditions is not met we have the possibility of change due to, respectively

1. Selection
2. Mutation
3. Random gene drift
4. Gene flow
5. Nonrandom mating

Each of these will be discussed briefly.

Selection

Natural selection is the factor in evolutionary change made famous by Charles Darwin. It is sometimes equated with all of evolution although, as you now know, this is an error. Selection is, however, extremely important and is, in many ways, the most interesting factor of change.

Selection occurs when the different kinds of genotypes do not, on the whole, reproduce equally well. The process of selection can be thought of as occurring in two ways, through differential mortality or differential fertility.

Differential fertility and differential mortality are usually combined in a single measure of selection. This measure is the differential representation among genotypes in the offspring generation after the effects of mutation, migration, and so on are eliminated. One convention for representing such differences is to let the most fit genotype or genotypes have a fitness value of unity and express the fitness of other genotypes as some fraction of this. For instance, we have

Genotypes	*AA*	*Aa*	*aa*
Fitness Values	1.00	0.90	0.80

where we are saying that the *Aa* genotype is intermediate in fitness between the two homozygotes. The definition of fitness just given means, in this example, that the ratio of *Aa* offspring to parents will be only 90 percent of the ratio of *AA* offspring to parents. Similarly the ratio of *aa* offspring to parents is only 80 percent of the *AA*.

The quantitative relations between genotypes, unless further qualified, represent true evolutionary change and not cyclic fluctuations within generations. The latter can occur if one allele is favored early in the life cycle and the other allele late in the life cycle with the net result of no change. In other words, the relative proportions of genotypes shown assume that the numbers of each genotype present are counted at comparable times in the life cycle for both parent and offspring generations.

To carry the example further assume that

p (the frequency of A) = 0.40
q (the frequency of a) = 0.60
N = 1000 individuals

then

Genotypes	AA	Aa	aa
Number	160	480	360
Fitness Values	1.00	0.90	0.80
	160	432	288

The last line, the products, represent the relative genotype numbers among offspring. This does not imply that selection decreases the absolute size of the population. The numbers in the offspring could be, just as easily

AA	Aa	aa
320	864	576

and the result would be the same as far as change in gene frequency is concerned. Counting genes gives us, for the offspring generation

$$q_1 = \frac{432 + 2(288)}{2(160 + 432 + 288)} = 0.57$$

Letting Δq represent change between generations,

$$\Delta q = q_1 - q_0 = 0.57 - 0.60 = -0.03$$

The frequency of a has decreased by this amount in one generation.

The rate of change in gene frequency, Δq, can be stated in general form by letting letters represent general quantities for fitness values. For purposes of understanding this chapter, only two or three specific results need be noted. We can represent dominance with respect to fitness as follows

Genotypes	AA	Aa	aa
Fitness Values	1	1	$1 - s$

where s, the coefficient of selection, represents the proportion by which aa is underrepresented in the next generation, relative to the other genotypes. In such a case the frequency of the a allele decreases at the rate

$$\Delta q = -\frac{spq^2}{1 - sq^2}$$

Selection against an allele does not necessarily mean it will be eliminated from the population. There are various ways in which an equilibrium can be achieved among the various forces causing gene frequency change. The "classic" example of a simple equilibrium exists when there is a balance between the rate at which mutation gives rise to a deleterious recessive allele, and the rate at which that allele is being eliminated by selection. The equilibrium frequency of the deleterious allele, in such a case, is given by

$$\text{equilibrium } q = \hat{q} = \sqrt{\frac{\mu}{s}}$$

where μ is the mutation rate and s is the coefficient of selection against the homozygous recessive.

Another important case of an equilibrium is that of heterosis or heterozygote advantage. This is the case in which the heterozygote is the most fit genotype. We can represent the selection coefficients in such a case as follows

Genotypes	AA	Aa	aa
Fitness Values	$1 - s_1$	1	$1 - s_2$

Neither allele will be lost completely and the equilibrium values for the alleles will be determined by the relative magnitude of selection on the two homozygous classes. Specifically

$$\hat{q} = \frac{s_1}{s_1 + s_2} \quad \text{and} \quad \hat{p} = \frac{s_2}{s_1 + s_2}$$

A rather dramatic example of heterosis, the sickle-cell trait, will be given in Chapter 5.

Mutation

In order to handle mutation in a quantitative fashion the assumption is made that the mutability of an allele is, to some extent, constant between populations or between generations. This assumption permits us to speak of a mutation rate characterizing a given allele. The mutation rate, μ, is the proportion of gametes per generation showing newly arisen mutants at a given locus. Since μ is a rate, the total number of newly arisen mutants is a function of both the mutation rate and the number of genes on which mutation operates. The case of mutation

$$A \xrightarrow{\mu} a$$

means that the A allele mutates to the a allele at a rate μ per generation. Unidirectional mutation will raise q, the frequency of a. But there may be a detectable rate of back mutation, represented by ν, such that change is proceeding both directions.

$$A \underset{\nu}{\overset{\mu}{\rightleftharpoons}} a$$

In this case an equilibrium will be reached where

$$\hat{p} = \frac{\nu}{\mu + \nu}$$

$$\hat{q} = \frac{\mu}{\mu + \nu}$$

But mutation rates are usually quite small and back mutation rates perhaps only a tenth as large. Therefore we usually do not expect an equilibrium on the basis of opposing mutation rates alone, as the other forces causing change will override such an equilibrium.

There are mutation rate estimates for alleles in a number of genetic systems in humans. These are not the same from system to system. Lumping them to get an average mutation rate has a number of methodological difficulties, but the figure of 1.2 times 10^{-5} is frequently cited and will be used when we need such an average. A major point in obtaining such a figure is to show that it is a small number. Some investigators feel that a correct average rate will be only 10^{-6} or less. Mutation must be viewed primarily as a source of genetic novelty, the raw material on which other evolutionary forces operate. Mutation can be shown to be important in determining the frequency of rare, deleterious genes but is generally of minor importance in determining the frequency of common genes.

Gene Drift

Random gene drift refers simply to chance deviations from what is expected. It is the same thing that the statistician calls sampling error. If, due to chance, genes are not transmitted in exactly the proportions expected this will result in gene frequency change. This is gene drift.

For a very simple and extreme example let us suppose we have a breeding population of only two individuals. Let us further suppose that the male and female are both heterozygous *Aa* individuals. They produce two offspring to replace themselves for the next generation. In terms of genotypes there are 6 possible pairs of offspring this couple could produce, these are (*AA*, *AA*), (*AA*, *Aa*), (*Aa*, *Aa*), (*Aa*, *aa*), (*aa*, *aa*) or (*AA*, *aa*).

Gene frequencies in the two-person parental generation are $\overline{A} = 0.5$ and $\overline{a} = 0.5$. But only two possible outcomes in the offspring generation maintain these gene frequencies, these are (*Aa*, *Aa*) and (*AA*, *aa*). But, by chance, the other outcomes can occur. Indeed, in such a small population, gene drift is highly likely.

Breeding populations are never so small as this but, as was noted, may get down to

50 interbreeding individuals. In a small, isolated population genetic drift may be expected to lead to fixation or loss of an allele, given enough time. Fixation refers to a gene frequency of 1.0. After fixation or loss has occurred there can be no more variation unless an allele is reintroduced. This factor has been referred to as the "decay" of variability. As a rule of thumb, we assume that random drift effects can be important in changing gene frequencies in groups where the breeding population is less than 200 individuals. This means that drift could be important in many preagricultural societies.

"NEUTRALISTS" VERSUS "SELECTIONISTS"

It appears that most morphological variation that we observe is molded quite closely by natural selection. Variation at the level of the molecule, however, seems to have a large component of randomness involved. Some amino acids, perhaps a large proportion, on the surface of a protein molecule serve only to maintain the structure of the molecule but have little or nothing else to do with the function of the molecule. They can be replaced by almost any other amino acid. They are, therefore, selectively neutral. The importance of random drift among selectively neutral alleles controlling molecule structure is so great as to have been labeled "non-Darwinian" evolution.

On the other hand, our best estimates indicate that no more than 10 percent of the mutation affecting molecular structure is, in fact, selectively neutral. How do we reconcile this with the observation that much, or even most, of the variation we see in a molecule is random variation? There is really no paradox in this. Some 90 percent or so of the mutations which occur are deleterious and are eliminated by natural selection. Therefore they contribute very little to the variation we can observe in a molecule at any given time. Similarly, the less than 1 percent which may be selectively advantageous mutants replace the preexisting forms and contribute to variation only during the period of replacement. The 1 to 10 percent of truly neutral alleles remain long in the population, drifting to various frequencies, and contribute a great deal to the variation we observe at the molecular level.

Gene Flow

Gene flow occurs when immigrants move from one population to another or when a visitor leaves offspring in a different population. Both processes result in the movement of genes from one population to another. If the two populations differ genetically, then gene frequencies change in the recipient population.

Isolation among human populations is a relative thing. They have been subject to gene flow at all times in the past. The mobility of individuals today makes gene flow the most important single factor in gene frequency change in most populations.

Nonrandom Mating

There are two major sources of nonrandom mating, *inbreeding* and *assortative mating*. Each will be discussed only briefly. As noted earlier some genetic systems manifest themselves in observable ways and can influence a person's choice of mate. Such features might be stature, skin color, eye color, general body build, or other features to which particular groups attach cosmetic or group-identification importance. Such deviations from random mating are termed assortative mating.

Assortative mating occurs on the basis of choice of phenotype. It can be divided into *negative assortative mating* and *positive assortative mating*. Negative assortative mating occurs when mates differ phenotypically more than would be expected by chance. But the old adage, unlikes attract, seems untrue for physical traits. Anthropologists have virtually no evidence of negative assortative mating in man. (The exception to this, of course, is gender itself where complete negative assortative mating is obligate.)

Positive assortative mating occurs where "like mates like." We do find positive assortative mating in human populations. But, again, not very strongly assortative. Perhaps a major reason for this is that strong positive assortative mating divides a breeding population into two populations. For example if, within a population, taller people start marrying only taller individuals and shorter people marry shorter people there will soon be two separate populations, one short and one tall. There can be, and often is, strong positive assortative mating on the basis of physical features which distinguish breeding populations. There are commonly labeled "racial" differences. They are more accurately labeled simply as genetic differences characterizing breeding populations or ethnic groups. This will be discussed further in Chapter 6.

The other major category of nonrandom mating is *inbreeding*. The genetic effects of inbreeding are similar to positive assortative mating. Both increase the frequency of homozygous genotypes at the expense of heterozygotes, relative to Hardy-Weinberg proportions.

Inbreeding is not determined by phenotypic similarity but by genealogical relationship. A person is said to be inbred if his or her parents are known to be related (i.e., a consanguineous marriage—a marriage between relatives). A *population* is said to be inbred if more people are the product of consanguineous marriages than would be expected given random mating.

Heavy inbreeding, like strong positive assortative mating, can in theory divide a population into a number of inbred lines or separate breeding populations. But even with preferential cousin marriage, a common cultural feature of many societies, inbreeding does not reach such high levels. The most important genetic effect of inbreeding is to increase homozygosity. Deleterious genes, if recessive, are "sheltered" from the action of natural selection when they are rare. For example, an allele of frequency $\overline{a} = 0.01$ will be in homozygous form with frequency $\overline{aa} = 0.0001$, given random mating. But among the offspring of first cousins the *aa* genotype, and the associated recessive trait, will be seven times as common. As can be seen from this example deleterious recessive alleles will be exposed to the action of selection far more frequently in the inbred portion of a population than in the noninbred portion.

PART TWO

HUMAN DIVERSITY AND ADAPTATION

Earlier textbooks in physical anthropology presented a picture of human diversity in quite a different manner than will be found here. Earlier books utilized racial taxonomies to characterize human diversity. The preferred approach was to first present the fossil record of evolution down to the time that skulls could be classified into the same set of racial categories as used for classifying living groups. From this point on the description of human diversity was a description of races.

Practitioners of this approach usually presented a list of measurements and observational characteristics (height, weight, skin color, and so forth) of peoples typifying each racial category. These were often accompanied by a set of pictures of people with interesting faces and, occasionally, more interesting body forms. Why this approach to human diversity has been dropped is the subject of Chapter 6.

Within these pages no attempt is made to describe or catalog all of human diversity today, even for a single trait. Specific examples of differences between individuals or between groups will be used to illustrate ways of studying and, hopefully, understanding human differences.

4

Microevolutionary Studies and Human Variability

The techniques used to compute gene frequencies and gene frequency change described in the last chapter make it possible for us to discuss what the anthropologist calls microevolutionary studies. The distinction between micro- and macroevolutionary studies is ostensibly the magnitude of the changes being investigated. However the critical variable is time. Evolution, which is almost always very slow change, results in macrochanges over a long period of time. We study such changes through the fossil record. Microchanges can, on the other hand, occur in a few generations or even a single generation. This means that the student of human evolution may witness evolutionary processes in action and validate hypotheses about such processes. Once validated these generalizations aid in understanding the larger, long-term changes that we see in the fossil record.

In this chapter some genetic systems that have proved useful in microevolutionary studies of man will be described. This does not mean that genetic studies are the only way to do microevolutionary studies. Measurable morphological change can in many cases be shown to occur in one or a few generations. Some of the earlier studies of this kind were carried out by Franz Boas. In 1894 he presented evidence that the offspring of French Canadians and Amerindian Canadians were taller, on the average, than either of the parental populations even though they were intermediate in some other characteristics (Boas, 1894). These unions were also apparently more fertile than others. These findings have been interpreted as evidence for overdominance (where greater heterozygosity is associated with greater size) and heterosis (heterozygote advantage with respect to fitness).

Another of Boas' findings was that the cephalic index—the ratio of head breadth to head length—of immigrants to the United States changed in the direction of the cephalic index of the settled Caucasian population of the country (Boas, 1911). This demonstration of the environmental changeability of the cephalic index sounds rather insignificant today but was important because this and a few other metric indices had been in use up to that time in setting up racial taxonomies. These taxonomies were used, in turn, to construct pseudohistories under the preevolutionary assumption that the traits or indices used had a "God-given" value that changed only with racial intermixture.

An interesting recent example of a morphological change that might be considered microevolutionary is the finding by Benoist that French-descended residents of Saint-Barthélemy (French West Indies) have, over the past 300 years, come to have a higher nasal index (due to great nasal width) than is found in other European or European-derived populations. The two small breeding isolates on Saint Barthélemy have also come to differ physically from one another in ways noticeable to themselves (Benoist, 1964).

This finding presents the possibility that natural selection has operated to cause a change in nasal dimensions in this group. The direction of change is that suggested by a world-wide correlation between nasal index and vapor pressure and hypotheses that have been advanced on respiratory physiology and climate. There are a number of obvious problems to be solved before a natural selection hypothesis can be confidently accepted. From population genetics we know what alterative hypotheses must be eliminated, such as mutation, gene flow, and so on. But the major problem is the old one of how much the morphological change is due to environmental change and how much to genetic change. Nasal breadth is highly correlated with palatal breadth. Nasal length is correlated with most linear measures of man, including head length, stature, and even middle finger length. Could a nutritional change result in differential growth rates to account for the change in nasal index without a genetic change?

This and other problems must be solved prior to dealing with selection hypotheses. These questions are worth answering and, unfortunately, there are far too few investigators working on them at present. But as a consequence of this situation, it is at present easier and more direct to work with traits whose variants are determined almost entirely by genotype and very little by environment. It is not that delimitable genetic systems are more interesting or are more important in microevolutionary studies. It is just that at the present time they promise more results per career expended in their investigation.

THE BLOOD GROUP SYSTEMS

The blood groups are sets of genetic features (phenotypes) that characterize red blood cells. Blood is composed of a dense suspension of individual cells (red blood cells or erythrocytes, white blood cells or leucocytes, and platelets) in a fluid portion called plasma. If blood is allowed to clot, the remaining straw-colored liquid is called blood serum. Serum is a solution of soluble proteins and salts but does not contain the fibrinogen of plasma that is important in blood clotting.

The ABO System

The first blood group system to be described was termed the ABO system by Karl Landsteiner in 1901. Before this it was known that attempts at blood transfusion were sometimes successful, sometimes disastrous. Landsteiner observed the reactions, *in vitro*, between the red blood cells and plasma of different persons. He found a regular set of agglutination (clumping) reactions and used the results of these reactions to classify persons as either A, B, or O. The serum of A persons would agglutinate the red blood cells of B persons but not of one another. The serum of B persons would agglutinate the red blood cells of A persons but, again, not of one another. Persons classified as O (for null) were so-called because their cells did not react with the serum of either A or B persons although the serum of type O persons would agglutinate the cells of both A and B persons. A fourth group, termed AB, was discovered in 1902. This latter and least common type of red cell was agglutinated by the serum of both A and B persons, but the serum of AB persons agglutinated no cells.

Blood typing, identifying the reaction types or phenotypes in the various blood group systems, depends on the immunological defense system of the body. When foreign proteins are introduced into the circulatory system, they stimulate certain cells of the system to produce small protein molecules, called *antibodies*, which form an important constituent of serum. The foreign particles that stimulate the production of antibodies are called *antigens*. Owing to their configuration, specific sites of the antibodies enter into molecular interactions with the antigens, combining with them and rendering them harmless to the system. The specificity by which antibodies recognize antigenic sites has proven to be a powerful analytic tool in protein chemistry and is the basis for typing the surface antigens of red blood cells.

The ABO system was the first to be discovered because it is the one system in which there are so-called naturally occurring antibodies. This means that a person who is type A always has anti-B antibodies. The person who is type B always has, conversely, anti-A antibodies. This occurs without the introduction of the foreign cells which is normally necessary to induce the production of antibodies. It has become apparent in recent years that A and B substances are very widespread in microorganisms and that the "natural" occurrence of anti-A and anti-B antisera is also an immune reaction. A specific anti-O antiserum is not available. The reactions that are the phenotypes of the ABO system are shown in Table 4.1.

Since no anti-O is used in routine blood

TABLE 4.1
Identifying the ABO Blood Group Types

	ANTISERUM USED		BLOOD TYPE	SERUM ANTIBODIES PRESENT
	Anti-A	*Anti-B*		
Reaction	−	−	O	anti-A, anti-B
	+	−	A	anti-B
	−	+	B	anti-A
	+	+	AB	−

Note: The blood type (red cell type) shown in column 3 is determined by the cell's reactions with the antisera shown in columns 1 and 2. For example, if cells give a positive (+) reaction with anti-A but a negative (−) reaction with anti-B we know that only A antigens are present on the surface and that the cells are, therefore, identified as type A. Column 4 shows the naturally occurring antibodies found in the serum of persons of given cell types.

typing, the heterozygotes involving O cannot be distinguished. If the locus of the genetic determinants of the ABO blood types is designated as *I*, then we can write the genotypes and their corresponding phenotypes as

Genotypes	*Phenotypes*
I^AI^A, I^AI^O	A
I^BI^B, I^BI^O	B
I^OI^O	O
I^AI^B	AB

There are now recognized a number of allelic subdivisions of both A and B. The subtypes A_1 and A_2 (with A_1 dominant to A_2) represent a common polymorphism in human populations today, although Amerindians and a number of small, relatively isolated populations in different parts of the world appear to lack A_2.

The MNS System

In 1927 Landsteiner and Levine discovered the MN blood group system. They produced a *heteroimmune* serum (one in which the antibody producer is of a different species than the antigen producer) as in Figure 4.1. The rabbit forms a mixture of antibodies against several antigens of the human red blood cells. This mixture can be rendered specific by adsorbing many of the antibodies to the surface of red blood cells of known antigenic type for removal from the serum. For example, by putting into the serum the red blood cells of another person who tests the same as the donor of the stimulus red blood cells, many antibodies will be attached to these red blood cells and removed. When this procedure no longer produces a reaction the donor cells themselves are introduced to the serum. If a reaction occurs there must be an antigen present in the donor cells which

FIGURE 4.1

is not present in the cells of the individual used in the adsorption procedure. A slightly more elaborate version of this procedure to identify genetic differences was used in isolating the MN system.

Anti-M and anti-N antisera are both available which permits identification of the heterozygote as in Table 4.2.

As shown in family studies, the ABO and MN blood groups are controlled by genetic loci that assort independently. Later the Ss system was discovered. These red cell antigens did not show assortment independent of the MN system. Both anti-S and anti-s are known, although the latter is not as com-

TABLE 4.2
MN Blood Group Types

	ANTISERUM		BLOOD	
	Anti-M	*Anti-N*	TYPE	GENOTYPE
	+	−	M	L^ML^M
Reaction	+	+	MN	L^ML^N
	−	+	N	L^NL^N

monly available. The best evidence now available indicates that the Ss types are controlled by genes at a different—but closely linked—locus to that of the MN types. Therefore there are four "chromosome types" *MS*, *Ms*, *NS*, and *Ns* that assort at meiosis.

The Rh System

By far the largest number of known blood group systems depend on *isoimmune* (antigen and antibody produced by the same species—in this case *Homo sapiens*) sera, discovered when pregnancy complications showed women to have circulating antibodies that reacted with the red blood cells of their own fetuses or newborns. Such an unexpected event can be due in some cases to prior blood transfusions that introduced unidentified antigens into maternal circulation. In some cases the fetal blood cells can themselves induce formation of antibodies in the mother. The latter case is represented by Rh incompatibility.

The Rh, or rhesus system, was named when it was discovered that experimentally produced heteroimmune serum to rhesus monkey (*M. mulatta*) red blood cells would also agglutinate the red blood cells of about 85 percent of the American Whites tested. It was then found that this system was the same as that identified earlier using the serum of a woman whose infant had been stillborn. This potent serum permitted the classification of persons as Rh-positive if their cells were agglutinated by the serum and Rh-negative if they were not. Letting *D* represent the gene for this trait and *d* its allele, we get

Phenotype	*Postulated Genotype*
Rh+	*DD, Dd*
Rh−	*dd*

This particular genetic system is best known to the public because of its connection with maternal-fetal incompatibility leading to hemolytic disease of the newborn. This condition, formerly called *erythroblastosis fetalis*, is a cause of morbidity and mortality among the newborn and provides an interesting kind of selection. A maternal-fetal incompatibility occurs when the fetus has an antigen against which the mother may produce antibodies. For this to occur the original zygote formed could only have received a gene for such an antigen from the paternal gamete. In other words, incompatible pregnancies are a result of incompatible matings.

In the Rh case the mother must be Rh-negative to be able to elaborate anti-D antibodies. The offspring must receive the *D* allele from the father who must then be Rh-positive (Figure 4.2). This means that there are two kinds of matings that can lead to hemolytic disease. These are shown in Figure 4.3. The frequency with which such matings occur and their possible evolutionary significance will be further explored in Chapter 5.

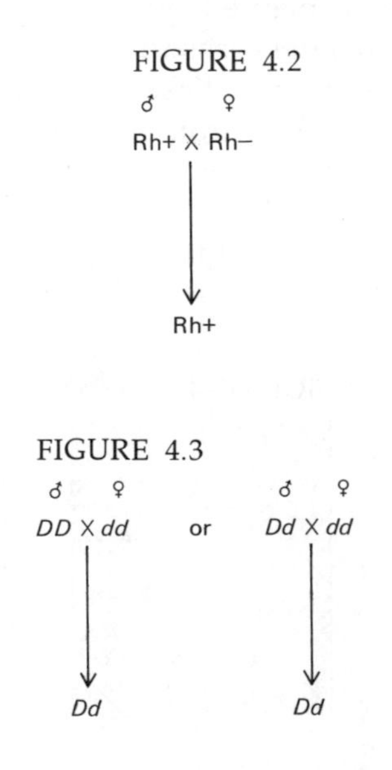

TABLE 4.3
Notational Equivalences Between the Fisher-Race and the Wiener Theories of the Rh Blood Groups

CHROMOSOME TYPE OF FISHER-RACE	GENE OF WIENER
CDE	R^z
CDe	R^1
cDE	R^2
cDe	R^o
CdE	r^y
Cde	r'
cdE	r''
cde	r

The Rh system as known today includes many more reaction types, and there are two competing theories on the genetics of the Rh system. Anthropologists commonly use the notation introduced by Fisher and Race (Race, 1944). In the original Fisher and Race theory, it was suggested that Rh reaction types could be explained by assuming three loci with two alleles at each locus. These loci were designated *C*, *D*, and *E*. Since the phenotypes explained by these loci did not show independent assortment, it was further proposed that these were linked loci. The homologous chromosomes would appear as in Figure 4.4 which means that there could be $2 \times 2 \times 2 = 8$ possible chromosome types that could combine to give 36 different Rh genotypes.

FIGURE 4.4

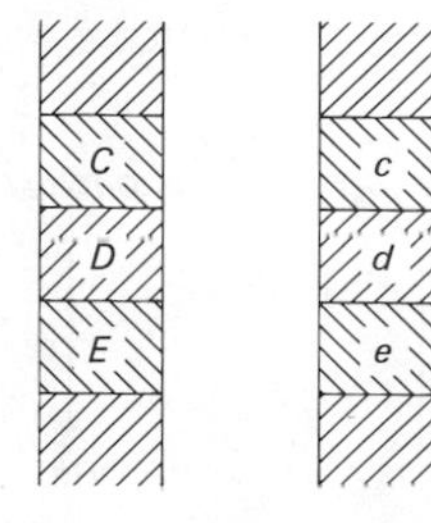

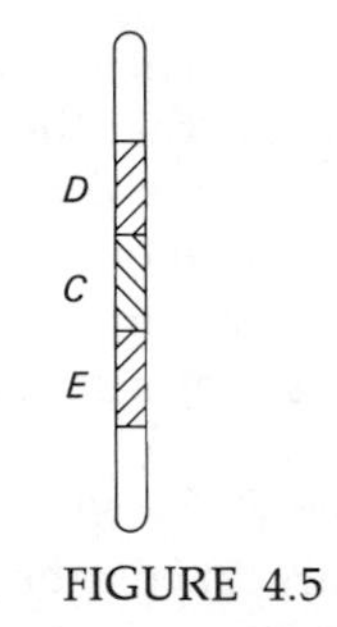

FIGURE 4.5

Since, after many families had been investigated, no unequivocal cases of cross-overs had been found, it was assumed that these loci were very, very closely linked. On the other hand there is apparently evidence for deletions, which indicates that the proper order should be that of Figure 4.5.

The theory of A. S. Wiener postulates a single genetic locus with multiple alleles (Wiener, 1943). But Wiener's theory gets much more complicated in that, although he posits a one-to-one relation between gene and antigen (agglutinogen), he also states that the antigens can have, in turn, multiple specificities or "factors." Another way of describing this situation is to say that one gene produces a site on the surface protein of the red cell to which more than one kind of antibody can attach. Table 4.3 compares the genetic notation of the Fisher-Race and the Wiener systems.

Since in most cases, these two theories predict similar results, there has been no critical test that could decide between them. Recently Race and Sanger have suggested that the *C* and *E* loci are within the same cistron (gene) (Race and Sanger, 1962). They seem to have evidence of a naturally occurring *cis trans* test (described in Chapter 2) in which an anti-ce antiserum will react with the red cells when the *c* and *e* genes are on the same chromosome (the *cis* position) but not when they are on opposite chromosomes (the *trans* position). A similar situation was found to hold for an anti-ce antiserum. This situation is shown in Figure 4.6.

If the "loci" of the Fisher and Race theory are within a cistron, the system approximates the theory of Wiener much more closely. There are many more difficulties to be worked out in the Rh system and there are many known variants of the Rh system that have not been mentioned in this discussion. Whether one takes this as discouraging news or as the promise of a future rich with new discoveries is a matter of personal choice. It is the case that the more we know about a number of blood group systems, the more their complexity comes to resemble that of the Rh system.

The Secretor System

Some persons show antigens of the ABO system in their saliva and other body fluids. Others do not. This "secretor status" of a person is controlled by alleles at a locus unlinked with the ABO locus. Since the ABO blood group substances are involved, this is also considered a blood group system. Secretor is dominant to nonsecretor as follows:

Genotypes	*Phenotypes*
SeSe	secretor
Sese	secretor
sese	nonsecretor

The person with at least one *Se* gene produces a water soluble ABO substance, whereas the *sese* homozygote produces only alcohol soluble ABO substance. This is the reason that ABO substance is found in many fluids of the secretor. It was noted earlier that the type O person has no identifiable O-antigen on his red cells against which antibodies are formed. If he is also a secretor, the type O person does, however, show the presence of a specific antigenic substance in his saliva. This is termed the H-substance. The H-antigen is identified by an anti-H found in saline extracts from seeds of the legume, *Ulex europaeus* (gorse).

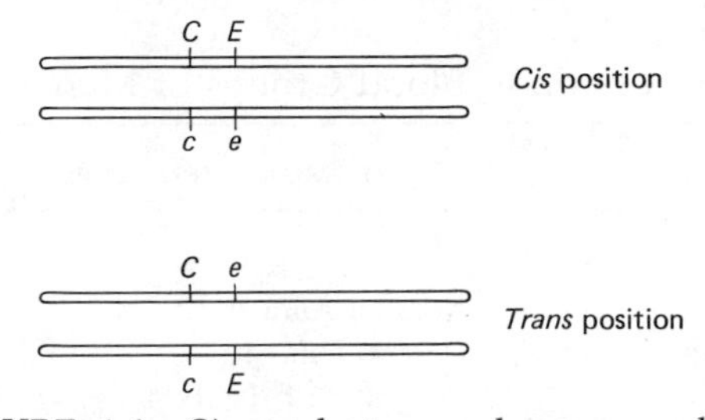

FIGURE 4.6 *Cis* and *trans* chromosomal arrangements.

The pace of discovery of new blood group systems, and new alleles in known systems, has continued to accelerate. Table 4.4 is an attempt to present, in summary form, the genetics of the best known of these systems. This table does not include the "private" blood group systems known for only one or two families. Each represents a significant polymorphism in at least one but usually most populations. As such, the existence and distribution of each system represents a problem or a set of problems in the dynamics of evolution.

HEMOGLOBINS OF MAN

The blood groups just discussed represent characteristics of the protein coat of the red cell or of proteins adsorbed to this protein coat. On a world-wide basis, we also find that the hemoglobin within the cells shows a great deal of genetic variety. Hemoglobin is a respiratory pigment (it gives the red color to erythrocytes) that transports oxygen to the capillary beds of body tissues and picks up carbon dioxide for return to the lungs. A heme group of hemoglobin is an organic molecule (protoporphyrin) containing an iron molecule. The globin fraction of

TABLE 4.4
Some Common Blood Groups of Man

SYSTEM	COMMON PHENOTYPES	POSTULATED GENES	ANTISERA
ABO	A_1, A_1B, A_2B, B, O	A_1, A_2, *B*, *O*	anti-A, anti-A_1, anti-B
Auberger	Au(a+), Au(a−)	Au^a, *Au*	anti-Au^a
Diego	Di(a+), Di(a−)	Di^a, *Di*	anti-Di^a
Duffy	Fy(a+B−), Fy(a−b+), Fy(a+b+), Fy(a−b−)	Fy^a, Fy^b, *Fy*	anti-Fy^a, anti-Fy^b
Kell	K k Kp^a K^b K k Kp^a Kp^b + + + + − − − − + + + − − + − + + − − + − + + + + − + + − + + − + − + − + + − +	K^a, K^b, K^o, k^a, k^b	anti-K, anti-k, anti-Kp^a, anti-Kp^b
Kidd	Jk(a+b−), Jk(a−b+), Jk(a+b+), Jk(a−b−)	Jk^a, KJk^b, *Jk*	anti-Jk^a, anti-Jk^b
Lutheran	Lu(a+b−), Lu(a+b+), Lu(a− b+)	Lu^a, Lu^b	anti-Lu^a, Lu^b
MNSs	MMSS, MNss, MMSs, NNSS, MMss, NNSs, MNSS, NNss, MNSs	*MS*, *Ms*, *NS*, *Ns* (closely linked)	anti-M, anti-N, anti-S, anti-s
P	P_1 P_2 + + − + − −	P_1, P_2, *p*	anti-P_1 anti-P_2 (also called anti-PP_1)
Rh	C c D E e C c D E e + − − − + + + −+ − + + − − + − + − + + − + − − + − + − + − + − + − + + − + + + + + + − + + − + + − − + + − + + + + + + + − − + + + + + + − + − − + − − + + + − + + − + + − + + + +	R^z *CDE* R^1 *CDe* R^2 *cDE* R^o *cDe* r^1 *CdE* r' *Cde* r'' *cdE* *r* *cde* (Wiener) (Fisher & Race)	
Xg	Xg(a+), Xg(a−)	Xg^a, *Xg*	anti-Xg^a

Note: Three ways of denoting phenotype are used. A phenotype given as MMSs indicates a positive reaction to anti-M, anti-S, anti-s, and a negative reaction with anti-N. A phenotype Jk(a+b−) indicates a positive reaction to anti-Jk^a, a negative reaction to anti-Jk^b. The Rh phenotypes are indicated by + or − symbols under letters denoting reactions with specific antisera, hence the inferred presence of the antigen.

hemoglobin is a large protein molecule of four polypeptide chains. The oxygen transport ability of hemoglobin depends on the ability of the iron atoms in the heme groups to combine reversibly with oxygen.

Hemoglobin is one of the proteins that has been studied well enough so that we know quite precisely its primary structure (the sequence of amino acids in each polypeptide chain and the number of chains), its secondary structure (the rotational symmetry of a chain about its axis), and its tertiary structure (the folding of the chains to produce a roughly spheroid molecular form). Such detailed knowledge of protein structure is due to recently developed analytical methods in protein chemistry. In particular, the chromatographic techniques have proven most useful in the identification of genetic variants in hemoglobin and other proteins.

The simplest illustration of chromatography is to place the nib of a fountain pen, containing washable black ink, on a wet tissue paper. Most such inks are mixtures of pigments that, on wet paper, will spread or migrate at different rates depending on their molecular size. This will produce bands of different color on the paper. A similar technique can be used to separate a mixture of proteins. They are applied to a wet filter paper, allowed to migrate, then stained for identification. This is called paper chromatography and can be used to separate proteins of different molecular size.

An extension of this technique is electrophoresis. Proteins may differ not only in molecular size but also in charge. By wetting the filter paper or other supporting medium in an electrolyte and passing a current from one edge of the filter paper to the other, proteins can be made to separate depending on the electrical charge differential between molecules as well as on the pH (hydrogen ion concentration) of the electrolyte. In hemoglobin work, filter paper strips or colloidal media such as starch gel slabs are the supporting medium. The use of this and similar techniques has permitted the identification of over 100 hemoglobins.

There are three genetically specified kinds of hemoglobin found in erythrocytes of "normal" individuals. These are hemoglobin A, A_2, and F. F stands for fetal hemoglobin. These hemoglobin variants differ in the polypeptide chains that join to form the completed molecule. Hemoglobin A (Hb A), the major component of normal adult hemoglobin, is a molecule of symmetrical halves, each half composed of one α and one β polypeptide chain. In other words, each Hb A molecule is made up of four chains, two designated as α chains, two designated as β chains. Each of these chains enfolds a heme group with Fe^{2+}. Each α chain is composed of 141 amino acid units. Each β chain is composed of 146 amino acid units.

Hemoglobin variants are designated in two ways: either by letters of the Roman alphabet, as was the initial practice, or by a structural description. Both are shown below:

	Letter Designation	*Structural Designation*
Adult	Hb A	$\alpha_2\beta_2$
Fetal	Hb F	$\alpha_2\gamma_2$
A_2	Hb A_2	$\alpha_2\delta_2$

The structural designation shows that all three variants have the same α chains and differ in whether they have β, γ, or δ chains.

Early in uterine life most of the hemoglobin in circulation is Hb F. It appears that Hb F has a higher affinity for oxygen than does Hb A *in utero* which should facilitate oxygenation of fetal blood across the placenta. About half-way through fetal life the production of γ chains begins to decrease at the same time that the production of β chains increases. One to two months after birth, the percentages of Hb A and Hb F are approximately equal. By the sixth or seventh month of life, Hb A has approached the adult proportion of 95 to 96 percent, while Hb A_2 may constitute 1 to 3 percent, and Hb F soon reaches a level of less than 0.5 percent of the total.

As was mentioned, although Hb A is the normal form of adult hemoglobin, there are now over 100 known "abnormal" hemoglobins that have a hereditary basis. Some of these varieties are known only for a single family but others, such as Hb C, Hb D, Hb E, Hb S, and the thalassemias occur in sufficiently high frequencies in some populations to be important in populational studies of human evolution.

SICKLE-CELL ANEMIA

Sickle-cell anemia, first described in 1910, is a condition in which the red blood cells take on irregular shapes, some reminiscent of the shape of a sickle blade. Such sickle-shaped cells do not flow as readily through fine capillary passages as do normal red cells and are easily broken open. This destruction of red cells results in chronic anemia plus occasional severe or massive destruction of red cells. It appears that up to the present many such "sicklers" died in infancy, and almost none reached adulthood to reproduce. In 1949 J. V. Neel showed that sickle-cell anemia was inherited as a simple autosomal recessive condition. The apparently normal heterozygote was referred to as the "carrier" or as having the sickle trait.

Genotype	*Phenotype*	*Referred to as*
Hb A/Hb A or *AA*	normal	normal
Hb A/Hb S or *AS*	normal	trait carrier
Hb S/Hb S or *SS*	sicklemia	sicklemia or sickle-cell disease

This provides another illustration of the statement that "dominant-recessive" refers to the phenotype, not to the genotype, and is relative to the method used to observe the phenotype. Clinically the *AS* individual is like the *AA* individual unless he happens to be at high altitude where the partial-pressure of oxygen is low. If, however, we place the red blood cells of *AA* and *AS* individuals

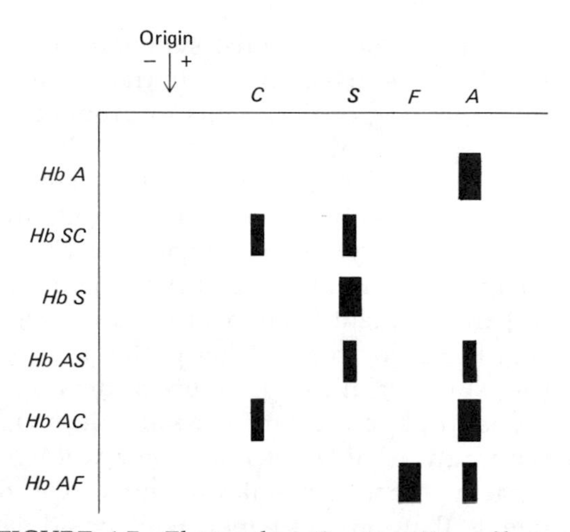

FIGURE 4.7 Electrophoretic separation of hemoglobin from individuals who are genotypically homozygous *A* and *S*, and heterozygous *SC*, *AS*, *AC* and fetal *AF*.

on a glass slide and seal them under cover plates with sodium metabisulfite, lowering the oxygen, some of the cells of the *AS* individual will assume sickle shapes; those of the *AA* individual will not. With this method of observing phenotypes the heterozygote is distinguishable and we no longer have "normal" dominant to "sickle." An even better way of observing the phenotypes of hemoglobin is by means of starch gel electrophoresis illustrated in Figure 4.7. As can be seen in that figure, the *AS* individual does not produce a new, hybrid protein but produces both Hb A and Hb S. Usually the latter hemoglobin constitutes 25 to 40 percent of the total in the *AS* individual.

In 1956 Vernon Ingram was able to show that Hb A differs from Hb S as a result of the substitution of a single amino acid in the sixth position of the β chain of hemoglobin (Figure 4.8). A single gene difference results in the substitution of valine for glutamic acid to produce Hb S. A different allelic gene re-

Hb A:	val –	his –	leu –	thr –	pro –	glu	– glu – lys . . .
Hb S:	val –	his –	leu –	thr –	pro –	val	– glu – lys . . .
Hb C:	val –	his –	leu –	thr –	pro –	lys	– glu – lys . . .

FIGURE 4.8 Amino acid sequences in the one peptide in which hemoglobins *A*, *S*, and *C* differ. (For a complete list of amino acids see Figure 2-16.)

sults in the substitution of lysine for glutamic acid producing Hb C. The majority of normal hemoglobins differ by a single amino acid substitution. These include α as well as β chain variants. There are also abnormal hemoglobins that represent unusual combinations of the polypeptide chains. Whether these are due to mutations of structural genes (genes that specify the amino acids in a protein) or to mutations of controller genes (genes that regulate other genes) is not yet settled. Four such unusual chain combinations are compared to Hb A in Table 4.5.

Abnormal hemoglobins usually reach appreciable population frequencies (one percent or higher) only in tropical or semitropical parts of the Old World. The relationship of this distribution to the distribution of certain blood parasites will be discussed in Chapter 5. The distribution of some of the most common hemoglobin variants is given in Figures 4.9, 4.10, and 4.11.

TABLE 4.5
Some Unusual Chain Combinations in Four Hemoglobins Compared to Normal Hemoglobin

HB TYPE	STRUCTURAL DESIGNATION
A	$\alpha_2\beta_2$
H	β_4
Barts	γ_4
α	α_4
δ	δ_4

THALASSEMIA

Thalassemia (from Greek *thalasses*, "the sea") is an inherited anemia first recognized as common in Greece, Italy, and other regions of the Mediterranean littoral. It is now known in Africa, Asia Minor, South Asia, and Oceania (Figure 4.11). Thalassemia has also been called Cooley's anemia. People with thalassemia major, assumed to be homozygotes, seldom live to adulthood. People with thalassemia minor, assumed to be heterozygotes, show a milder anemia and are able to live fairly normal lives. Indeed, many thalassemia heterozygotes may show no symptoms of anemia.

We now know that there is not one, but a number of thalassemias. The various thalassemias have in common a number of characteristic changes in the erythrocytes, the most important of which are the occurrence of microcytic (very small) and hypochromic (deficient pigment) erythrocytes. These symptoms reflect a deficiency in the amount of hemoglobin in the cells. The polypeptide chains present in the various thalassemias are electrophoretically normal. The genetic block or blocks involved affect the rate of synthesis of hemoglobin.

In thalassemia and, to a lesser extent, sicklemia, there are changes in bone marrow, hyperplasia – abnormal increase in the number of cells – which is accompanied by radial striation detectable in radiograms of the skull, thinning of the cortical bone, and osteoporosis – decreased density of the bones. Osteoporosis occurs even with thalassemia minor. This leads to the possibility that changes in the frequency of thalassemia over long periods of time can be

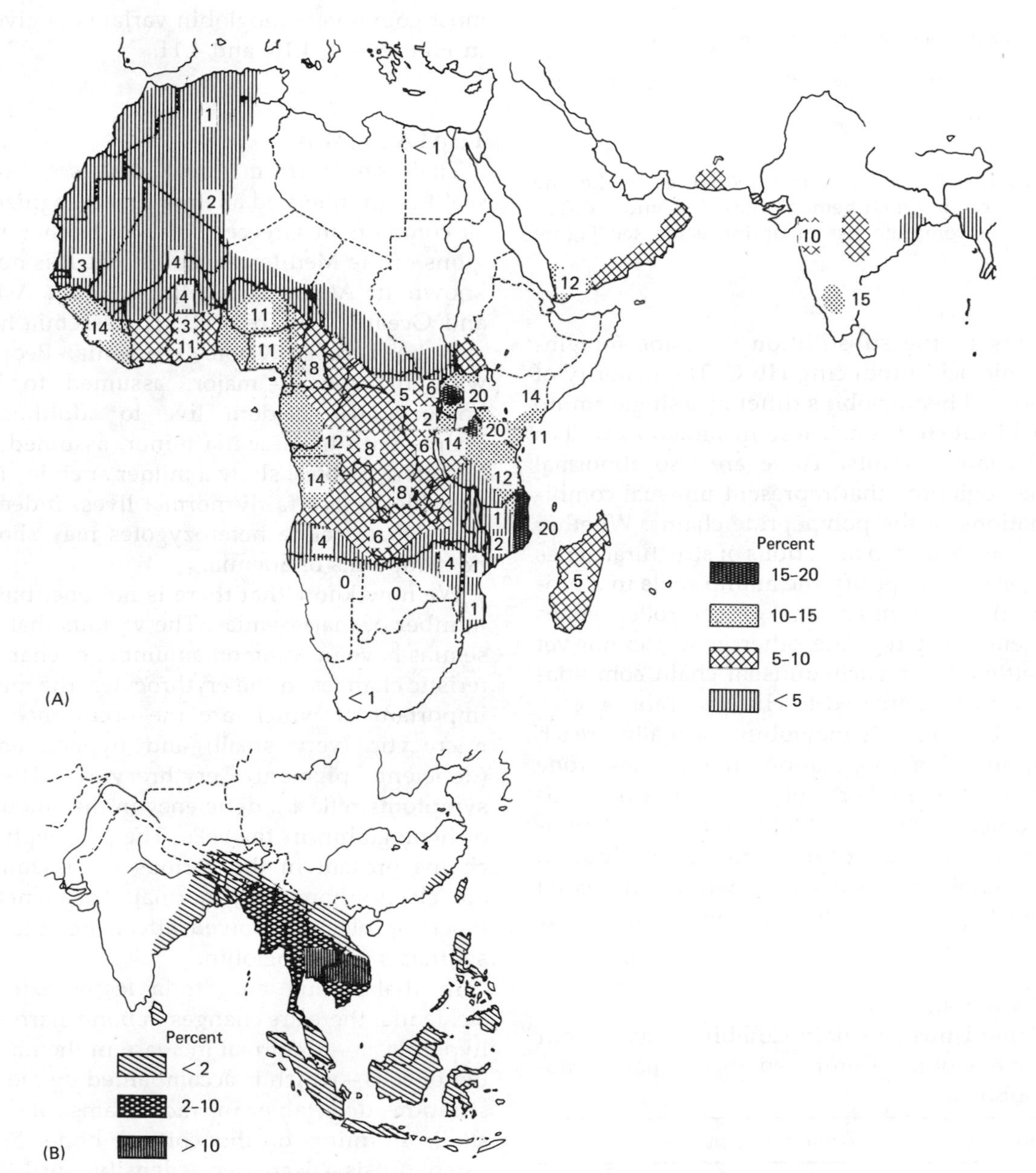

FIGURE 4.9 Known Old World distributions of hemoglobins *S* and *E* frequency of (A) sickle-cell gene in Africa, Arabia, and southern Europe, (B) hemoglobin *E* gene in southeast Asia. (*Source:* Allison, 1961. Used with permission.)

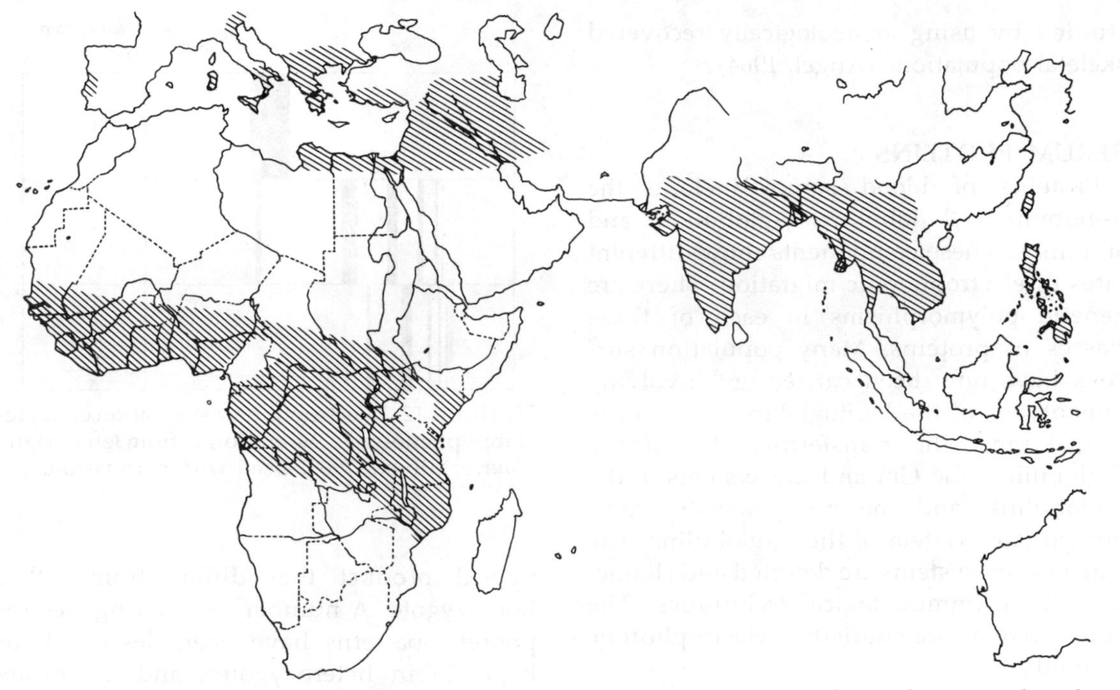

FIGURE 4.10 Known Old World distribution of greater than 2 percent incidence glucose-6-phosphate dehydrogenase deficiency. (*Source:* Allison, 1961. Used with permission.)

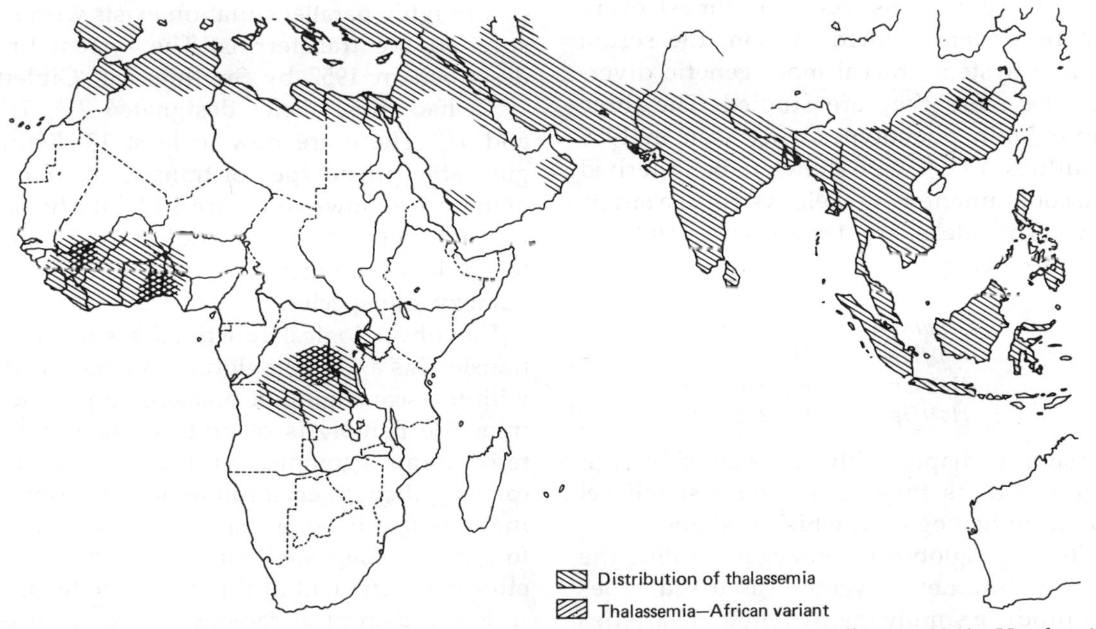

FIGURE 4.11 Known Old World distribution of Thalassemia. (*Source:* Allison, 1961. Used with permission.)

studied by using archeologically recovered skeletal populations (Angel, 1964).

SERUM PROTEINS

Proteins of blood serum include the α-globulins, β-globulins, γ-globulins, and albumins. These components show different rates of electrophoretic migration. There are genetic polymorphisms in each of these classes of proteins. Many population surveys have now been carried out involving phenotypes of the haptoglobins (Hp) of the α_2-globulins, the transferrins (Tf) of the β-globulins, the Gm and Inv systems of the γ-globulins, and the group-specific component (Gc) system of the α_2-globulins. The Gm and Inv systems are detected and characterized by immunological techniques. The other systems are studied by electrophoretic methods.

The haptoglobins and transferrins have been studied more extensively than have the other three systems. As with almost every genetic system studied in man, the serum protein systems reveal more genetic diversity the more they are studied. The three haptoglobin phenotypes, first seen in 1955 (Smithies, 1955), were shown to be inherited as a codominant, two-allele system at a single locus. The alleles are designated as Hp^1 and Hp^2 producing

Genotype	*Phenotype*
Hp^1/Hp^1	Hp 1-1
Hp^2/Hp^1	Hp 2-1
Hp^2/Hp^2	Hp 2-2

These three haptoglobin types are shown in Figure 4.12 as they appear on a starch gel bound to hemoglobin which is stained.

The haptoglobin heterozygote, unlike the hemoglobin heterozygotes discussed, does not produce simply the two protein varieties of the homozygotes but instead produces a hybrid product that differs from either homozygote. A number of differing electrophoretic patterns have been described for haptoglobin heterozygotes, and it appears that a number of alleles exist for the haptoglobin system.

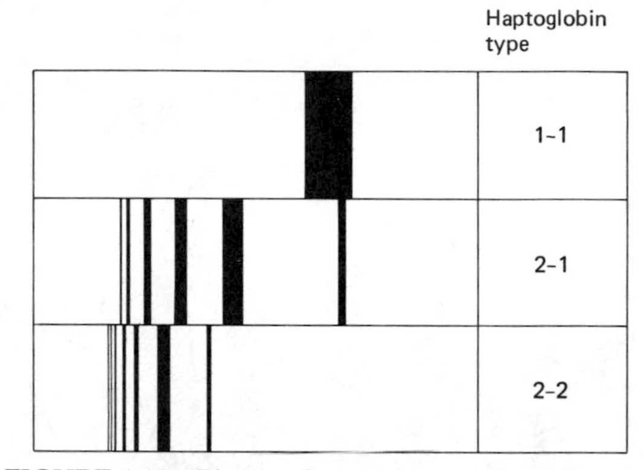

FIGURE 4.12 Electrophoretic separation of haptoglobin phenotypes. Migration is from left to right. (*Source:* Giblett, 1969. Used with permission.)

A roughly parallel situation exists with respect to the transferrins. The system first described in 1957 by Smithies (cf. Giblett, 1962) had three alleles designated Tf^B, Tf^C, and Tf^D. There are now at least 18 distinguishable phenotypes of transferrins. Some of these are shown in Figure 4.13. Like hemoglobin—and unlike haptoglobin—the transferrin heterozygotes are a mixture of the homozygote species.

The physiological functions assumed for transferrins and haptoglobin both have to do with the scavenging of breakdown products from the hemolysis of erythrocytes. Transferrin forms a complex with ferric iron ions, carrying them to erythropoietic (blood forming) tissues in bone marrow. It has, therefore, been suggested that transferrin varieties more efficient at this might be favored in lowland tropical regions where acquired or genetic anemias are common and the de-

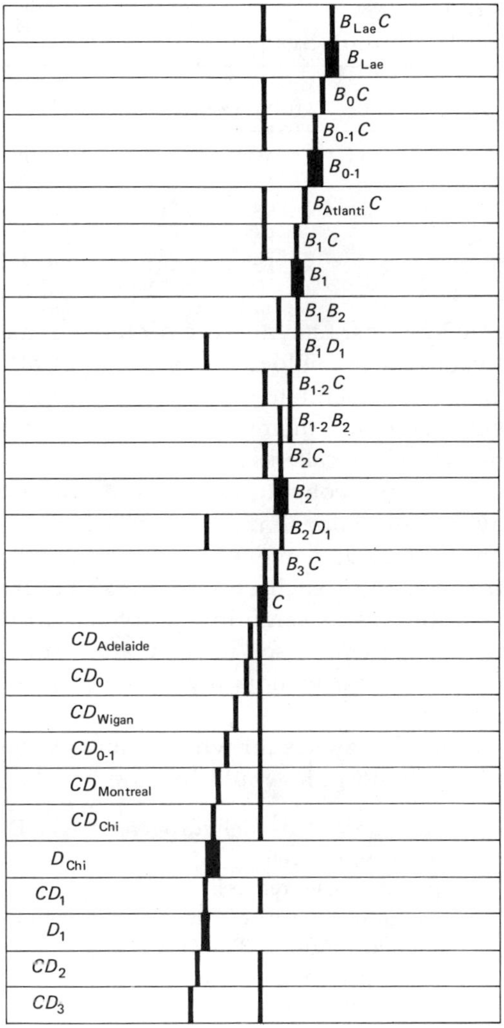

FIGURE 4.13 Electrophoretic separation of transferring phenotypes. (*Source:* Giblett, 1969. Used with permission.)

creased life span of red blood cells might result in iron deficiency. So far no one has been able to demonstrate a difference in iron-binding ability among the transferrin types. Most populations of the world show a frequency of 0.90 to 1.00 of Tf^C.

Haptoglobin combines with free hemoglobin after the lysis (breaking open) of red blood cells. This prevents the passage of free hemoglobin into the tubules of the kidneys. Continued passage of such large molecules as hemoglobin (mol. wt. approx. 67,000) can produce kidney damage. This, as in the reasoning applied to transferrin, suggests that haptoglobin varieties that sequester free hemoglobin most readily should be favored in lowland tropical regions. Haptoglobins do vary in this respect, and Hp 1-1 has the greatest hemoglobin-binding capacity.

In Africa the distribution of the Hp^1 allele is in rough correspondence with an expected frequency gradient of hemolytic episodes. In the tropical rain forest areas of central West Africa, Hp^1 is found in frequencies 0.70–0.90. This decreases to frequencies of 0.20–0.40 in North and South Africa. Buettner-Janusch and Buettner-Janusch (1964) also found a higher frequency of Hp^1 in the lowland coast populations of Madagascar than in the inland plateau-dwelling populations. The latter case may, however, be explained in terms of gene flow alone, since we have other anthropological evidence that seafaring peoples from southeast Asia settled prehistorically in central Madagascar. Outside of Africa we find no obvious relation between the tropics and the frequency of Hp^1.

The Gm and Inv systems reflect differences in the γ-globulin fraction of the serum. The phenotypes in each of these systems are identified by serological techniques in which properly sensitized red cells react with and are coated by the specific γ-globulin components designated the Gm and the Inv types. (See Steinberg, 1962, for a description of the procedures involved.) In their original and simple form, these two systems could be described as is done in Table 4.6. A great many other Gm reaction types (phenotypes) have now been described, and whether they

TABLE 4.6
The Gm and Inv Systems Shown with Two Alleles Each

	PHENOTYPE	GENOTYPE
	Gm (a+b−)	Gm^a/Gm^a
Gm system	Gm (a+b+)	Gm^a/Gm^b
	Gm (a−b+)	Gm^b/Gm^b
	Inv (a+b−)	Inv^a/Inv^a
Inv system	Inv (a+b+)	Inv^a/Inv^b
	Inv (a−b+)	Inv^b/Inv^b

TABLE 4.7
The Gc System

	PHENOTYPES	GENOTYPES
	Gc 1-1	Gc^1/Gc^1
Gc system	Gc 2-1	Gc^2/Gc^1
	Gc 2-2	Gc^2/Gc^2

are all dependent on simple allelic differences or not is unclear at present.

The Gc system, the "group specific component," is an α_2-globulin polymorphism described by Hirschfield in 1959. Three electrophoretically distinguishable phenotypes are determined by two alleles as shown in Table 4.7. There are now known a number of rare alleles in addition to two alleles common in isolated populations. The latter are $Gc^{Chippewa}$ among the Chippewa Indians and $Gc^{Aborigine}$ among Australian aborigines and some New Guinea populations.

Other serum protein polymorphisms include the prealbumins, lipoproteins, and a number of enzyme systems. The prealbumins are so named because they migrate faster than the albumins. The genetics of this system is shown as Table 4.8.

TABLE 4.8
The Prealbumin System

ALLELES	PHENOTYPES	GENOTYPES
Pr^F, Pr^M, Pr^S	MM	Pr^M/Pr^M
	MS	Pr^M/Pr^S
	SS	Pr^S/Pr^S
	FM	Pr^F/Pr^M
	FS	Pr^F/Pr^S

ENZYME POLYMORPHISMS

The most rapidly developing area of new genetic polymorphisms in man concerns enzyme systems. Enzymes are proteins that act as catalysts in biological reactions. In other words, they facilitate such reactions without being used up themselves in the process. They could be called isozyme systems, since each system constitutes a set of proteins having the same enzymatic function, and hence the same name, but differing in structure. Although such enzymes may occur in many tissues, a sampling problem dictates that the enzymes used in population studies must be primarily those identifiable in blood samples.

A list of enzymes for which man is known to be polymorphic would include

- glucose-6-phosphate dehydrogenase (G6PD) deficiency, red cell
- acid phosphatase, red cell
- serum cholinesterase or "Pseudocholinesterase"
- phosphoglucomutase (PGM), red cell
- carbonic anhydrase, red cell
- 6-phosphogluconate dehydrogenase (6GPD), red cell
- serum alkaline phosphatase
- adenylate kinase, red cell
- lactate dehydrogenase
- phosphohexose isomerase (PHI)
- B-aminoisobutyricaciduria (BAIB)

Not all of these have been studied extensively. The best known are the first five or six listed. The last listed is not an enzyme but a urinary excretion product that is a genetic polymorphism and appears to reflect a simple enzymatic difference.

G6PD Deficiency

Red cell glucose-6-phosphate dehydrogenase (G6PD) deficiency is a sex-linked trait occurring in the Old World coextensive with the sickle-cell gene and thalassemia. G6PD deficiency has been shown to be connected with a disease called "favism," which seems to have been present in the eastern Mediterranean region prior to the Christian era and which continues to account for an appreciable mortality today. The term favism describes a mild to severe hemolytic episode that takes place when the victim eats the fava bean (*Vicia fava*) or even inhales the pollen of the plant. Such hemolytic reactions also occur in persons with G6PD deficiency in response to certain drugs, including some antimalarials (termed "primaquine sensitivity," etc.) and even to some bacterial infections.

Since the allele, or alleles, producing G6PD deficiency is deleterious from the standpoint of producing anemia and neonatal jaundice, it is suggested that, as in the case of some of the hemoglobin varieties, it somehow confers an advantage in a malarious environment. Glucose-6-phosphate dehydrogenase, as its name implies, is involved in the metabolism of red cell glucose, specifically in the production of reduced glutathione. Since the falciparum malaria parasite (*Plasmodium falciparum*) also utilizes reduced glutathione, it is suggested that G6PD deficiency may slow the spread of the parasite in the red blood cells.

Our knowledge of this system is expanding rapidly as new techniques are developed and new alleles are identified. That more than one allele might be producing G6PD deficiency was first suggested when African varieties of the condition were shown to be less deleterious than those of southern Europe. There now appears to be a number of alleles that can produce the deficiency in addition to more than one "normal" allele identifiable by electrophoresis.

Red Cell Acid Phosphatase

Red cell acid phosphatase differences were described for Seattle populations in 1963 by Hopkinson, Spencer, and Harris. Five phenotypes were distinguished by starch gel electrophoresis. Family studies showed these to be determined by allelic autosomal genes. The system was described as is done in Table 4.9.

The five phenotypes listed proved on investigation to differ in enzyme activity level, and these fit a hypothesis of additive gene effect in which activity level increases in the order $P^a < P^b < P^c$. Since the P^c allele is least common, occurring primarily in European-derived populations, the homozygote, *CC*, was not identified until 1964. It proved to have the highest activity as predicted from an additive-effect hypothesis. A fourth allele has now been identified and designated P^r. In order of activity $P^r < P^a < P^b < P^c$.

An interesting aspect of the study of red cell acid phosphatase is that the system is a quantitative trait when looked at through the measurement of enzyme activity level; when looked at by means of starch gel electrophoresis, however, it is a qualitative polymorphism (Figure 4.14).

Enzyme activity level, measured in a population sample, shows a continuous, uni-

TABLE 4.9
Phenotypes and Genotypes of the Red Cell Acid Phosphatase System

PHENOTYPE	GENOTYPE
A	P^a/P^a
BA	P^b/P^a
B	P^b/P^b
CA	P^c/P^a
CB	P^c/P^b

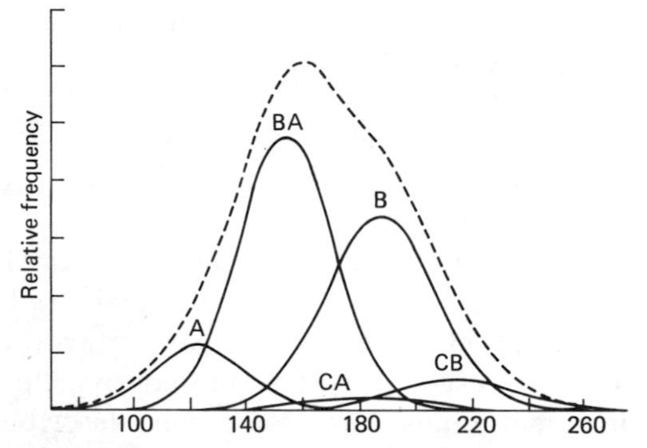

FIGURE 4.14 Red cell acid phosphatase activity: solid lines represent enzyme activity found in the different phenotypes after they have been identified electrophoretically; dashed line represents the unimodal distribution that would be observed if types were not first identified by electrophoresis. (*Source:* Harris, 1966. Used with permission.)

modal distribution (Figure 4.14). The usual explanation of continuous variation that has a pronounced hereditary component is that the variation is "polygenic" or "multifactorial." This means that alleles at more than one locus affect the trait. For example, if three alleles at each of three loci affect a trait there can be, genotypically, 216 kinds of individuals with respect to this trait; at each locus, that is, there will be six possible combinations and there will be $6^3 = 216$ combinations overall. If the total range of variation of the trait is divided up among many genotypes, then successive genotypes will differ by small increments. A small environmental variance superposed on the successive genotype categories will produce, from the observer's point of view, continuous variation from which individual genotype categories cannot be identified.

But the number of genotypic categories need not be great to get this effect. Red cell acid phosphatase provides a good illustration of the fact that alleles at a single locus can produce a "continuous" trait in the manner described. To be exact, we cannot be sure at present that other genetic modifications contributing to the "environmental" variation in acid phosphatase activity are not present. Nevertheless this case suggests that some apparently complex continuous variables in man may prove to have a very simple genetic foundation.

OTHER GENETIC POLYMORPHISMS

Alleles producing defective color vision are found in low frequencies in all large population samples in man. Such alleles reach a frequency of 8 percent in European populations. Color vision defects are apparently controlled by alleles at two loci on the *X* chromosome. These color vision defects are described in terms of the trichromatic theory of color vision. Normal individuals can match any given color by mixing various proportions of three monochromatic light beams—red, blue, and green—on a screen, or perhaps better stated, they require these three colors to be able to match the color discriminations they can normally make. Such a person is therefore referred to as a normal trichomat.

Persons with color vision defects will be unable to distinguish some of the mixtures distinguished by normal trichromats. Persons are classified by whether they can make distinctions based on one, two, or three colors and on the basis of which color they are unable to utilize in making distinctions. Therefore a person can be classified as having trichromatic, dichromatic, or monochromatic vision as well as "protan" (red recognition) deficiency or "deuterans" (green recognition) deficiency. This classification system is shown in Table 4.10.

A good discussion of color vision and a

TABLE 4.10
A Classification of Color Vision Responses

TRICHROMATIC DISTINCTIONS	DICHROMATIC DISTINCTIONS	MONOCHROMATIC DISTINCTIONS
Normal vision		Monochromats
Protanomaly	Protanopia	
Deuteranomaly	Deuteranopia	

more extensive classification of color vision defects is given by Kalmus (1965). The color vision defects listed in Table 4.10 are best described by reference to Figure 4.15.

Protanopes react little if at all to long (red) wavelengths. Protanomalous individuals have alleles that produce a less severe form of the same defect. Deuteranopes perceive all points along the red-green axis as though restricted to yellow. Again, deuteranomalous individuals have an allele of less severe effect. Monochromats react only to brightness differentials and not to wavelength of light. Protan and deuteran defects are lumped together in the lay term "red-green color blindness." The majority of cases of reported "red-green color blindness" constitute deuteranomaly. Family studies indicate that protan and deuteran defects are controlled by alleles at two separate loci, close to the locus for G6PD deficiency, on the *X* chromosome.

Post (1962) has suggested that the alleles for defective color vision are increasing today as a result of a relaxation of selection against such alleles. Post tabulated data on the populations of the world in which red-green color blindness had been investigated. He classified these populations into three categories depending, presumptively, on how long it had been since their ancestors were hunters and gatherers. Averaging the frequency of red-green color blindness in each of these categories he found the following:

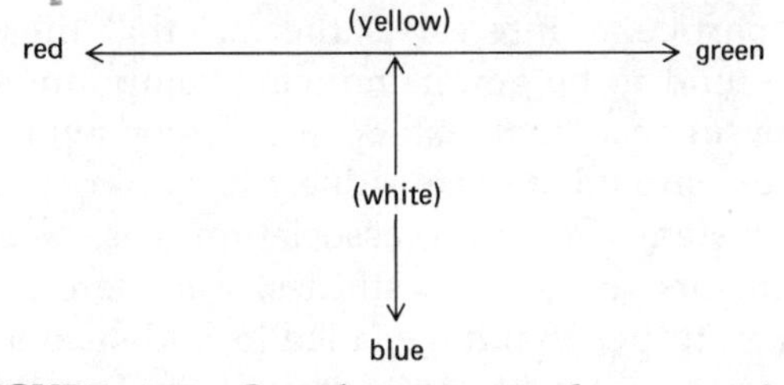

FIGURE 4.15 Complementary colors according to Hering. Red light plus green light gives yellow light. Yellow light plus blue light in proper proportions gives white light. (*Source:* Adapted from Kalmus, 1965. Used with permission.)

	Color Vision Defect Frequency
Hunters and Gatherers	0.020
Intermediates	0.033
Long-Standing Agriculturalists	0.080

To explain these data Post suggests that color defect alleles are disadvantageous in hunting societies but not in agricultural societies. The adoption of agriculture would therefore bring about a relaxation of selection against color vision defects. The primary difficulty with this hypothesis is that it requires, for this amount of change to occur in three to four thousand years, a fairly high mutation rate. Conversely, an equilibrium at $\hat{q} = 0.02$ in hunting groups, in the face of such a mutation rate, requires that we suppose a disadvantage of $s = 0.06$ against color blind individuals (Neel and Post, 1963; Post, 1965).

Another genetic system in man which has been investigated in many parts of the world concerns the ability to taste PTC (phenylthiocarbimide). The genetics of this trait was described in Chapter 2. Tasters or nontasters of PTC are also tasters or nontasters, respectively, of a number of thiourea compounds.

Of particular interest is the fact that these also tend to be goiter-inducing compounds. This suggests that tasters, due to food avoidances, should be less subject to goiter than nontasters. Such an association has been confirmed by several studies. But there are other studies that have failed to find such an association. Such results may depend on the distribution of goitrogen-producing plants versus iodine deficiency in the induction of goiter.

In this chapter we have reviewed some of the genetic systems known and now being studied in man. The 37 or so loci involved must represent only a small part of the genetic diversity in man. Undoubtedly most genetic systems express themselves in continuous variation and, as a consequence, we have no Mendelian analysis of such systems. This means that the number of possible genotypes in man is unimaginably large. Considering only 37 loci, if there are only three alleles at each locus there can be six possible genotypes at each locus and a total number of possible genotype combinations of 6^{37}. How large is this number? It is approximately 6×10^{29}. The present population of the earth is approximately 3×10^{9}, which means that if every person alive today gave rise to another three billion persons, each having a different genotype, this would utilize less than one ten-billionth of the genetic possibilities available at these 37 loci.

Obviously there is no shortage of genetic diversity in man. Consequently there is great potential for evolutionary change. In fact, the problem becomes one of how, in the face of a bewildering number of possibilities, an investigator goes about establishing the existence of fitness differentials or evolutionary change at all. Another difficulty is that the subjects of study are rather long lived, as long lived as the investigator. In spite of these difficulties a small beginning has been made in the study of natural selection in living populations. The results of some of these studies will be presented in the next chapter.

REFERENCES CITED

Allison, A. C. 1961 Abnormal Haemoglobin and Erythrocyte Enzyme-Deficiency Traits. *In* G. A. Harrison (ed.), *Genetical Variation in Human Populations,* pp. 16–40. Oxford: Pergamon Press.

Angel, J. L. 1964 Osteoporosis: Thalassemia? *American Journal of Physical Anthropology* 22:369–374.

Benoist, J. 1964 Saint-Barthélemy Physical Anthropology of an Isolate. *American Journal of Physical Anthropology* 22:473–488.

Boas, F. 1894 The Half-Blood Indian. *Popular Science Monthly* 45:761–770.

Boas, F. 1911 *Changes in the Bodily Form of Descendants of Immigrants. Reports of the Immigration Commission, Senate Document No. 208.* Washington, D.C.: Government Printing Office.

Brues, A. M. 1954 Selection and Polymorphism in the ABO Blood Groups. *American Journal of Physical Anthropology* n.s. 12:559–597.

Buettner-Janusch, J. and V. Buettner-Janusch 1964 Hemoglobins, Haptoglobins, and Transferrins in the Peoples of Madagascar. *American Journal of Physical Anthropology* 22:163–169.

Giblett, E. R. 1962 The Plasma Transferrins. *In* Arthur G. Steinberg and Alexander G. Bearn (eds.), *Progress in Medical Genetics,* vol. 2, pp. 34–63. New York: Grune and Stratton.

Giblett, E. R. 1969 *Genetic Markers in Human Blood.* Oxford: Blackwell.

Harris, H. 1966 Enzyme Polymorphism in Man. *Proceedings of the Royal Society, Series B.* 164: 298–310.

Hirschfield, J. 1959 Immune-Electrophoretic Demonstration of Qualitative Differences in Human Sera and Their Relation to the Haptoglobins. *Acta Pathologica et Microbiologica Scandinavia.* 47:160–168.

Hopkinson, D. A., N. Spencer, and H. Harris 1963 Red Cell Acid Phosphatase Variants: A New Human Polymorphism. *Nature* 199:969–971.

Ingram, V. M. 1965 A Specific Chemical Difference Between the Globins of Normal Human and Sickle Cell Anaemia Haemoglobin. *Nature* 178:792–793.

Ingram, V. M. 1963 *The Hemoglobins in Genetics and Evolution*. New York: Columbia University Press.

Kalmus, H. 1965 *Diagnosis and Genetics of Defective Colour Vision*. Oxford: Pergamon Press.

Landsteiner, K. 1901 Über Agglutinationserscheinungen Normales Menschlichen Blutes. *Wiener Klinische Wochenschrift* 14:1132–1134.

Landsteiner, K. and P. Levine 1927 A New Agglutinable Factor Differentiating Individual Human Bloods. *Proceedings of the Society for Experimental Biology and Medicine, New York.* 24:600–602.

Neel, J. V. and R. H. Post 1963 Transitory "Positive" Selection or Colorblindness? *Eugenics Quarterly* 10:33–35, 84–85.

Post, R. H. 1962 Population Differences in Red and Green Color Vision Deficiency: A Review, and a Query on Selection Relaxation. *Eugenics Quarterly* 9:131–146.

Post, R. H. 1965 Selection Against "Colorblindness" Among "Primitive" Populations. *Eugenics Quarterly* 12:28–29.

Race, R. R. 1944 An "Incomplete" Antibody in Human Serum. *Nature* 153:771–772.

Race, R. R. and R. Sanger 1962 *Blood Groups in Man*, 4th ed. Oxford: Blackwell.

Smithies, O. 1955 Grouped Variations in the Occurrence of New Protein Components in Normal Human Serum. *Nature* 175:307–308.

Steinberg, A. G. 1962 Progress in the Study of Genetically Determined Human Gamma Globulin Types (the Gm and Inv Groups). *In* A. G. Steinberg and A. G. Bearn (eds.), *Progress in Medical Genetics* , vol 2, pp. 1–21. New York and London: Grune and Stratton.

Wiener, A. S. 1943 Genetic Theory of the Rh Blood Types. *Proceedings of the Society for Experimental Biology and Medicine, New York,* 54: 316–319.

5

The Search for Selection in Man

No attempt has been made, as yet, to codify the various approaches to the study of natural selection in man. The primary reason for this is that the data used in studying selection came from many sources and were originally collected for unrelated purposes. This means that the student of human microevolution must use these data by applying any intellectual and methodological device he can find. Only in the past few years have large studies been initiated specifically to investigate human adaptations at the population level.

As a beginning, we may use the following four headings as a classification of kinds of microevolutionary studies: clinical studies, correlational studies, distributional studies, and deductive approaches. By clinical studies I refer to cases where certain conditions are an object of investigation because they result in pathology or have a direct effect on fitness and, in the course of investigation, prove to be genetically based. In the large majority of cases, these conditions will represent rare alleles. They have been kept rare by selection. But in some cases—a well-known example is Rh incompatibility—they are common polymorphisms.

Correlational studies attempt to establish an association between the presence of a particular allele and prevalence of a particular disease. Other correlational studies have attempted to associate presence of an allele with more direct measures of fitness such as

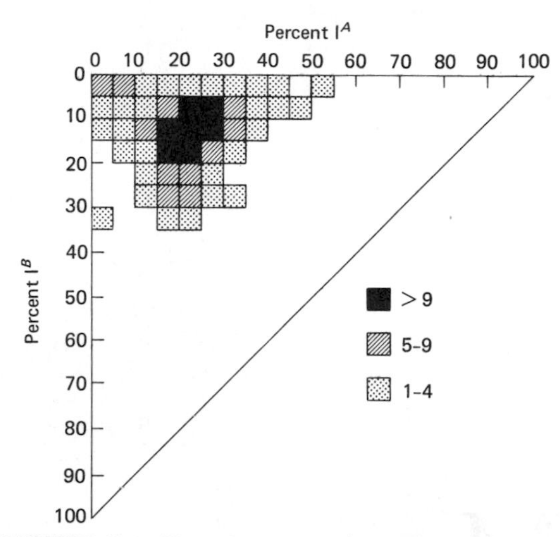

FIGURE 5.1 Distribution of ABO allele frequencies in 215 human populations around the world. The frequency of *O* is simply 1-*A*-*B*.

fertility rates, mortality rates, or general morbidity rates. A little has been done on the effect of parental genotypes.

Distributional studies is a category in which I include both geographic distributions and theoretical distributions based on a priori expectations. Any regular geographic distribution, as shown by some of the charts in Chapter Four, represents problems that must be explained. Migrational theories are occasionally involved and investigated to explain such distributions. Ecological correlates of gene frequency distributions are often sought. Certainly the most successful case of the latter kind of study concerns the sickle-cell gene. A certain hypothesis may lead to a theoretical distribution of gene frequencies. If this distribution is not found, the hypothesis is rejected. As an example of the latter, Brues (1954) reasoned that if selection did not limit the range of ABO blood group gene frequencies, then random gene drift would have led, in the long run, to populations having all possible frequencies of the ABO alleles. This expectation is far from being realized (Figure 5.1.)

In Figure 5.1 the values of the allele frequencies of I^A and I^B are given along the axes. The frequency of I^O is, of course, the difference between unity and the sum of $I^A + I^B$ frequencies. From this Brues makes the following generalizations.

> The *O* gene is rarely reduced below 50%, and then only by a small amount. In the absence or near-absence of the *A* gene, the *B* gene usually remains low. In the absence or near-absence of the *B* gene, the percentages of the genes *A* and *O* are rather widely distributed, from zero to 50% *A*. Where the frequency of the *B* gene is high, on the other hand, the range in respect to *O* and *A* is more restricted, centering around 20% gene *A*. (Brues, 1954, p. 561)

Neither mutation nor drift can explain these regularities. "Admixture," or gene flow hypotheses are inadequate as ABO polymorphisms are ancient, occurring in other primates as well as man. We are left with selective factors as the only serious possible explanation for this distribution.

Finally there are deductive approaches. These rest on inferences concerning the presumed function of a trait or its deduced mode of interaction with selective factors of the environment. An example of this approach might be found in the study of haptoglobin varieties. As mentioned in the previous chapter, Hp 1-1 has a higher avidity for free hemoglobin than do other haptoglobins. Assuming this to be the primary function of haptoglobin, an investigator can set up deductive hypotheses about ecological correlates and proceed to test them.

Deductive approaches are appealing but are unlikely to be fruitful. We can assume that the number of "functional" interactions involving a given trait will generally be large. This means that hypotheses deduced on the assumption of a simple one-to-one relationship of cause and effect will seldom be confirmable even if valid under an "all other things being equal" principle.

A good example of a deduction that is rather ingenious, but which has given equivocal results, concerns the ABO system. Vogel and others (1960) have suggested that some epidemic diseases that have periodically devastated large populations in the past, may alter gene frequencies in a very simple fashion. If the infectious organism has an ABO-like antigen then the host individuals who have this antigen on their red blood cells will not be able to elaborate antibodies to this particular antigen and will therefore be less able to resist infection than other individuals who have the antibody present. Vogel and his associates have pointed to the bubonic plague bacillus as having an H-specific antigen and to smallpox (variola) virus as having an A-specific antigen. From this they deduce that type O individuals (who, if secretors, elaborate H-substance) are selected against by bubonic plague and type A individuals by smallpox.

Attempts to relate the history of plague and smallpox epidemics to the distribution of the ABO blood group alleles have so far given equivocal results. The problem is still being actively investigated, but we lack knowledge of the "prehistory" of these diseases as well as the exact specificity of surface antigens of the disease organisms (Otten, 1967).

Before considering the findings of some studies on natural selection in living populations, we must first consider, and perhaps reject, an aphorism often repeated in textbooks and articles on evolutionary biology. This is the statement that mortality, in order to be effective in selection, must occur prior to or during the reproductive period of the individual. This is based on the quite reasonable idea that two different genotypes, once they have completed their reproductive periods of life, have already made their contribution to the next generation. Differing life spans after this, therefore, are of no genetic consequence. This is generally true. But it is less true for man than it is for other animals.

Let us look at a form of life that is quite different from man in this respect, the Pacific salmon. The salmon is a pelagic fish that returns to a fresh-water stream, which is its spawning ground, after five years in the ocean. It deposits eggs only once and as many as 10,000. Shortly after depositing eggs and milt, the female and male salmon die without returning to the sea. Indeed they are physiologically adapted to just achieve the spawning trip, and no more. If we look at animals that reproduce more than once, we find that they too do not, in general, live for any length of time in a postreproductive period. The reason for this is that postreproductive differential mortality, since it has no genetic effect, cannot select for resistance to those diseases or disabilities whose onset is late in life.

In sexually reproducing species there is a genetic correlation of $r = 0.5$, for autosomal loci, between parent and offspring. This means that the offspring gets half of his genes from one parent. If the postreproductive viability of the parent contributes to the viability or fertility of the offspring, then, as a result of the genetic correlation between them, indirect selection for parental longevity takes place.

Man's normal condition in his evolutionary past was one in which he had no access to what is now considered proper medical and health care. It has not been sufficiently noticed or stressed that populations living under these normal conditions show reduced female reproduction after the age of 35. But even in societies living under very primitive conditions most of the adults alive at age 35 will reach the 45–50 age class. At this age, however, "degenerative disease" appears and life expectancy from this point on, especially in males, is very short. It appears that such a vital history profile is related, through the evolutionary process just discussed, to the length of dependency of the human child.

Having stated this we can now state the obverse where we began the discussion: Mortality, even in the postreproductive period, to the extent that it prejudices the reproductive chances of the offspring, constitutes selection against parental genes. Therefore, especially in man, we cannot accept as a truism the assertion that postreproductive mortality is not selective. By the same token we must be suspicious of the similarly common statement that nonlethal disease is not selective. Due to the economic nature of the family unit any burden placed on it, as a result of chronic disease, can affect the reproductive potential of the offspring. Again this affects the fitness of parental genes. Therefore it seems best, as a general principle, to assume that morbidity itself is a continuum with respect to its effect on fitness even though only one end of this continuum mortality, may have an obvious and direct fitness effect. In the following pages, then, direct fitness effects will be cited where known, but a demonstrated connection between genotype and disease will be taken as indirect evidence of fitness differentials.

BLOOD GROUP INCOMPATIBILITY

The Rh blood group system was described in 1940, and it was only shortly after this that the Rh system was shown to be connected with hemolytic disease of the newborn due to isoimmunization of the mother (Levine, 1943). This can occur when an incompatible pregnancy results from an incompatible mating as described in Chapter 4. The consequences of this for a population depend in part on the frequency of the *D* and *d* alleles leading to the Rh-positive or Rh-negative classification. In an Asian population there may be few or no Rh-negative females, as *d* is rare in much of Asia. Such populations will have very few Rh-incompatible pregnancies.

In Europeans and American Whites we find approximately 15 percent Rh-negative individuals and 85 percent Rh-positive. We can compute the frequency of $d = \sqrt{0.15} = 0.39$. From this we can estimate the expected frequency of genotypes and incompatible matings (Figure 5.2). Incompatible matings of type A occur with frequency = 0.06. Incompatible matings of type B occur with frequency = 0.07. But only half of the pregnancies of mating type B result in incompatible pregnancies, as no antibody can be elaborated in maternal circulation against red cells from the *dd* fetus. We therefore would expect nine percent of all pregnancies in European populations to be incompatible for Rh and potential cases of hemolytic disease.

Hospital records of such hemolytic disease have shown it to occur in approximately one of each 150 births. There are a number of reasons why hemolytic disease does not occur with each Rh-negative incompatible pregnancy. Since anti-D antibodies are not "naturally occurring" as are the antibodies of

FIGURE 5.2

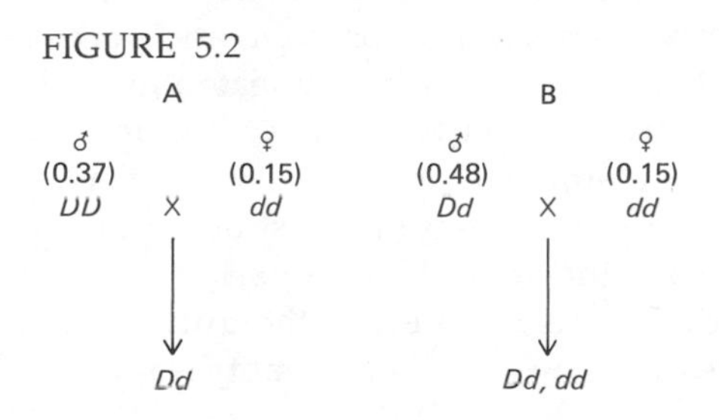

the ABO system, they will be produced only after Rh-positive red cells are introduced into the circulatory system of the Rh-negative female. But antibody strength probably will not reach levels damaging to the fetus (through transplacental immunization with fetal Rh-negative cells) on the first and second pregnancy. The danger of this happening increases with subsequent pregnancies. Moreover antibody strength does not increase invariably with each successive pregnancy. The reasons for this are not fully understood, but a number of factors seem to be involved, including the ABO antigens of the fetal red cells. This latter factor will be mentioned again later.

What are the possible evolutionary effects of Rh incompatibility? Selection occurs against the *Dd* genotype. Selection against the heterozygote eliminates each allele in equal numbers but, percentagewise, this means that the least common allele drops in relative frequency. On the face of it, this suggests that in almost all populations Rh-negative should be decreasing. But if this is all that is involved how could the *d* allele have reached the frequencies in which it is found today?

One of the first suggestions was, of course, migration. If a population is either 100 percent *D* or 100 percent *d* there is no Rh incompatibility in matings. The migration hypothesis was that aboriginal European populations had only the *d* allele. In China, Japan, and most of Asia the *d* allele is rare. Had migration or gene flow from Asia into the West occurred on a large scale, this would have created Rh incompatibility and a transient polymorphism in which the least common allele (now *d*) would have been eliminated slowly. As with most migration hypotheses, there is little that can be done to confirm or reject the hypothesis.

Another factor involved in Rh incompatibility, which may alter the picture considerably, is reproductive compensation. Reproductive compensation occurs when a couple conceives to replace one or more of their offspring that have died. The degree of reproductive compensation can be measured on a scale from zero to one. What effect has reproductive compensation on the effects of Rh incompatibility? When the father is homozygous *DD* there is no effect, as each additional conception results in a *Dd* zygote. The situation is altered for the *Dd* father. The mating *Dd* (♂) × *dd* (♀) will, in the long run, produce equal numbers of *Dd* and *dd* zygotes. But the *dd* individuals will not suffer hemolytic disease. Any child lost to Rh hemolytic disease will be a *Dd*. And for each loss, if compensation occurs, there is a 50 percent chance that the lost *Dd* zygote will be replaced by a *dd* zygote. The obvious result is that this mating will produce more living *dd* offspring than *Dd*. Reproductive compensation may therefore increase the frequency of the *d* allele.

Another factor that influences the prevalence of Rh hemolytic disease is the occurrence of ABO incompatibility. Quite early it was noted that the mating A (♂) × O (♀) produced a smaller proportion of A offspring than did the reciprocal mating of O (♂) × A (♀) (Hirszfeld and Zborowski, 1926). Although jaundice, neonatal death, and other symptoms of severe hemolytic anemia have been shown to be attributable to such ABO incompatibility, it is a much less common clinical problem than is Rh incompatibility. This seems surprising in view of the consistently found and marked effect of A × O incompatible matings in reducing the incidence of A offspring. The resolution of this seeming paradox is that the effect of ABO incompatibility is expressed quite early. This could occur by antibody action on sperm prior to their reaching the ovum (prezygotic selection) or by early spontaneous abortion which can occur without the female being aware of a pregnancy.

Many studies have now been carried out showing an interaction between ABO incompatibility and Rh incompatibility. The Rh incompatible fetus is protected to a certain extent by ABO incompatibility. The findings can be summarized as follows:

1. Among Rh-negative mothers who become Rh sensitized (produce antibodies), the frequency of the ABO compatible matings is much higher than would be expected from the population as a whole.

2. Among Rh affected infants, a greater proportion are from ABO compatible matings than from the reciprocal ABO incompatible matings. This is especially marked in the AB × O mating (Levine, 1958).

3. Among Rh-sensitized mothers those with high titer anti-Rh antibodies tend to be from ABO compatible matings (Vos, 1965).

The explanation of these findings is that the naturally occurring anti-A and anti-B antibodies are able to hemolyze (break open, releasing hemoglobin) red cells. They can therefore, initiate the removal of red blood cells crossing the placenta before the production of antibodies against other antigens (including Rh) on the fetal red cells is induced.

That this is the mechanism by which ABO incompatibility confers protection against Rh incompatibility effects seems confirmed by an experiment carried out by Stern (Stern et al., 1956). They administered Rh-positive red blood cells, through a sequence of injections, into 39 Rh-negative males. Some of the recipients (17) were compatible with the injected cells with respect to ABO antigens; some (22) were incompatible. All were of course incompatible for Rh and potentially capable of producing antibodies against Rh-positive cells. Results are shown in Table 5.1.

TABLE 5.1
ABO Interaction in Experimentally Induced Rh Antibody Production

	TOTAL NUMBER DEVELOPING ANTIBODIES	TOTAL NUMBER WITHOUT ANTIBODIES
ABO Compatible Blood (17)	10	7
ABO Incompatible Blood (22)	2	20

Source: Adapted from K. Stern, I. Davidsohn, and L. Masaitis, "Experimental Studies on Rh Immunization," *American Journal of Clinical Pathology,* 26 (1956), 833–843.

Although the sample was not large the results are clear. A far smaller proportion—9 percent—developed antibodies to the Rh antigen when ABO incompatible than when ABO compatible (59 percent).

In the past there has been a reluctance on the part of many persons to believe that natural selection is important today in human populations. The ABO system provides a case where rather heavy selection may occur without being apparent to the casual observer. In Japan there is little Rh incompatibility since the frequency of the Rh-negative phenotype is less than 1 percent. But approximately 44 percent of Japanese matings are ABO incompatible as compared to a figure of approximately 36 percent among Europeans. Matsunaga and Itoh (1958), on the basis of a fairly large survey, found that couples who were ABO incompatible had a higher frequency of infertility, more spontaneous abortion, and a smaller mean number of living children per pregnancy than in the sample of ABO compatible couples. They estimated the overall mortality due to maternal-fetal incompatibility to be one in every five of the pregnancies in which maternal-fetal incompatibility is present.

This is marked differential mortality against the heterozygotes and, therefore, tends to decrease the frequency of the I^A and I^B alleles relative to I^O. But counter to this they found differential fertility tending to a decrease in I^O. The latter occurred in that the mean number of pregnancies as well as mean number of living offspring was lower when the father was type O than when neither parent was type O. This effect is not explicable in terms of ABO incompatibility.

At least some of the effects of ABO incompatibility occur as prezygotic selection. Females who have anti-A or anti-B antibodies in blood serum have been shown to have the same antibodies in the reproductive tract. Conversely A and B antigens, when present in spermatic fluid, adsorb to the surface of spermatozoa. This led to the suggestion that sperm of secretor males might be inhibited in their ability to reach and fertilize the ovum in ABO incompatible matings (Behrman et al., 1960).

That there are many kinds of evidence for selection at the ABO locus is probably due to the fact that this is the most extensively tested system. Another, more indirect, kind of evidence suggesting the operation of natural selection in the ABO system comes from studies of the association between disease prevalence and the presence of particular alleles. A number of diseases have shown a positive association with a blood group allele in one or more studies. Some of these may be fortuitous associations. Some have been confirmed consistently in studies from different areas, and the likelihood that the associations shown are due to chance is incalculably small. Three such associations, the three most widely confirmed, are shown in Table 5.2. These correlations suggest that these diseases constitute selective factors but they do not measure selection nor do they provide a causative connection. They merely say, in short: Here is an important area of research.

Studies of disease and blood group associations are carried out, typically, on data from hospital records. Records are pulled for all patients being admitted for a given condition, rheumatic fever for example, during a specific period of time. Gene frequencies are tabulated on the blood group systems for which the patients have been typed. These frequencies are then compared to those of a control group, perhaps donors at a local blood bank, to see if they differ. If in the rheumatic fever group a given allele is found in much higher or lower frequency than can be expected by chance, when compared to the control group, this is accepted as an association between the gene and the disease.

The preceding simplified explanation may irritate some of the practitioners of the art,

TABLE 5.2
Some Associations Between Blood Groups and Disease

DISEASE	POSITIVELY ASSOCIATED WITH BLOOD TYPE	NUMBER OF STUDIES	TOTAL CHI-SQUARE (APPROX.)	P (APPROX.)
Duodenal ulcer	O	12	222	0
Cancer of the stomach	A	15	78	1/10,000,000
Pernicious anemia	A	9	28	1/1000

Note: The column heading "P" indicates the probability of such an association, measured by chi-square, that could occur as a result of pure chance.

but it does summarize the essentials of the method. It also points to some of the difficulties encountered in these studies. Since the data are usually assembled from records kept not for this purpose, there are often difficulties in knowing if more than one breeding population is represented in the hospital sample and, moreover, whether the control group represents the same populations in the same proportions. This has been called the problem of a "stratified" sample or the problem of ethnic and racial heterogeneity. Failure to account for such heterogeneities can lead to spurious conjunctions or correlations as shown in Figure 5.3. A large ethnic group, Group A, may be sampled many times and no association between, say, rheumatic fever and R^o will show up. Similarly Group B, sampled many times, may show no association between rheumatic fever and R^o. But if samples from A and B differ in disease prevalence and gene frequency a spurious association can arise from merging the samples as shown in Figure 5.3.

FIGURE 5.3 Hypothetical spurious association (dashed trend line) produced by merging two groups (+ and o) when no within-group association exists.

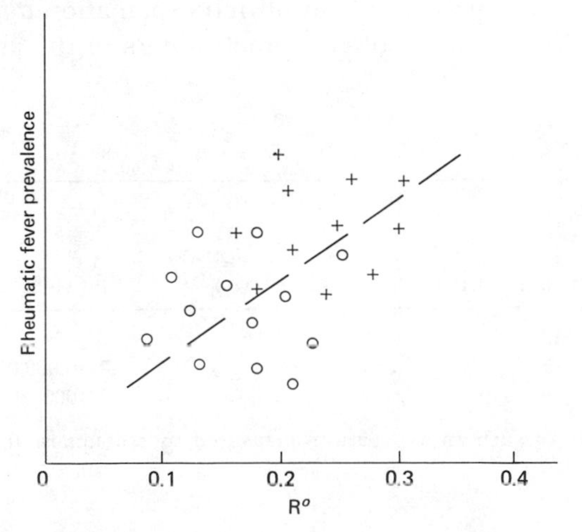

SICKLE CELLS AND MALARIA

The genetics of sickle-cell anemia was described in Chapter 4. This anemia was first described in American Negro populations in 1910. As late as 1951, however, it was thought that this condition was rare or lacking in African populations. This supposition turned out to be simple lack of knowledge resulting primarily from the lack of adequate diagnostic services in most of central Africa and, secondarily, from early mortality among homozygous sicklers. We now know that many Central African populations exceed 20 percent *AS*, and that some reach 40 percent. A population with 20 percent *AS* (sickle-cell trait) at birth should have approximately 1.3 percent *SS* (sickle-cell anemia) at birth. Better studies made in recent years have shown sicklemia to exist roughly as expected in these populations.

Two possible explanations of these frequencies were eventually rejected. These were the migration and mutation hypotheses. Migration, as an explanation, was advanced because Hb S was known to exist in tropical Old World populations outside of Africa, particularly in India. Again better data showed African populations to have as high or higher frequencies of the Hb S gene as any populations outside of Africa. Migration, or gene flow, alone cannot explain a higher frequency than existed in the supposed ancestral population. Furthermore migration, as always, begs the question of how such differences originated in any parental population.

The mutation hypothesis was investigated by Vandepitte and others. (1955). Since *SS* individuals seldom live to reproduce, any *SS* child can be expected to be the offspring of an *AS* × *AS* mating. Of 235 mothers of such sickler children in Kinshasa (formerly Léopoldville) there were two mothers found to be normal *AA*. Assuming that these cases, *AS* (♂) × *AA* (♀) → *SS*, represent mutation of $A \rightarrow S$ a mutation rate of $\mu = 1.7 \times 10^{-3}$ was calculated. Gene frequency equilibrium due to a balance between mutation and selection was given in Chapter 4 as

$$\hat{q} = \sqrt{\mu/s}$$

In the Kinshasa area

$$q \simeq 0.125$$

Therefore

$$\hat{q} \simeq 0.125 = \sqrt{\mu/1.0}$$
$$\overline{\mu} \simeq 1.6 \times 10^{-2}$$

In other words, the calculated mutation rate is only one-tenth that necessary to produce equilibrium.

It should be noted, further, that Vandepitte et al. followed a procedure that produced a maximal estimate of mutation. This is so because child substitution or unreported adoption will increase the frequency of $AS \times AA \rightarrow SS$ cases that are scored as mutations when they are not mutations. Later studies have shown that the mutation rate of $A \rightarrow S$ must be very much lower than this maximal estimate.

As early as 1946 Beet published observations showing that, in Zambia (then Northern Rhodesia), persons with Hb S showed a lower frequency of malaria parasites in blood cells than did persons with normal hemoglobin. The difference, in this and in later studies, is most apparent during the season when the reinfection rate is lowest. Allison (1954), on the basis of this and his own data, advanced the hypothesis that the *S* allele was maintained by a fitness advantage of the *AS* individual as compared to the *AA* individual in a malarious environment. He suggested this particularly for the malaria parasite *Plasmodium falciparum* transmitted by the mosquito *Anopheles gambiae*. Allison showed that there was a correlation between the geographic distribution of malaria and that of sickle cells. He confirmed Beet's observation

TABLE 5.3
Number of Individuals of *AS* and *AA* Genotypes in Which Malaria Parasites Became Established, After Exposure, in a 40-Day Observation Period

GENOTYPE	INFECTION ESTABLISHED	INFECTION NOT ESTABLISHED	TOTAL
AS	2	13	15
AA	14	1	15

Source: Adapted from A. C. Alleson, "Protection Afforded by Sickle-Cell Trait Against Subtertian Malarial Infection," *British Medical Journal,* 1 (1954), 290–294.

that the incidence of parasitemia was less in *AS* individuals than it was in *AA* individuals. Further he experimentally introduced *P. falciparum* infection into 30 Luo volunteers by inoculation and mosquito bites. One-half of the volunteers had the sickle trait, one-half were normal. Results are shown in Table 5.3.

If the sickle gene is maintained by heterozygote advantage, we should be able to state the magnitude of the fitness differential involved. For example, in an equilibrium population where no *SS* adults are living and $AS = 0.25$, $q = 0.125$. We can compute the relative fitness value that will yield this equilibrium as

Genotype	*AA*	*AS*	*SS*
Fitness	0.86	1.00	0

In other words, the "normal" homozygote must have a fitness that is 14 percent less than that of the heterozygote in order to maintain this equilibrium.

The data cited thus far suggest that a fitness difference exists in the direction expected but more direct evidence is desirable. One kind of evidence for differential mortality is the change in frequency of sickle-cell trait in various age groups. Rucknagel and

Neel (1961) summarized the data from seven such studies. These data show a general trend of increasing frequency of the trait. They found 1.15 times the proportion of *AS* individuals in adults as in infants. The proportion of *AS* individuals in these seven studies is 0.281. Therefore we can see that the value of 1.15, which can be viewed as the selective advantage of *AS* relative to *AA* due to differential mortality between infancy and adulthood, is approximately what is necessary to maintain a balance.

Most mortality due to malaria appears to occur in infancy and early childhood. A few studies have been done of children dying of malaria. These show that *falciparum* malaria deaths, and especially the involvement known as "cerebral malaria," include a smaller proportion of *AS* individuals than would be expected on a random basis. One such study, for example, showed that only one of 100 malarial deaths was *AS*, where the expected number was 22.6.

Burke et al. (1958) showed that, in overall mortality, *AS* infants and children in the Congo had a decided advantage over *AA* infants and children. In a sample of 3059 *AS*, there was 24 percent mortality; in 6588 *AA*, there was 27.4 percent mortality.

Roberts and Boyo (1960; 1962) found a similar age trend in the Yoruba of Nigeria as reported for a number of other groups. They found that the *AS* frequency tended to rise between the ages of six months to four years, at which time adult frequencies were approximated. In this connection it should be noted that the newborn are relatively resistant to malaria, perhaps due to fetal hemoglobin and also to the fact that maternal antibodies are still in circulation and confer some immunity. By the age of four to seven years, children in malarial areas have developed immunity of their own. This leaves early childhood as a critical period. But Roberts and Boyo, in reviewing data on 43 populations, conclude that differential mortality alone is not enough to maintain the *S* allele at its present level and suggest that differential fertility is also involved.

A number of studies in Africa have shown a slight increase in the number of children from *AS* × *AA* matings and a decreased number of children from *AS* × *AS* matings compared to *AA* × *AA* matings. The results have not been as striking as mortality differences. Firschein (1961) has, on the other hand, shown that among the Black Caribs of British Honduras, *AS* females have 45 percent more children than *AA* females. The suggestion has been made that this may be due to spontaneous abortion from malarial infection of the placenta.

If malaria does maintain the *S* allele, then elimination of malaria should result in at least equal fitness of *AA* compared to *AS*. If *SS* remains at a disadvantage we can predict that a decrease in the prevalence of malaria will result in a decrease in frequency of the *S* allele. The American Negro may be a case in point. On the basis of blood group antigens it has been estimated that the gene pool of the American Negro on the average represents approximately 20 to 30 percent European derived genes, and approximately 70 to 80 percent West African derived genes. Occasional small populations classed socially as American Negro have a significant genetic component from Amerindian populations but overall this contribution is insignificant. American Negro populations have overall approximately 10 percent *AS*. West African populations have overall approximately 20 percent *AS*. This suggests that the frequency of the *S* allele in the American Negro has gone down faster than can be explained by gene flow alone.

The sickle-cell gene in Africa provides a rather striking example of the interaction of cultural and biological evolution. This has been explored primarily by Livingstone (1958; 1961). Livingstone notes that some groups on the West Coast of Africa, living in highly malarious areas, have a lower frequency of the *S* allele than expected. These are the same populations considered by anthropologists to have been the last to adopt agriculture. What relation has agriculture to the frequency of an inherited trait? Livingstone lists the ecological conditions favoring the *A. gambiae* mosquito as

1. Fresh water, not brackish
2. Standing pools, not flowing
3. Sunlit water
4. Resting places near the host population (e.g., hut walls and eaves)

The clay soils found in many tropical areas, after prolonged cultivation, become almost impervious to water. The spread of more intensive forms of agriculture resulted in clearings, standing pools of water, and villages that provided the ideal conditions for an increase in *A. gambiae*. This resulted in an increased exposure to *P. falciparum* malaria and selection for higher frequencies of the *S* allele.

The other side of the story has not been investigated. Agriculture-based kingdoms such as the early kingdoms of West Africa must have required a certain productivity and population density to maintain the social superstructure of a kingdom. What are the necessary values of these parameters and to what extent did the spread of the sickle gene create these conditions? Unfortunately this equally interesting question has not been investigated.

The story of the sickle-cell gene has been given in some detail as a paradigm of microevolutionary change in man. It is by no means typical of evolutionary change, in that the selective differentials involved are quite large. Obviously an animal with the reproductive physiology of man cannot afford many such systems with high selective pressures against common genotypes. Such a situation would mean extinction of the species. It is, however, a useful illustration because the various levels of investigation have produced an interconnected and concordant picture: formal genetics, specification of the protein, the mechanical basis of sickling (Murayama, 1966), population dynamics involved in the spread of the gene, and finally, cultural factors in ecological change. Fuller data would exemplify another important point in the study of microevolutionary change, and that is their difficulty. Only a few studies have been cited. Many studies not cited confirm those given. Although no studies have produced contradictory results, a number of reasonably well-conceived studies have produced no results, for example, not showing greater parasitemia in *AA* than in *AS*. If results are so hard won where selective differentials are high, we can expect that a great deal of time and effort will be involved in understanding ongoing evolutionary change in more "typical" cases where differentials are small.

REFERENCES CITED

Allison, A. C. 1954 Protection Afforded by Sickle-Cell Trait Against Subtertian Malarial Infection. *British Medical Journal* 1:290–294.

Beet, E. A. 1946 Sickle Cell Disease in the Balovale District of Northern Rhodesia. *East African Medical Journal* 23:75–86.

Behrman, S. J., J. Buettner-Janusch, R. Heglar, H. Gershowitz, and W. L. Tew 1960 ABO(H) Blood Incompatibility as a Cause of Infertility: A New Concept. *American Journal of Obstetrics and Gynecology* 79:847–855.

Brues, A. M. 1954 Selection and Polymorphism in Physical Anthropology 12:559–597.

Burke, J., G. de Book, and O. de Wulf 1958 La Drepanocytemie Simple et L'anemie Drepanocytaire au Kwango (Congo Belge). *Academie Royale des Sciences Coloniales. Classe des Sciences Naturelles et Medicales* 7:1–128.

Clarke, C. A. 1961 Blood Groups and Disease. *In* Arthur G. Steinberg (ed.), *Progress in Medical Genetics,* vol. 1, pp. 81-119. New York: Grune and Stratton.

Firschein, I. L. 1961 Population Dynamics of the Sickle-Cell Trait in the Black Caribs of British Honduras, Central America. *American Journal of Human Genetics* 13:233–254.

Hirszfeld, L. and H. Zborowski 1926 Über die Grundlagen des Serologischen Zusammenlebens Zwischen Mutter und Frucht. *Klinische Wochenschrift* 5:741–744.

Levine, P. 1943 Serological Factors as Possible Causes of Spontaneous Abortion. *Journal of Heredity* 34:71–80.

Levine, P. 1958 The Influence of the ABO System on Ph Hemolytic Disease. *Human Biology* 30:14–27.

Livingstone, F. B. 1958 Anthropological Implications of Sickle Cell Gene Distribution in West Africa. *American Anthropologist* 60:533–562.

Livingstone, F. B. 1961 Balancing the Human Hemoglobin Polymorphisms. *Human Biology* 33:205–219.

Matsunaga, E. and S. Itoh 1958 Blood Groups and Fertility in a Japanese Population, with Special Reference to Intrauterine Selection due to Maternal-Faetal Incompatibility. *Annals of Human Genetics* 22:111–131.

Murayama, M. 1966 Molecular Mechanism of Red Cell "Sickling." *Science* 153:145–149.

Otten, C. M. 1967 On Pestilence, Diet, Natural Selection, and the Distribution of Microbial and Human Blood Group Antigens and Antibodies. *Current Anthropology* 8:209–226.

Roberts, D. F. and A. E. Boyo 1960 On the Stability of Haemoglobin Gene Frequencies in West Africa. *Annals of Human Genetics* 24: 375–387.

Roberts, D. F. and A. E. Boyo 1962 Abnormal Haemoglobins in Childhood Among the Yoruba. *Human Biology* 34:20–37.

Rucknagel, D. L. and J. V. Neel 1961 The Hemoglobinopathies. *In* A. G. Steinberg (ed.) *Progress in Medical Genetics,* vol. 1, pp. 158–260. New York: Grune and Stratton.

Stern, K., I. Davidsohn, and L. Masaitis 1956 Experimental Studies on Rh Immunization. *American Journal of Clinical Patholology* 26:833–843.

Vandepitte, J. M., W. W. Zeulzer, J. V. Neel, and J. Colaert 1955 Evidence Concerning the Inadequacy of Mutation as an Explanation of the Frequency of the Sickle Cell Gene in the Belgian Congo. *Blood* 10:341–350.

Vogel, F., H. J. Pettenkofer, and W. Helmbold 1960 Über die Populationsgenetik der ABO-Blutgruppen, 2. Mitteilung. Gehäufigkeit und epidemische Erkrankungen. *Acta Genetica et Statistica Medica* 10:267–294.

Vos, G. H. 1965 The Frequency of ABO-Incompatible Combinations in Relation to Maternal Rhesus Antibody Values in Rh Immunized Women. *American Journal of Human Genetics* 17:202–211.

6

Race Studies and Human Variation

In the preceding chapters we have gone in great detail into the genetic approach to the study of human diversity. A second approach deals not with variation in gene frequencies but with variation in traits. Such traits are height, weight, skin color, blood pressure, the frequency of infectious diseases, and many other characteristics of populations. This second approach will be illustrated in Chapter 7. A third approach, to be discussed now, is by far the oldest. It consists of classifying groups of humankind into different races and then asking questions about the origins and distributions of these races. This is the study of race.

This is a chapter on race studies but one which contains no racial classifications. The study of racial classifications is an activity that belongs to an era now ending. To show why this is so, the entire chapter is devoted to the *concept* of race. As the analysis of a concept this chapter must be somewhat different from other chapters in this book. And if the analysis seems controversial this is because, although the racial approach to the study of human variation belongs to an era that is ending, that era is not completely over.

The concept of race and the use of racial classifications in the study of human variation were prominent features of the natural-history phase of anthropology. The early taxonomists not only classified plant and animal species but also classified different

MAMMALIA.

ORDER I. PRIMATES.

Fore-teeth cutting; upper 4, parallel; teats 2 pectoral.

1. HOMO.

Sapiens. Diurnal; varying by education and ſituation.

2. Four-footed, mute, hairy. *Wild Man.*

3. Copper-coloured, choleric, erect. *American.*
Hair black, ſtraight, thick; *noſtrils* wide, *face* harſh; *beard* ſcanty; *obſtinate*, content free. *Paints* himſelf with fine red lines. *Regulated* by cuſtoms.

4. Fair, ſanguine, brawny. *European.*
Hair yellow, brown, flowing; *eyes* blue; *gentle*, acute, inventive. *Covered* with cloſe veſtments. *Governed* by laws.

5. Sooty, melancholy, rigid. *Aſiatic.*
Hair black; *eyes* dark; *ſevere*, haughty, covetous. *Covered* with looſe garments. *Governed* by opinions.

6. Black, phlegmatic, relaxed. *African.*
Hair black, frizzled; *ſkin* ſilky; *noſe* flat; *lips* tumid; *crafty*, indolent, negligent. *Anoints* himſelf with greaſe. *Governed* by caprice.

Monſtroſus Varying by climate or art.

1. Small, active, timid. *Mountaineer.*
2. Large, indolent. *Patagonian.*
3. Leſs fertile. *Hottentot.*
4. Beardleſs. *American.*
5. Head conic. *Chineſe.*
6. Head flattened. *Canadian.*

The anatomical, phyſiological, natural, moral, civil and ſocial hiſtories of man, are beſt deſcribed by their reſpective writers.

Vol. I.—C 2. SIMIA.

FIGURE 6.1 The classification of man as given by Carolus Linnaeus. (A facsimile reproduction from Linne, 1806.)

groups of man. Such a classification of human races from Linnaeus is shown as Figure 6.1. It is important to note that racial studies became a part of "natural history" or "natural science" in this period. This means that the original concept of race was both pre-Mendelian and pre-Darwinian in its formulation. Keeping this in mind will help us understand some of the shortcomings of race studies up to the present.

Georges Cuvier recognized three major races of man: *Caucasian* or white, *Mongolian* or yellow, and *Ethiopian* or black. Johann F. Blumenbach even earlier (in 1776) recognized five "principal varieties" of *Homo sapiens.* Blumenbach's five varieties (races) were named *Mongolian, Ethiopian, American, Malay,* and *Caucasian.* These were based on skin color and on easily observed morphological features of the head and face. Many racial classifications have been proposed since that time. Some classifications involve over 200 categories but most have approximated the divisions set up by Cuvier or Blumenbach. The European and American layman's concept of human races has also continued to reflect a three- to fivefold division and this has been incorporated into many Western social usages.

It was noted that the conception of race represented by these early attempts at subspecific classification date from a pre-Darwinian period of natural science. They were therefore preevolutionary and certainly pregenetic. The same is true of the species concept, which in the post-Mendel period was redefined as what we have called the biospecies in order to make it serve as a useful concept in evolutionary theory. Attempts to redefine race have been somewhat less successful than the attempts to redefine species and the reasons for this will be explored in the following paragraphs.

Over the years a great many definitions of the term *race* have been advanced. A useful collection of excerpted writings on the concept of race has been presented by Count (1950). Three definitions of race, representing slightly differing concepts, are given below.

> Race may be defined as a great division of mankind, the members of which show similar or identical combinations of physical features which they owe to their common heredity (Seltzer, 1939, quoted in Count, 1950, p. 608).

> A race is a great division of mankind, the members of which, though individually varying, are characterized as a group by a certain combination of morphological and metrical features, principally non-adaptive, which have been derived from their common descent (Hooten, 1946, p. 448).

> We may define a human race as a population which differs significantly from other human populations in regard to the frequency of one or more of the genes it possesses (Boyd, 1950, p. 207).

These definitions represent steps in the evolution of the race concept. They were put forward by men well trained in evolutionary and genetic theory in differing attempts to make race a useful evolutionary concept. But these attempts have not been wholly successful.

As noted in the discussion of the history of evolutionary theory, classification is a central activity in the natural-history phase of science. The procedure, greatly simplified, is to classify differences into nominal (named) categories and then to ask: What are the relationships between these categories? Why do these categories exist? The latter question is rather intriguing. One possible answer is simply that the investigator created them. But this is an overly facile answer that begs

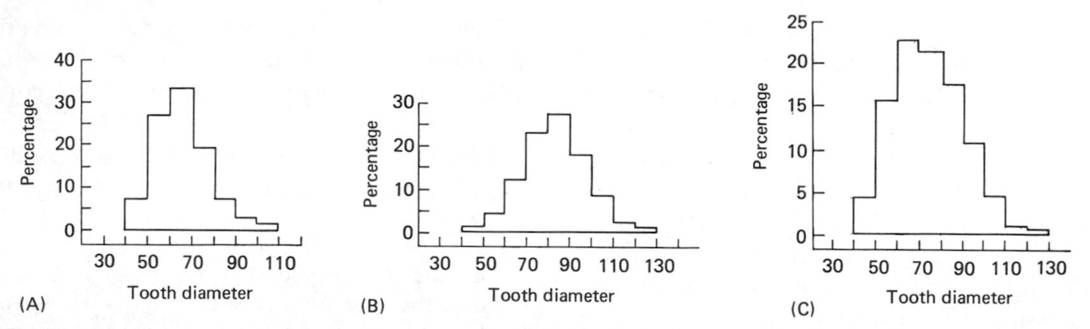

FIGURE 6.2 The error of an archetypal approach: (A) and (B) frequency polygons of diameters of teeth separated into two groups according to an a priori concept of taxonomic difference, (C) single frequency polygon of a merged unimodal distribution of diameters.

the question really being asked, which is, Why do differences exist? It is important to note that all people, when dealing with phenomena in terms of nominal categories, tend to lose sight of the fact that they are asking, Why do differences exist? and think in terms of the question, Why do categories exist? or Why do differences between categories exist?

This universal tendency, if not frequently scrutinized, can lead to *archetypal thinking*. An archetype is an original form or model from which present representatives are patterned. In archetypal thinking the named category comes to be thought of as the archetype. Once this mode of thinking is adopted any individual member of the named category is then treated as a *representative* of the archetype. If the individual member is not exactly the same as the description of the archetype then it is treated as an *imperfect representative* of the archetype. In other words the description of the category is no longer just a summary statement about the individuals who compose the category. It takes on a reality of its own. This is the error of archetypal thinking.

To render the preceding paragraphs more intelligible I will illustrate how archetypal thinking can mislead. The example oversimplifies the actual procedure and is of course entirely hypothetical. Suppose we have a sample of teeth that shows variation in size. Let us further suppose that this variation, if plotted as a frequency polygon, shows a graph that reflects a continuous, unimodal distribution of sizes (Figure 6.2).

Suppose however that the approach of our hypothetical investigator is to create two classes. He separates the sample into large teeth and small teeth, and he suspects the sample to be representative of a large and small species. He now has categories that he may name and treat as taxonomic divisions. But what about the "reality" of this taxonomic division? Once archetypal thinking has set in this question translates as, Can we show that there are real differences between the taxa? Our investigator answers this by measuring the teeth in the two taxa, applying the appropriate statistical test, and showing that the means of the two groups of teeth differ significantly. Our investigator now considers that he has established a real difference between taxa that exists or existed regardless of the presence of the investigator. Having come this far he now seeks names for these two taxa that will be valid under the International Rules of Zoological Nomenclature. He is now in a position to compound his error by asking questions about the phylogenetic relationships between these groups.

This sounds like a parody and it is. No one engages in the error of archetypal thinking in such transparent form. But in only slightly more involved form this sequence of *non sequiturs* has occurred in discussions of fossils, stone tools, races, and, indeed, can be found in every field of inquiry.

Although the error of archetypal thinking has been a pernicious tendency in all fields in which a taxonomy of continuous variation is important, it has been most tenacious in racial taxonomies. *Races* have seldom remained descriptive categories but have tended to take on a reality of their own. Once races are set up people do not ask why variation exists but why races are distributed as they are and how they have been distributed in the past. These questions imply that there are some unchanging and essential features of races. Such an implication means that races are no longer summary descriptions of variation as it exists.

Such archetypal thinking has further consequences. Populations that do not fit neatly into one of a limited number of races are explained as being the result of an intermixture of two or more races. Related to this is the concept of the *pure race*. The postulated existence of formerly pure races is, in fact, the most extreme form of archetypal thinking. In any racial classification it can be shown that almost no individual conforms precisely to the *type* of the race. There is usually a great deal of variation within each race. Furthermore there are always intermediate populations that do not fit easily into any one of a finite number of races. The archetypal thinker solves this dilemma by assuming that there was once a time in which human varieties did indeed come in neater categories that would have corresponded to the racial taxonomy. These are the hypothetical pure races.

Once pure races are postulated, whether labeled as such or not, it is then a simple matter to explain the variation actually seen today by postulating the necessary migrations and intermixtures of populations. For this reason many early books on ethnic prehistory would include a section on racial migrations. It required only an outline map of the world and some sinuous arrows winding their way across the map to represent racial migrations from the homelands of the pure races.

Logically this approach requires that one postulated race be the shortest known, one the tallest known, one the lightest, one the darkest, and so on for all variables. This kind of division would permit the necessary migrations and mixing to get any combination of traits observed today. As this art was practiced, only a few traits or variables were used. Roland Dixon's *The Racial History of Man* (1923) is a lucid example of this approach. This book was thoroughly criticized for the racist overtones of its conclusions and for the use of ideal types not represented by any living population. What the critics did not recognize was that Dixon was taking the usual archetypal approach, carrying it to its logical extreme, and being quite explicit about the process.

Dixon used three traits, the cephalic index (a ratio of head width to head length), the length-height index of the head, and the nasal index (a ratio of nasal width to nasal length). Dixon divided each of these indices into three categories: narrow, medium, and broad. This could give $3 \times 3 \times 3 = 27$ combinations of traits and therefore, potentially, 27 races or types. Dixon quickly concluded "that groups without medial factors may be regarded as definite types, now rarely found in their pure state, and that all the groups comprising one or more medial factors are the result of the blending and fusion of these" (Dixon, 1923, pp. 15–16). This gave him $2 \times 2 \times 2 = 8$ fundamental types which he named Caspian, Mediterranean, Proto-Negroid, Proto-Australoid, Alpine, Ural Palae-Alpine, and Mongoloid.

Since Dixon was quite aware of what he was doing I will quote him at length. He says:

> We may then, I believe, regard these 'types' as, so to speak, archetypes, fundamental patterns, more or less perfectly evolved in the process of development of the human species, and, like other animal varieties and species, having had a definite origin in both time and space. From the complex fusion between these archetypes, or fundamental races, . . . the existing, actual races which might be described as stable blends, have been derived (Dixon, 1923, p. 502).

From this point Dixon goes on to assume that these archetypal races existed in more or less pure form until relatively recent times and as far back as the Pliocene. Accordingly he relegates the then known fossil forms of man to his archetypal races. A recent example of this same kind of approach, differing only in details and in the amount of fossil material surveyed, is found in *The Origin of Races* (Coon, 1962). Archetypal thinking, although no longer present in the straightforward manner of Dixon, is far from dead. Archetypal races continue their hypothetical journeys and migrations in the minds of many today.

All of this is not to say that migrations do not occur. We know that they do. We further know that large-scale movements of people or continued gene flow between regions will be reflected in genetic similarities. But the pure race approach presented large-scale migrational schemes as explanations, not as hypotheses. These were in effect pseudo-histories of man.

It is obvious at this point that racial taxonomies have certainly been misused due to the trap of archetypal thinking. We must now ask if this is only a case of imperfect application of the concept of race. Can race be useful in the study of human variation? To answer this let us again look at the evolution of the race concept. In the Linnaean classification of human races, forms of dress, speech, and manners were noted as relevant distinguishing features. These were soon dropped, not on grounds of subjectivity alone, but because these features were cultural variables that could change with differences in enculturation or training.

The next step was to use only biological characteristics. These were not as changeable as cultural characteristics. Furthermore only those biological characteristics less influenced by environmental differences were those most desired. It was a blow to many racial studies when Franz Boas (1911) showed that the cephalic index was strongly influenced by environment. The cephalic index has been thought to be one of the most *stable* traits as well as one that could be measured on both living and dead.

The definition of race characteristic of this period can be labeled as representing *typological* concepts of race. In the typological concept of race only the clustering of traits need be demonstrated to validate the existence of a race. Some authors have equated the typological approach with archetypal thinking. This is a mistake. It is true that, as was pointed out in previous pages, there is too often a close psychological connection between a typological approach and archetypal thinking. But there is no necessary logical connection, and indeed, what many "numerical taxonomists" are attempting today is a typological approach to classification that manages, at the same time, to avoid archetypal overtones.

The definitions of race of both Seltzer and of Hooten are modified typological definitions. They are modified from earlier typological definitions by the stipulation that the similarities among members of a race are due to common descent. But, as I will show, the traditional use of racial taxonomies based on such definitions is the reconstruction of a kind of *phylogeny* within the species. This means that the stipulation of common descent and the reconstruction of descent relationships involve a logical circularity. Such added phrases hardly improve these typological definitions.

The next step in evolution of the concept of race can be seen in Hooten's definition of race. Morphological and metrical features are specified, but those used should be *principally nonadaptive*. A further step was to go from morphology to the gene itself. But it skould be noted that when blood group genetic studies were first introduced in the study of race it was assumed that these systems too were nonadaptive. Why was it thought important that nonadaptive traits be used? The underlying reason was the same as that which caused language and manners, tools of early racial diagnostics, to be dropped. The whole trend represents an effort to find traits that do not change through time; an attempt to find traits that are *racial markers* and that can be used to unravel the supposed intertwined racial threads in the prehistory of our species. If natural selection could alter a trait this could interfere with the process of unravelling racial prehistory. Consequently it was hoped that nonadaptive traits could be found.

We see from this that the first effect of evolutionary theory was not to change the goal of racial studies but only to modify the requirements for "good" racial traits. The goal of such studies was, in a most important sense, antievolutionary. To be able to trace races back through time implies, if this is to be more than the prehistory of a single gene, that genetic reassortment does not occur, that other traits characterizing a race occur whenever one of the market traits occurs. This is genetically unsound and is yet another form of archetypal thinking. The approach is further antievolutionary in that it avoids the problem of how differences have developed. Instead the assumption is made that differences have existed and persisted from an unspecified early time that is not the object of investigation. This approach continues to have some spokesmen even to the present.

A large number of investigators have attempted a more drastic modification of the concept of race in order to bring it into line with genetic and evolutionary theory. But why not discard the term and the concept completely? The answer would go something like this. We can specify the factors that cause gene frequency (evolutionary) change: selection, mutation, migration, drift, and nonrandom mating. But what if we ask the question, what determines the gene frequencies in a given population at a given time? The answer to this question is that there are six factors involved, the five factors listed above plus the gene frequencies of the parental population. And when we are concerned with one or several tens of generations it will be the parental gene frequencies that are the most important factor. The argument is then that we need to preserve a classificatory (racial) approach because we need to label the genetic continua that are such important factors in determining gene frequencies over a significant time range.

Persons taking this position generally have defined race much like Boyd did. The two major elements in this view of race are:

1. Races are breeding populations and not ideal types.
2. Races differ genetically from one another.

These and some related concepts I will call the *populational* definition of race. This kind of change in the race concept attempts to parallel the change that occurred in the species concept. Linnaeus used a purely typological concept of species. The biospecies of today represents a redefinition of the Linnaean concept of species and is a central concept in evolutionary theory today. But a somewhat different history has befallen the term race. The field of population genetics developed using the term *breeding population* or *Mendelian population.* These were fairly well-defined concepts and became central to population genetic theory. Only after this usage was well established was it proposed that race be redefined as, so to speak, a breeding population with a difference.

But the term race itself has for centuries carried with it troublesome connotations that persist despite the fact that these connotations are not part of the attempts at a scientific definition of race. This has led many today to advocate dropping the term entirely. If race denotes a breeding population with a difference why not simply use the term *breeding population.* If such a breeding population needs to be delineated and labeled for a particular study this can be done as easily. This approach has become common in anthropology today.

The concepts that have been labeled *geographical race* and *ecological race* are also included in the populational concept of race. These terms are simply intended to designate clusters of breeding populations that are more similar to one another than to other breeding populations because of similar selective factors or gene flow within a common ecological zone or a common major geographic region.

In addition to the typological and the populational concepts of race there is another concept that deserves some attention for the sake of a complete view of possibilities in categorizing subspecific differentiation. For lack of a commonly agreed upon name I will call this the *biological subspecies.* This concept has been labeled by others as the *incipient species* (Dobzhansky, 1966) or the *semispecies* (Mayr, 1963, p. 501). The biological subspecies concept is analogous to the biological species concept in that the degree of obligate genetic isolation is the fundamental variable behind each.

If speciation is a result of the slow acquisition of genetically isolating mechanisms then it can be useful to recognize intermediate stages in this process. Biological subspecies are then subspecies among whom partial genetic isolation has evolved. This does not mean that such biological subspecies will necessarily continue to differentiate until they become separate species. It does mean that biospecies that recently evolved were, at one time, biological subspecies.

Let us now ask under what circumstances, or in what kind of species, would we expect the concept of biological subspecies to be useful? Minimum conditions would be where a species is widespread and its total range includes ecologically diverse regions. If these regions are large relative to the average migrational distance of the individual then there will be more gene flow within than between regions and between-region differences could be expected to evolve. The process of speciation requires something more. It requires the buildup of reproductive barriers. These reproductive barriers may be in the form of lowered fertility of between-region matings, they may be the lowered fertility of the offspring of such matings, or they may be an increased mortality among offspring of such matings. Such situations are in fact found in a number of plant and animal populations. In such species biological subspecies can be designated legitimately and usefully.

We now ask, is *Homo sapiens* such a species? Certainly *sapiens* is regionally differentiated in a number of morphological and physiological traits having a genetic basis. But such differences are not directly relevant as critera of biological subspecies. Using the speciation model as our guide, we can claim to be able to validly designate biological subspecies, but only to the extent that partial reproductive barriers have evolved.

We have no evidence for the existence of partial reproductive barriers within *sapiens.* Most studies of "race mixture" or "hybridization" in man have not controlled cultural factors in fertility and mortality differentials successfully. (For examples of such studies see Boas, 1894; Penrose, 1955; Morton, Chung, and Mi, 1967.) Therefore it is presumptuous at this point to say that no differentials will be found. But at present there is as much or more evidence for heterosis effects in fertility as for the opposite. Therefore we must conclude that the biological subspecies as I have described it here does not appear to be a valid model for human populations.

I have presented three major approaches to the race concept: the typological race, the populational race, and the biological subspecies. The typological and, to an extent, the population approaches have been used most in the construction of pseudohistories of man. This entire attempt was preevolutionary in origin and outlook. But we must also ask if racial taxonomies may have more legitimate uses in the study of human variation and human evolution.

The answer to this is that racial taxonomies have extremely limited scientific uses if they have any at all today. Typological approaches, as was noted, all too often lead to an archetype mental set in framing questions and in interpreting the data on human variability. Further, typological classifications can have no utility unless knowledge of the value of one trait observed in a population tells you something about a trait not observed. Such predictability is generally low in typological classifications. It is a function of how closely associated or how highly correlated the traits might be. Many genetic traits in man show very low associations in their distributions. In other words, on the world-wide basis to which racial taxonomies apply you can not predict stature from skin color, nor skin color from ABO gene frequency, and so on. In only a few traits or genetic systems are there high correlations on a global scale. For this reason the predictive power of typological categories is very low, and indeed, if racial categories are set up empirically—on the basis of trait clusters—most populations fall into no race at all.

The populational definition of race does nothing for us that is not better done by the concept of the breeding population. It adds an emotionally loaded term that continues too often to convey covert social meanings. Under such circumstances a clear gain in analytic ability must be demonstrable to justify the intellectual price paid for the continued use of the term race. No such gain is demonstrable.

Finally, the biological subspecies as I have defined it here could have utility in the study of species that have genetically coadapted subgroups. Such coadaptation can, in theory, be maintained or elaborated by the evolution of partial genetic reproductive barriers. There is at present no evidence of such a situation in man.

This brief survey of the concept of race and

its employment has not done justice to any single point of view. It has been too succinct to represent any one person's views. It has summarized the major concepts of race, the trends of change in these concepts, and the intellectual problems involved. The overall conclusion must be that studies of human variation that are conceptualized in terms of races have very little utility today. And they had something less than scientific utility in the past.

This does not mean that the names of various races will cease to appear in the literature of either the social or biological sciences. We can expect the literature on race *per se* to cease to occupy the attention of investigators whose concerns are strictly biological. But we can expect the names of racial categories of various types to continue in use simply because they frequently designate some social and historical reality important in human societies today. This means that racial terms, no matter how derived, frequently indicate *culturally* conditioned mating barriers. These are, then, *ethnic* groups that are sometimes designated by nationality and religion names, as well as race names. These groups may differ in some biological variable of interest in a particular investigation. Such categories will continue to have some utility as sampling categories. This occurs primarily in three kinds of studies.

First, and more important to the social scientist than to the biologist, if someone advances a generalization about a race, then in order to reply to or investigate the validity of the generalization, the racial classification must, in many cases, be accepted to provide a comparable sample. This is obviously what is done in replying to some kinds of racist statements.

Secondly, racial categories may be employed as sampling units in biological surveys where the problem of *stratified samples* may exist. This was discussed as a problem in establishing a correlation between blood groups and disease in Chapter 5. In Figure 5.1 we saw a scattergram of a spurious correlation between R^0 and heart disease. In such a case a racial breakdown of the data would have shown that, within the two populations considered separately, there was no correlation between R^0 and heart disease. There are many ways other than the use of an ethnic category, of course, in which a sample may be stratified. There may be significant stratification (or heterogeneity) by economic level, sex, occupation, and so forth. What may be a relevant category to include in a study is a matter of the investigator's judgment and his resources, since there are usually a large number of ways in which a sample can be cross-partitioned. And it should be noted that such categories of race can obscure, with equal ease, significant heterogenetities. For example, in physiological studies the category *Caucasian* has often proven to be a markedly heterogeneous class.

A third area in which racial designations will continue to have some minor utility is in the area of climatic adaptation studies. Again this involves a question of sampling. If an investigator wishes to investigate adaptational differences in the ability to tolerate heat stress or cold stress he maximizes his chances of finding such differences by comparing samples, the majority of whose ancestors were long resident in a hot or a cold climate. Racial categories that are, in fact, ethnic categories can be utilized frequently in selecting such samples.

In summary we see that the concept of race as a biological phenomenon has but little utility in present-day studies of human variation. The early typological concept of race provided what might be called preevolutionary definitions of race. Classifications of man were set up in terms of these definitions. But these definitions had little or no utility to an evolutionary approach in understanding the origin of biological differences. They represent an early natural-history stage in the study of human variation. The resultant classifications were frequently less than useful as they usually involved archetypal thinking, the construction of pseudohistories, and the avoidance of evolutionary problems. This early phase lasted through the 1940s and appears occasionally today. Raciation as partial speciation is valid and useful for some species. This I have spoken of as the biological subspecies. There is no evidence at present for the validity of such a model for *Homo sapiens*. Race therefore has ceased to be an object of study. Race names, more accurately called ethnic names, however obtained, continue to be used as convenient sampling categories, as they may designate heterogeneities in both biological variables and in the breeding structure of a society.

REFERENCES CITED

Boas, F. 1894 The Half-Blood Indian. *Popular Science Monthly* 45:761–770. 1911 *Changes in the Bodily Form of Descendants of Immigrants. Reports of the Immigration Commission, Senate Document No. 208*. Washington, D.C.: Government Printing Office.

Boyd, W. C. 1950 *Genetics and the Races of Man*. Boston: Little, Brown.

Coon, C. S. 1962 *The Origin of Races*. New York: Knopf.

Count, E. W. (ed.) 1950 *This Is Race*. New York: H. Schuman.

Dixon, Roland B. 1923 *The Racial History of Man*. New York: Scribner's.

Dobzhansky, T. 1966 Spontaneous Origin of an Incipient Species in the *Drosophila paulistorum* Complex. *Proceedings of the National Academy of Sciences* 55:727–733.

Hooten, E. A. 1946 *Up from the Ape*, rev. ed. New York: Macmillan.

Linne, C. Von 1806 *A General System of Nature, Through the Three Grand Kingdoms of Animals, Vegetables, and Minerals, Systematically Divided into their Several Classes, Orders, Genera, Species, and Varieties, with their Habitations, Manners, Economy, Structure, and Peculiarities*, Translated by William Turton. London: Lackington, Allen and Co.

Mayr, E. 1963 *Animal Species and Evolution*. Cambridge, Mass.: Belknap Press.

Morton, N. E., C. S. Chung, and M.-P. MI 1967 Genetics of Interracial Crosses in Hawaii. In *Monographs in Human Genetics*, vol. 3. New York: S. Karger.

Penrose, L. S. 1955 Evidence of Heterosis in Man. *Proceedings of the Royal Society of London, Series B* 144:203–213.

7

Climatic Adaptation

In the previous chapter it was shown that in the human species races—to the extent that they are something more than breeding populations are in no sense integral biological systems. Attempts to develop more analytic approaches in the study of human variation and microevolutionary change have led students of man to largely abandon racial taxonomies as a source of hypotheses and problems on human adaptations.

The old racial history approach has been supplanted by two major approaches that do have a systematic integrity and that are currently found to be highly promising and productive in the study of microevolutionary change. The first of these is exemplified by the genetic analysis described in Chapter 5. In such studies the dependent variables of interest are gene frequencies at a defined locus, or a group of interacting loci in a breeding population. The second approach is the study of whole-organism responses to environmental variables. The dependent variables of interest are physiological responses, growth responses, or behavioral responses.

The independent variables involved in this second approach obviously could be quite diverse and could include disease vectors as described in Chapter 5. The independent variables usually dealt with, however, are a more limited set such as nutritional differences, temperature differences, altitude differences, and intensity of radiant energy in different parts of the solar spectrum. Many studies of whole-organism responses to these variables can be subsumed under the heading *climatic adaptation*. This is the subject of the present chapter.

Climatic adaptation in man, as in other animals, involves at least four different modes of response:

Genetic changes
Growth changes
Physiological changes
Behavioral changes

These are listed in order of increasing speed of response to changing conditions. Genetic changes are slow, intergenerational changes occurring not within individuals but between successive generations of a population. Growth changes, as the phrase is used here, refer to ontogenetic changes, changes that occur within the lifetime of the individual and which are relatively irreversible. Physiological changes are, for the most part, reversible. And behavioral change is in theory quickly responsive to environmental change.

The latter three modes of response are not mutually exclusive categories. The demarkation between what is growth response and what is a physiological response can be quite hazy. The same is true for the dividing line between physiological response and behavioral response. Despite this overlap these remain useful categories for discussing the adaptations expected or found to be made by an organism.

In studies of man the behavioral response takes on greater importance than is the case with any other animal. The presence of culturally patterned ways of dealing with the environment means that much of man's behavior, though in theory rapidly changeable, usually shows quite stable patterns that are transmitted nongenetically from generation to generation. This has proven to be a highly efficient mode of response, since it involves a kind of natural selection among behavioral forms and attitudes and not necessarily selection among individuals. This kind of "hereditary" change is less costly to a population and has permitted man to explore many exploitative possibilities for moving into new environments. For this reason man today inhabits a very diverse range of environments. In fact, the only other animal species enjoying an equal dispersion are those that are parasitic to, or otherwise dependent on, man.

Due to his great range of dispersal, it appears that man would be a good subject for the study of climatic adaptation, and he is. But due to reasons given in the preceding paragraph, these are primarily cultural adaptations to climatic differences. This has led some to suggest that cultural adaptations have *replaced* genetic adaptations and that we can expect to find no differences that are a result of climatic adaptation among human populations. But this conclusion is based on a false dichotomy. The occurrence of one kind of adaptation does not exclude the other. Human groups on the edges of the inhabited regions of the world have always tended to push or to be pushed into the less hospitable uninhabited regions. The limit on their ability to expand into such areas, for any given technology, has been biological—the ability to reproduce their numbers. This would lead us to believe that there may be microevolutionary changes in man based on climatic differences.

A far more important factor in cutting down microevolutionary differentiation in man in his great mobility. Whole populations may move from one ecological or climatic zone to another or, more commonly, gene flow may be continuous between populations in different zones. And as geological time is measured, man is a quite recent invader of the cold regions of the world.

A major problem in studying the genetic component of climatic adaptation in man is the frequent fact of unequal acclimatization in the samples compared. Acclimatization falls primarily within the realm of physiological adaptations. These are usually considered to be rapid changes. For some purposes a person may be considered acclimatized after he has spent six days in a new climate. But just as physiological changes shade imperceptibly into growth changes, there is no fixed point at which one can be said to be "fully acclimatized."

Frequently the control group has not been long resident in the region of climatic extreme where studies of local populations are conducted, or it is the group of investigators. Under these circumstances differential acclimatization must be considered. Further, the individual's microenvironment influences acclimatization.

Scientists working in Antarctica were found to differ greatly in their ability to tolerate cold stress, depending on whether their normal tasks were out-of-doors or within the base camp huts. One implication of such findings is that minor cultural differences can be mistaken, in their effects, for genetic differences in adaptation. If the control group has been in the same area as the sample group, but has been leading a different life style, there can still be differential acclimatization. On the other hand, minor cultural differences can result in unnoticed differences in exposure to climatic variables that can also lead to performance differences.

Another source of difficulty in these studies in the past has been that, too often, very small samples were compared. Studies of performance physiology usually require that a number of variables be measured and recorded. This means that much effort goes into the study of one individual. Therefore investigators have often had to be content with results from a small number of subjects. Generalizations based on such small numbers are in many cases not very secure.

There is another minor problem with the selection of suitable samples and controls which should be mentioned. If the search is for microevolutionary differences an effort must be made to compare groups whose ancestors have long resided under differing conditions. But finally the investigator works with the available subjects, and these may not provide ideal comparisions. For example, in cold-tolerance studies a common comparison has been between Eskimo and European or North American populations of European extraction. European and European derived populations are convenient for this purpose. But there is every reason to expect that the north Europeans would themselves be cold-adapted and would therefore show less difference in such a comparison than might exist otherwise. Indeed, if classic Neandertal is ancestral to modern Europeans, they may be as highly adapted to cold as anyone.

Despite the numerous methodological problems, there are studies whose results indicate that microevolutionary differences in climatic adaptation are real. One kind of evidence brought forward showed that certain so-called *rules* of ecological or climatic adaptation fit man as well as the species for which they were originally formulated. The three oldest and most commonly known statements of environmental correlation are Gloger's rule, Bergmann's rule, and Allen's rule. A brief paraphrase of each will be given.

Gloger's rule states that within a given species of bird or mammal lighter coloration occurs in the population occupying drier and cooler regions. Darker coloration occurs in populations occupying warm and humid regions. Bergmann's rule states that, within a given species, individuals of populations in colder regions tend to be larger than individ-

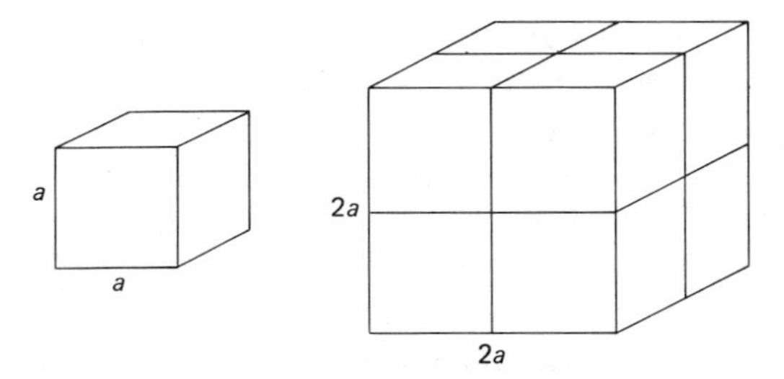

FIGURE 7.1 A change in size without a change in shape changes the ratio of surface area to volume. (The large cube differs from the smaller by a factor of two in linear dimensions and eight in volume.)

uals of populations in warmer regions. Allen's rule states that within a given species appendages tend to be lengthened in populations in warm regions, shortened in populations inhabiting cold regions.

Gloger's rule as originally stated (Gloger, 1833) had to do with the pigmentation of fur and feathers. But man, with his exposed skin, seems to conform roughly to Gloger's rule in skin pigmentation. Figure 7.5 on pp. 108–109 illustrates this. Although imperfect in that it does not show minor within region variation, it is a reasonable representation of broad pigment differences in man. Some possible evolutionary reasons why man should show this distribution will be discussed in the section on skin color later in this chapter.

Bergmann's and Allen's rules can be discussed together, as they both involve the same set of physical relationships. Loss of heat in a body of any shape depends primarily on the surface area of the body. The generation of heat in mammals and birds at rest (basal metabolism) depends primarily on lean body weight. Therefore one possible way in which heat load can be regulated is by changing the ratio of surface area to volume (assuming that weight per unit of volume is unchanged).

The relationship of Bergmann's rule to the preceding statements can be shown by a simple illustration. If we double the linear dimensions of a cube, as is done in Figure 7.1, the surface area increases as the square of the linear dimension, but the volume goes up as the cube of this dimension. Given our generalization about the physical basis of heat regulation, it is then obvious that of two animals having the same shape and body composition the larger individual can maintain thermal equilibrium more easily in a cooler climate; the smaller, in a warmer climate.

Allen's rule is concerned with changes in shape to achieve an optimal volume-to-surface-area ratio. In Figure 7.2 two cylinders are shown having the same volume, but the longer cylinder has twice the surface area. The appendages of birds and mammals can be thought of as systems of cylinders that can vary in the ratio of volume to surface area.

There is less evidence for Allen's rule than there is for the others in human populations. There is evidence for shorter lower extremities in cold-climate people, but not especially short upper extremities. For example, the ratio of sitting height to stature is higher in

FIGURE 7.2 A change in proportions of a shape changes the ratio of surface area to volume. (The longer cylinder has twice the circumferential area of the shorter cylinder but an equal volume.)

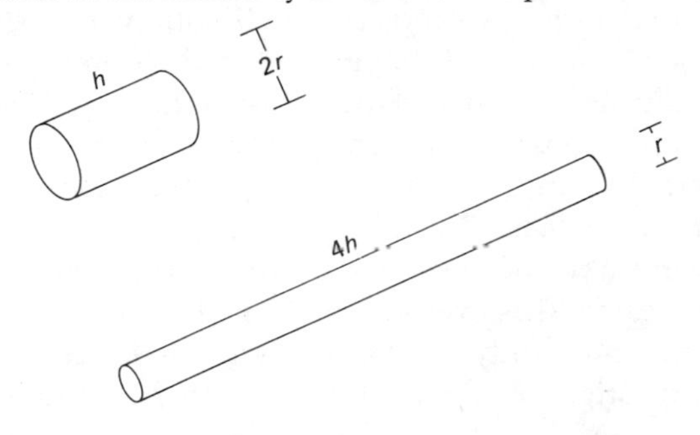

TABLE 7.1
Correlation Coefficients Between Body Sizes and Temperature

	MEAN ANNUAL TEMPERATURE	JULY NOON "EFFECTIVE" TEMPERATURE	MEAN JULY TEMPERATURE	MEAN JANUARY TEMPERATURE
Weight/surface area	−0.535	−0.404	−0.384	−0.587
Weight	−0.460	−0.319	−0.310	−0.528
Surface area	−0.359	−0.222	−0.223	−0.436
Stature	−0.044	+0.043	+0.007	−0.111

Source: R. W. Newman and E. H. Munro, "The Relation of Climate and Body Size in U.S. Males," *American Journal of Physical Anthropology*, 13 (1955), 7.

Eskimo populations, several indigenous Siberian groups, Koreans, and North Chinese than in most other populations of the world. There is no evidence, on the other hand, that arm length, relative to sitting height, is reduced in these cold-climate peoples.

The applicability of Bergmann's rule in man is fairly well established. One of the best studies of the relationship between weight and mean annual temperature was reported by D. F. Roberts (1953). Roberts examined the data for 116 male samples and 33 female samples from both the Old and the New World. Findings were comparable for males and females so only the results for the males, which were the larger number of samples, will be cited. Roberts found an overall correlation coefficient $r = -0.600$ between mean weight per sample and mean annual temperature. Since stature is highly correlated with weight ($r = -0.735$ in this study), he controlled for stature (by partial correlations) and found the correlation between weight and temperature still highly significant at $r = -0.538$. When controlling for weight he found the correlation between stature and temperature to be insignificant.

When Roberts classified his samples into ten ethnic groups he found there was a highly significant correlation between weight and mean annual temperature within groups as well as a difference between groups. Roberts also noted that where the data were good enough, as in Britain, the negative correlation between mean annual temperature and weight was evident between regions and between counties.

The latter observation suggests that body build differences are in part differential growth responses to environmental differences. Newman and Munro (1955) studied U.S. Army inductees and found the body-build variables to conform to Bergmann's rule within the United States. The body-build variables showed the highest negative correlations with mean January temperature, as shown in Table 7.1. This suggests that the annual temperature minimum is the most important climatic stress in this series. Indeed, when the winter temperature measure is held constant (by partial correlations), the summer temperature correlations become insignificant.

The United States is not genetically homogeneous, and some small part of this gradient in body build may be due to genetic gradients, but it is reasonable to assume that most if not all of the correlation is due to direct growth responses to climate. That man is generally quite plastic in response to environmental differences has been amply documented by Kaplan (1954). Reduction in growth rate has been shown in studies of migrants from temperate to tropical climates.

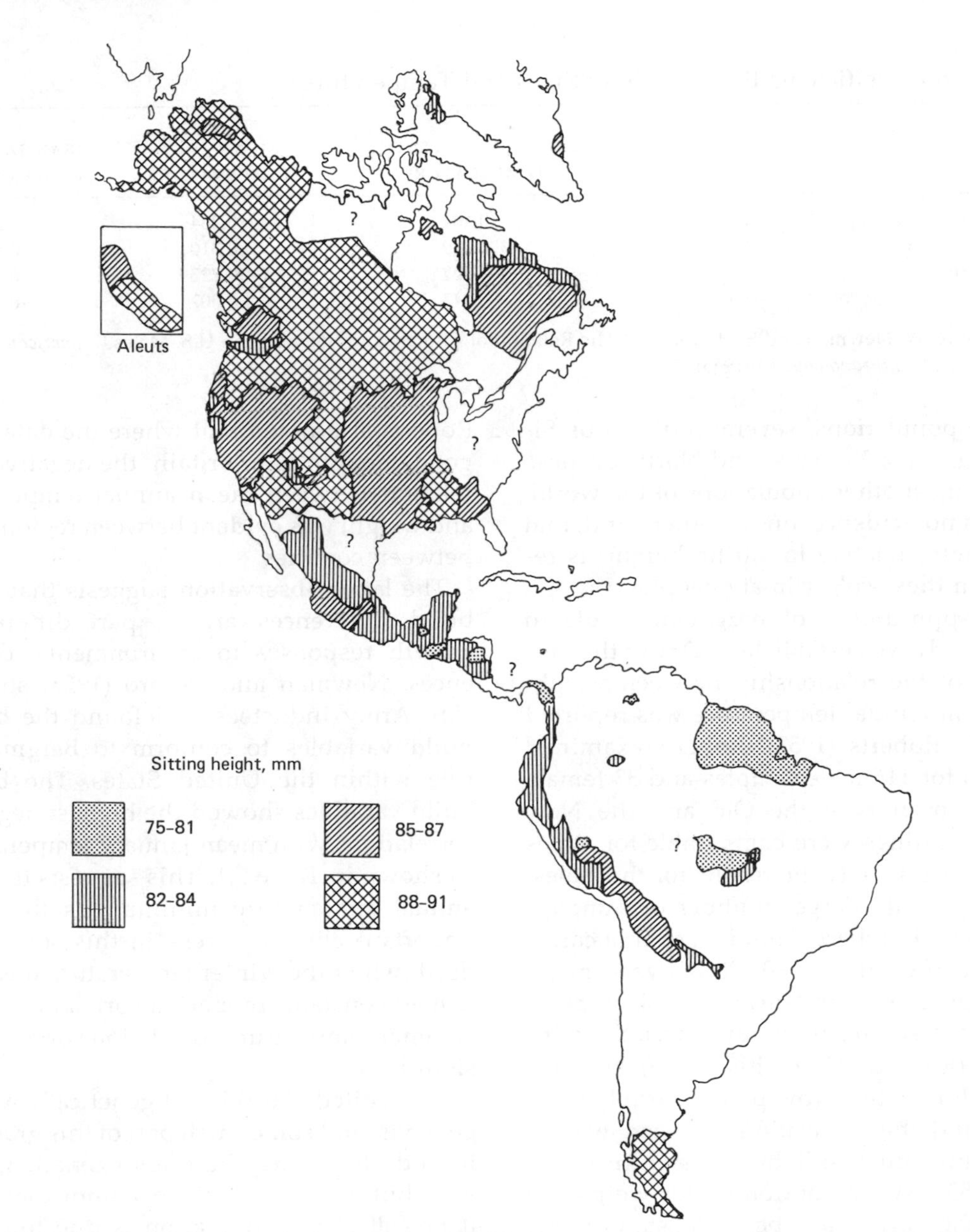

FIGURE 7.3 Sitting heights of living New World aborigines. (*Source:* Newman, 1953. Used with permission.)

Eveleth (1966) has shown that American children raised in Rio de Janeiro, Brazil, weigh less and are more linear in build than than their midwestern U.S. contemporaries even though food habits and rearing were comparable. They did not show differences in the proportion of limb length to body length. It seems clear from this and similar studies that growth rate in man is affected by climate, and the result is in a direction promoting thermal equilibrium.

Newman has investigated the applicability of Bergmann's and Allen's rules to Amerindian populations (Newman, 1953). Again Bergmann's and Allen's rules appear to describe the data well. Indeed they may fit man better than they do most animals. One of several distribution maps given by Newman is shown in Figure 7.3.

Other approaches are required to get to the "why" type of questions concerning the correlations expressed in the ecological rules, and concerning other known correlations between climate and morphology. Two general approaches have been used and these have been applied in both field and experimental situations. The first approach is to focus on a single *dependent* variable and to ask what are the independent variables influencing it. And further, what is the magnitude of effect of each independent variable? This is the approach used when we ask what determines skin color, or more precisely, human pigmentation.

The second approach involves the selection of a single *independent* variable and then asking what are the dependent variables, in human physiological response, that are influenced by this variable. This approach, since it does not focus on a single resultant, does not conform to a straightforward or syllogistic model of scientific method, as could be used to describe the first approach. But most studies in climatic adaptation fall into this category. This is the approach used when we ask what are the effects of heat stress, cold stress, or high altitude.

HUMAN PIGMENTATION

Human skin color is determined by a number of factors. The most important of these are melanin, occurring in granules called *melanosomes* and the red color of hemoglobin of the dermis. In dark-skinned persons the melanin is dense enough to mask the color of the underlying hemoglobin of red blood cells when viewed with the unaided eye. In light-skinned persons the hemoglobin gives a pinkish cast to the skin. In this section we will view some of the suggestions about the adaptive significance of differences in melanin concentration.

The oldest evolutionary explanation for Gloger's rule in birds, and mammals other than man, is that this represents cryptic coloration. A light color should be of advantage in prey-predator relationships in northern latitudes, and a dark color in tropical latitudes. Although such a suggestion has also been advanced for man (Cowles, 1959) it has not been taken seriously. Most ideas on human pigmentary differences revolve about the regulation of the amount of ultraviolet light energy reaching the dermal layer of the skin. To understand these ideas we need to know a little about the structure of the skin.

The skin is composed of two distinct layers, the underlying dermis or corium and the epidermis. The dermis contains living, dividing cells served by nerves, venules, arterioles, and capillaries. The upper surface of the dermis is convoluted by papillae (conical protuberances). The density of the papillae varies directly with the sensitivity of the skin region to touch. On the contact surfaces of the hands and feet of primates, these

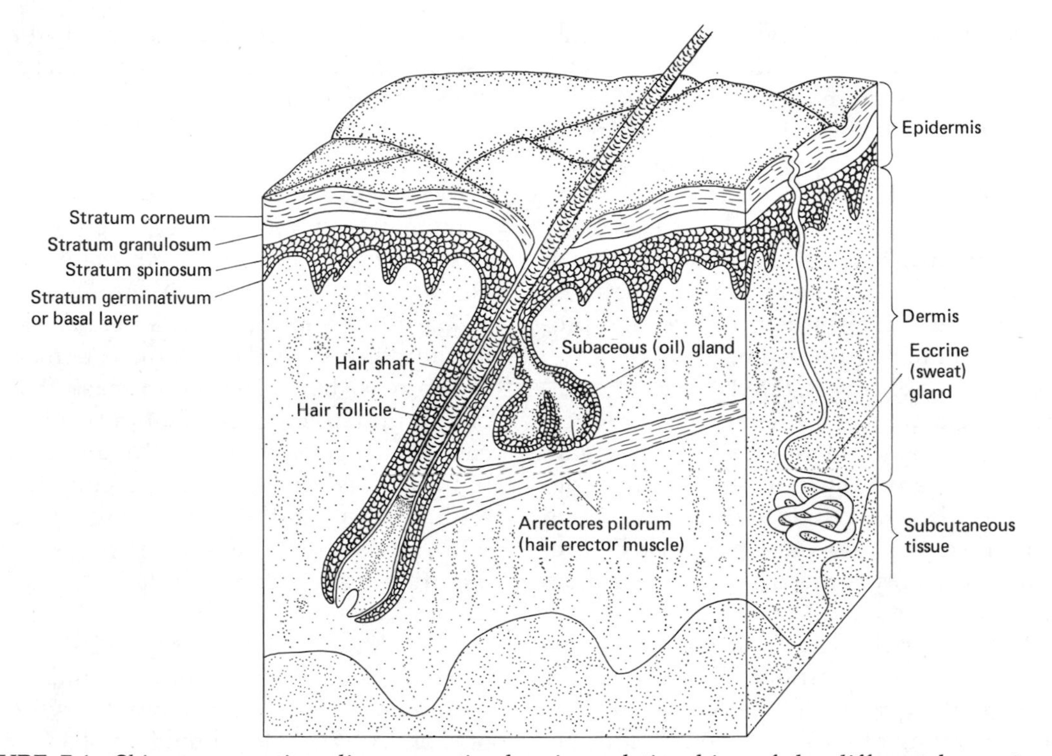

FIGURE 7.4 Skin cross section diagrammatic showing relationships of the different layers to each other.

papillae are aligned in closely packed double rows under the dermal ridges that produce finger, palm, and sole prints.

The epidermis is divided into a number of strata (Figure 7.4). It can be thought of as consisting of two layers, the lower malpighian layer and the outer horny layer (*stratum corneum*). The malpighian layer is composed of living cells, but cells above the basal layer are no longer undergoing mitotic division and are, in a sense, dying as they migrate upward. The horny layer consists of dead cells having no nucleii, flattened and scale-like in shape (squamosal), and containing a high percentage of keratin, a tough, insoluble protein. The malpighian layer can be further divided into a basal cell layer (*stratum germinativum*), a spinous layer (*stratum spinosum*), a granular layer (*stratum granulosum*), and, on palms and soles, a clear layer (*stratum lucidum*) that is transitional to the epidermis.

The melanin-producing cell, the melanocyte, occurs at the interface between the dermis and epidermis with its dendritic projections rising into the columnar malpighian cells, to which it supplies melanin. Melanin granules vary from yellow-brown to red to black in color. Light reflected by melanin through a thick epidermal layer can appear bluish.

Groups of people who differ greatly in skin color do not differ in the number and distribution of melanocytes. They differ in the rate of production of melanin within the melanocytes. Differences in this rate of production of malanin between individuals are primarily genetic differences. But the rate is also influenced by nutritional and hormonal differences and by the absorption of radiant energy.

This brings us back to the problem of why genetic differences exist in skin color and why a similar cline with latitude exists in all parts of the world. In 1919 it was discovered that artificially produced ultraviolet irradiation would cure rickets. Between this date and 1922 the ultraviolet portion of direct sunlight and cod liver oil were also found to be antirachitic in action. Up to this time rickets had been a serious disease in Europe. Rickets manifests itself in soft and easily deformed bones. And in rachitic women with deformed pelves the chance of death at birth, of both mother and infant, was high.

In 1934 Frederick Murray combined these factors into a hypothesis to explain selective advantage in skin color (Murray, 1934). He noted that reports indicated that Negro children in the United States were more prone to rickets than were white children. In the absence of a fish diet, vitamin D (calciferol) could be produced only by ultraviolet irradiation of the skin. Vitamin D was essential to the normal calcium and phosphate metabolism that prevented rickets. Heavily pigmented skin would absorb rather than transmit ultraviolet rays more than lightly pigmented skin. The intensity of ultraviolet in the solar spectrum drops rapidly in higher latitudes. Therefore there would be selection for lightly pigmented individuals in higher latitudes, as they could better utilize the ultraviolet rays available. Murray further documented the fact that light skinned persons in the tropics suffer a variety of difficulties and skin diseases which he assumed were related to depigmentation. These conditions would favor dark-skinned persons in lower latitudes.

With many modifications, this is the currently dominant hypothesis—or set of hypotheses—on selective differentials in skin color. We will now review in a bit more detail some of the evidence and arguments for and against this hypothesis.

The distribution of skin color in the Old and New Worlds is shown in Figure 7.5. These distribution maps are admittedly not ideal since they rely on earlier studies which, for the most part, judged skin color by visual comparison to a standard set of colored tiles rather than by the more recent use of reflectance photometry. Also, they involve interpolations across areas where no data are available. But overall they give a reasonable picture of the distribution of grades of pigmentation in man. The distribution of ultraviolet radiation is shown in Figure 7.6. The general correspondence between the distributions of pigmentation and ultraviolet intensity can be seen by a comparison of the maps.

Short wavelength radiation is generally destructive to metabolic processes. Radiant energy above 320 nm is not absorbed by molecules of protoplasm as are wavelengths shorter than this. Shorter wavelengths that are absorbed can provide the energy to break chemical bonds. This may lead to a change in the structure of the molecule. Perhaps even more important is the production of "free radicals," which are positively or negatively charged particles that can react with other molecules. Ozone in the upper atmosphere absorbs solar radiation shorter than 290 nm. This means that the biologically significant part of the ultraviolet spectrum is that short range between 290 and 320 nm. Intensity of radiation in the ultraviolet range cannot be judged visually, as the visible spectrum ranges from 400 to 700 nm. The intensity of ultraviolet radiation drops off much more rapidly from the equator than does visible light. At a latitude where the sun is 35° or

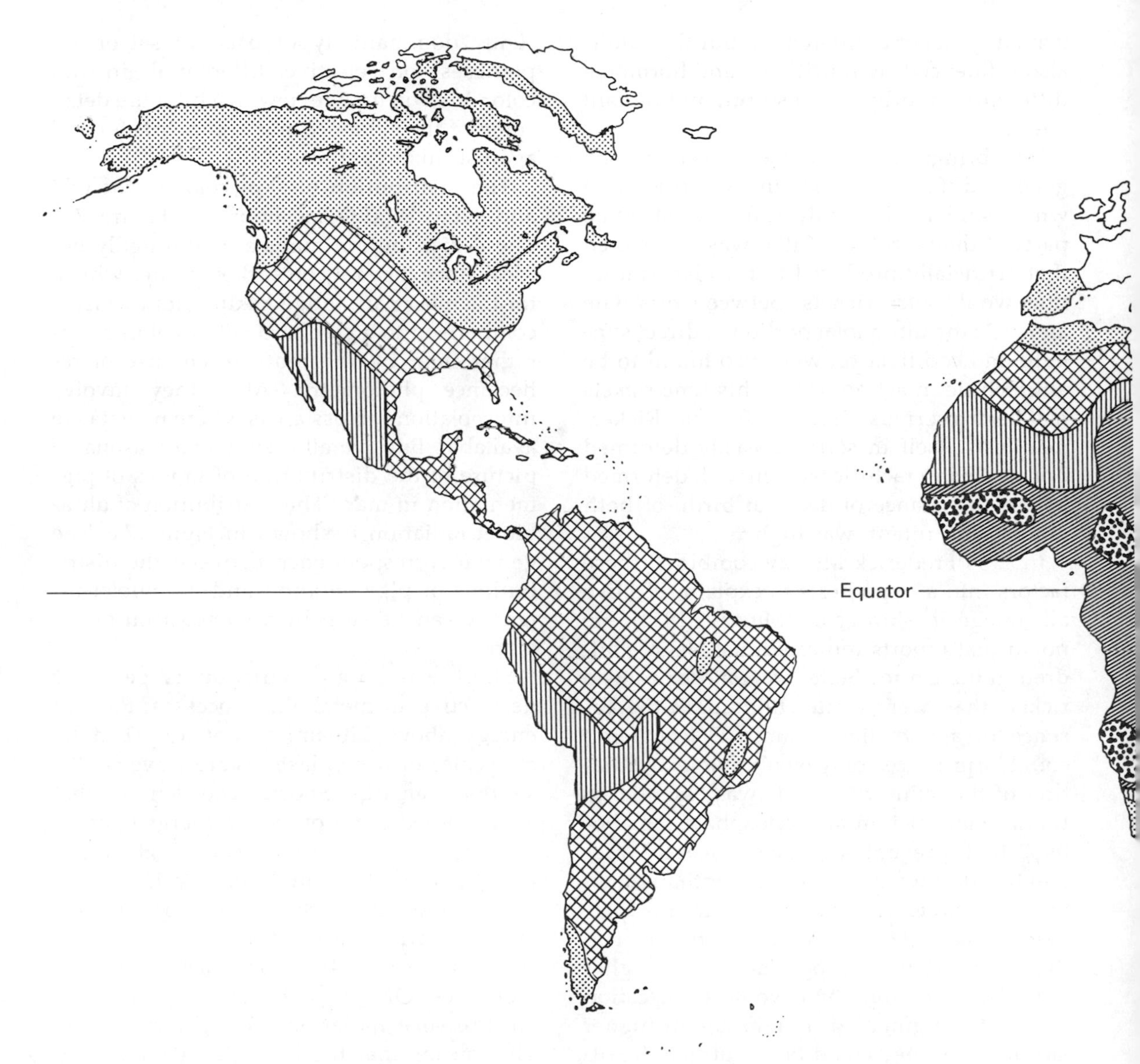

FIGURE 7.5 Skin color distribution before 1400 A.D.

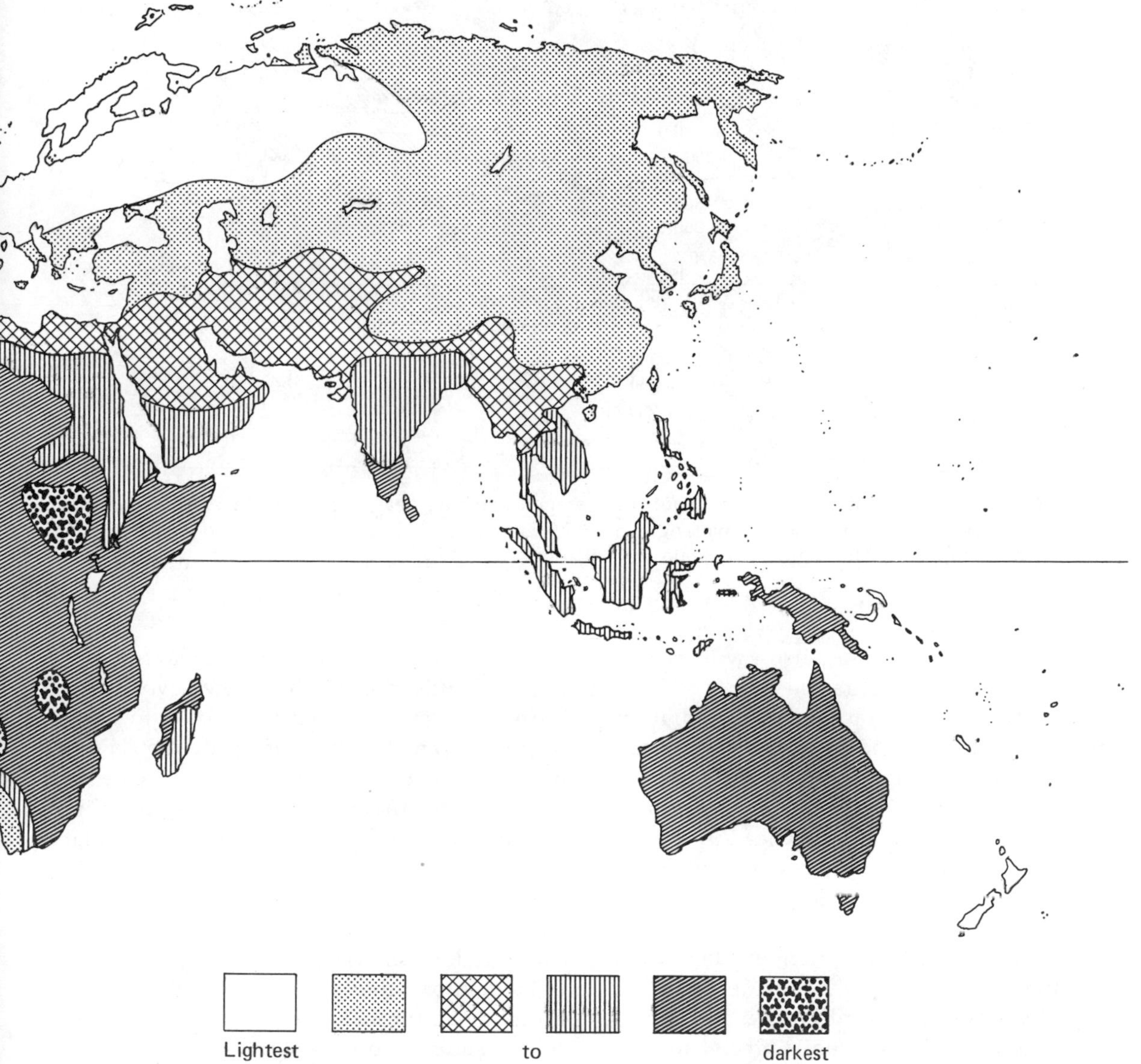

Lightest
to
darkest

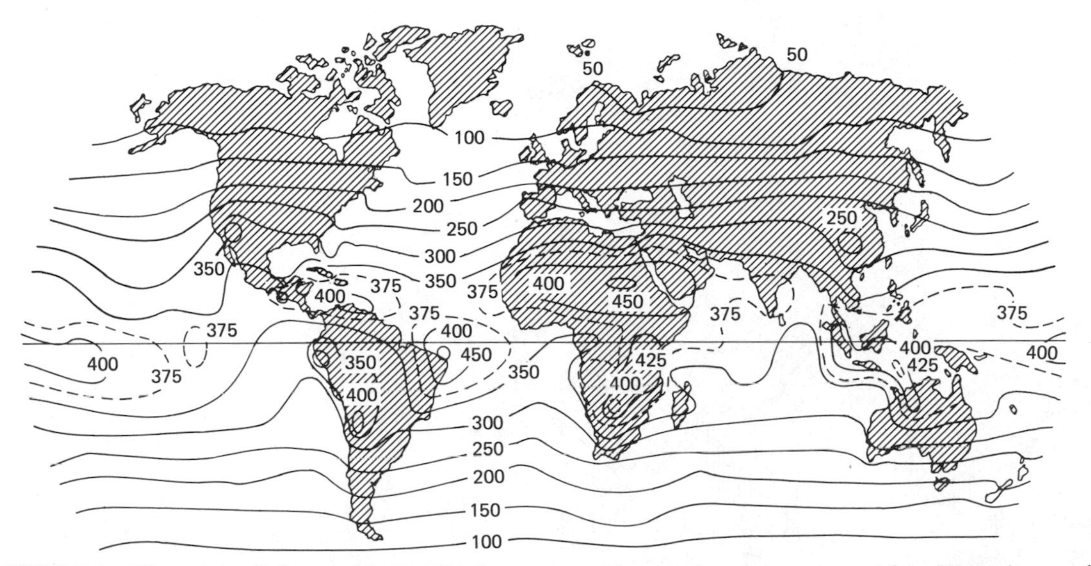

FIGURE 7.6 Ultraviolet light world distribution averaged over the year, measured in W-sec/sq cm for a 10 nm bandwidth centered on a wavelength of 307.5 nm. (*Source:* Urbach, Davis, and Forbes, 1966. Used with permission. This figure uses data from Schultz, 1966.)

less above the horizon a person will receive no effective ultraviolet rays.

The amount of ultraviolet (UV) radiation impinging on the individual is influenced by many factors other than latitude of course. There is greater UV radiation at higher altitudes. There are great differences in the reflectivity of various surfaces to UV. Mist, smoke, smog, or overall cloud cover cuts out UV rays entirely.

The effects of UV radiation on the human skin have been generally considered under three headings: (1) sunburning effects, (2) carcinogenic and mutagenic effects, and (3) promotion of the endogenous production of vitamin D. All of these effects seem confined to the UV wavelengths. Therefore it has generally been assumed, as it was assumed by Murray, that melanin concentration is a result, evolutionarily, of selection that optimizes fitness in these three vectors simultaneously.

Mild sunburn involves erythema (reddening) and mild discomfort. More severe sunburn involves increased sensitivity to pain, blistering, and peeling of epidermal layers with loss of melanized tissue and lowered resistance to further sunburn, susceptibility to secondary infection, damage to vascular beds in the dermis, and loss of sweating in the damaged areas. Thompson (1951) has shown that even mild sunburn can significantly reduce the rate of sweating. This can, in turn, impair the individual's ability to lose excess heat through evaporation. Repeated severe sunburn obviously would be a factor in the fitness of individuals.

Before going on to consider carcinogenesis and vitamin D production in response to UV radiation, let us see what kind of physiological acclimatizations occur in the skin as a response to UV. The two major changes that we need consider are, first, thickening of the *stratum corneum* and, second, darkening of the skin. Both of these responses reduce further penetration of UV rays into the skin.

Increased cell division at the base of the epidermis and the rapid migration of cells to the surface produce a thicker epidermis. A thicker epidermis reduces the penetration of UV rays in at least three ways. First, many proteins, including keratin of the epidermis, absorb UV. This absorbed energy is probably then reradiated as heat. Second, in dark-skinned persons at least, a thicker epidermis means a thicker layer of melanoid particles to absorb UV. And third, the scalelike structure of the corneum is an efficient way to defract and reflect UV rays. The defraction and reflection increase the length of the path that they travel and therefore increase the probability that they will be absorbed before reaching lower strata.

The skin darkening or "tanning" with exposure to ultraviolet can be thought of as a three-part process: immediate pigment darkening, migration of preformed melanin, and the formation of new melanin. Immediate pigment darkening is a rapid change resulting from the oxidation of melanin already present in the malpighian layer. As cells in the malpighian layer migrate upward from the basal layer, the melanin becomes reduced and is lighter or even colorless. In the presence of oxygen, UV radiation—or even radiation through the range of visible light—oxidizes this reduced melanin to a darker form. Immediate pigment darkening occurs at the moment of exposure and fades away completely within two to three hours (Pathak, 1967). Another effect associated with immediate pigment darkening, and which may account for some of the apparent color change, is that the melanosomes (the melanin granules) of the malpighian cells become condensed into a tightly organized cloud or shell around the nucleus of each cell.

The migration of preformed melanin is a simple consequence of the accelerated rate of cell migration mentioned under sunburn effects. The formation of new melanin can be observed as darkening that occurs 48 to 72 hours after exposure to ultraviolet. In some manner UV radiation seems to serve as an inducer for heightened tyrosinase activity. Tyrosinase is the enzyme important in the first two steps of the metabolic pathway leading from tyrosine to melanin.

The mutagenic and carcinogenic effects of UV radiation will be treated here as a single phenomenon manifesting itself in the form of skin cancer. There are three kinds of associations that point to the role of UV light in the genesis of skin cancer. There is a positive correlation between ultraviolet intensity and the prevalence of skin cancer among light-skinned persons. There is a negative correlation between pigmentation and skin cancer for a given region. And finally, among light-skinned individuals, skin cancer occurs primarily on areas of the body most exposed to sunlight and is roughly proportional to the frequency and length of exposure. These findings in man are supported by experimental induction of skin cancer in mice by ultraviolet irradiation.

It is a moot point at present as to how important skin cancer might be as a selective agent. Blum has argued that skin cancer has had no significant evolutionary effect (Blum, 1959). Most skin cancers do not represent a serious disability. Since they are due to long, repeated exposures they do not occur with appreciable frequency before middle age (except among albinos). Some have argued that even benign skin cancer can provide a site of secondary infection. Another suggestion is that the disfiguring effects of skin cancer might lead to sexual selection (Daniels, 1964, p. 981).

Blum has also argued, quite aside from the problem of whether or not skin cancer affects

Darwinian fitness, that melanin is unimportant in screening ultraviolet radiation. He argues that the thickness of the epidermis is the major variable controlling ultraviolet penetration (Blum, 1945; 1961). It is true that the thickness of the epidermis strongly affects UV transmission. The thickening of the epidermis subsequent to UV exposure can be considered an adaptive response that decreases the likelihood of further damage.

But the amount of melanin is an important between-group variable in regulating UV transmission. M. L. Thompson compared the thickness and UV transmission of the epidermis of 22 European and 29 African subjects in Nigeria. The epidermis was separated by a blistering agent applied to the side of the hip where clothing protected the skin from the sun's rays in all subjects. Thompson found no significant difference between means for European and African epidermis thickness. He found great variation in the amount of pigmentation in the African subjects. He found that the transmission of ultraviolet through the excised pieces of epidermis varied inversely with the amount of pigmentation and that the one albino African transmitted approximately as much UV radiation as did the Europeans. In his words, "These measurements are, therefore, consistent with the theory that differences between the two races in transmission per unit thickness, as well as scatter about the regression line, are caused by differences in pigmentation" (Thompson, 1955, p. 243).

What are the genetic systems involved in pigmentary differences in man? We simply don't know. Estimates of the number of loci involved range from 3 to 6.

Does melanin act only as a barrier to ultraviolet light? Probably not. It is more reasonable to assume that it has a number of "functions" in the sense of regulatory relations to other physiological processes. But these may be related, in turn, to photodynamic effects. Commoner and others (1954) showed that melanin has a stable free radical structure. In other words it contains unpaired electrons capable of pairing with other electrons. The implication of this is well stated by Mason and others (1960, p. 229) who say that

> melanin may act in *some* organisms as a biological electron exchange polymer able, by means of its capacity for oxidation and reduction, and its stable free radical state, to protect a melanin-containing tissue or associated tissues against reducing or oxidizing conditions which might otherwise set free within living cells reactive free radicals capable of disrupting metabolism.

And, as was mentioned earlier, an immediate effect of UV exposure is the reorganization of melanosomes in the malpighian cells into a perinuclear sphere. This suggests that the melanosomes do function as a protective device for the nucleus against disruption by free radicals.

The other side of the question is "Why light skin?" In the theory put forward by Murray, light skin was an adaptation to the need for vitamin D in regions where populations get very little ultraviolet during large parts of the year. Vitamin D_3 (cholecalciferol) is produced in the malpighian layer of the skin. Ultraviolet radiation must reach this stratum to convert the precursor substance (7-dehydrocholesterol) to vitamin D_3. Vitamin D can also be ingested but the only natural foods having significant amounts of vitamin D are the bony fishes. Archeologically we find no evidence of marine fishing prior to the Upper Paleolithic. This means that man, up until relatively recent times, had to get by with endogenous production of vitamin D. Insufficient production of vitamin D over a number of months can result in decalcification of bone. The consequent distortion of the bones of an individual is recognized as rickets in infants, as osteomalacia in adults.

In the absence of dietary supplements, skin color has been a critical variable in many recorded clinical cases of vitamin D deficiency leading to rickets. Such reports, from both North America and Europe, indicate that Blacks are more susceptible to rickets than are Whites. Even Whites living in the smoky early industrial cities or living in a sufficiently northerly latitude suffer from lack of ultraviolet. The early Norse colony on Greenland, which finally became extinct, suffered heavily from rickets. And Ronge has shown that Swedish children, when supplied with ultraviolet in the school during the winter, do not show the seasonal decrease in physical fitness found otherwise (Ronge, 1948).

W. Farnsworth Loomis has added another piece to this picture by suggesting that hypervitaminosis D can have severe deleterious effects. In other words, excess vitamin D can produce excessive calcification: the calcification of soft tissues, and the production of kidney stones that are frequently fatal if surgical removal is not possible (Loomis, 1967). This leads Loomis to suggest that the amount of melanin regulating vitamin D production is not only important in higher latitudes but also in the tropics. Studies have yet to appear that will show whether or not light-skinned individuals are subject to hypervitaminosis D in the tropics.

Few anthropologists doubt that the distribution of skin color as it is known in historic times is adaptive to climatic variables. And undoubtedly a major climatic variable, if not the major variable, affecting this distribution is the distribution of ultraviolet radiation. Ultraviolet radiation has other systemic effects that may also be important but that have not been reviewed here. Melanization in the skin has effects on other traits, such as eye color. It has been suggested that visual acuity under differing conditions is influenced by eye color (Daniels, 1964). All of this suggests that while we certainly do not have the whole story, we can be equally certain that we do have an important part of the story of human skin color.

COLD ADAPTATION

Men have long been fascinated with the ability of other groups of people to tolerate climatic extremes. Tierra del Fuego at the tip of South America is such a place. The climate is wet, cloudy, and cool, and temperatures are often little above freezing. The tolerance to cold of the local inhabitants was noted by Charles Darwin, who said in an often quoted paragraph:

> The climate is certainly wretched; the summer solstice was now passed, yet every day snow fell on the hills, and in the valleys there was rain, accompanied by sleet. The thermometer generally stood about 45°, but in the night fell to 38° or 40°. . . . While going on shore we pulled alongside a canoe with six Fuegians. These were the most abject and miserable creatures I anywhere beheld. On the East coast the natives, as we have seen, have guanaco cloaks, and on the West, they possess sealskins. Amongst these central tribes the men generally have an otterskin, or some small scrap about as large as a pocket-handkerchief, which is barely sufficient to cover their backs as low down as their loins. It is laced across the breast by strings, and according as the wind blows, it is shifted from side to side. But the Fuegians in the canoe were quite naked, and even one full-grown woman was absolutely so. It was raining heavily, and the fresh water, together with the spray, trickled down her body. In another harbour not far distant, a woman, who was suckling a recently-born

> child, came one day alongside the vessel, and remained there whilst the sleet fell and thawed on her naked bosom, and on the skin of her naked child. . . . At night, five or six human beings, naked and scarcely protected from the wind and rain of this tempestuous climate, sleep on the wet ground coiled up like animals (Darwin, 1839, pp. 234–236).

Even more than the Yahgan or Alacaluf of Tierra del Fuego, the Eskimo and other far-northern people have excited the interest of temperate-climate man by their ability to survive under conditions of extreme cold. Man evolved as an Old World tropical primate. Man's closest primate relatives remain confined to narrow ecological zones of the humid tropics. Man, on the other hand, has extended his range enormously. This expansion has been due to developments in technology. But this in turn has placed groups of men under conditions of marked cold stress. Most populations living in circumpolar regions today are descendants of populations that have been north of 45° latitude since at least the beginning of the Upper Paleolithic and perhaps for the entire Upper Pleistocene.

Has man in his 1000 or more generations in this nontropical environment been able to adapt, in part, because of his biological responses? And if biological adaptations are detectable do they, in part, represent microevolutionary change? We cannot answer such questions with finality. But we do have some intriguing bits of evidence, and these will be briefly reviewed in the following pages of this chapter.

At the broadest level we can see that any animal has possibly two major adaptive strategies available in the management of cold stress. With lowered ambient temperature it can either produce more heat or it can save heat more effectively. In cold-adapted arctic mammals the second strategy, that of heat saving, most differentiates them from closely related temperate or tropical forms. The primary way of saving heat is by means of increased insulation. This is usually achieved by a thickened fur and by thickened subcutaneous deposits of fat. Some arctic animals hibernate during the coldest part of the year as a way of saving heat. Heat exchange mechanisms also conserve deep body heat while steeply dropping superficial temperatures. Such heat exchangers are the *venae comitantes*, the veins of the deep tissues of the body and the extremities. They lie parallel to the arteries and in some places form a simple network around the artery. Such close juxtaposition of the incoming and cooled venous blood with the outgoing and warm arterial blood provides an opportunity for heat exchange. This warms the venous blood prior to its reaching the body cavity and helps maintain the critical core temperature at the price of dropping the surface temperature of the extremities even further.

Man must maintain the temperature of his brain, thoracic viscera, and abdominal viscera at approximately 37°C. Other tissues may have a slightly lower "normal" temperature. The temperate-climate individual produces at rest approximately 39 Calories of heat per square meter of surface area per hour. This is the "basal" metabolic rate. This basal rate is sufficient to maintain normal body temperature at temperatures only as low as 27°C for the nude individual resting in still air and with no proximate cold objects to absorb radiant energy. Any worsening of these conditions and additional heat-producing or heat-saving mechanisms must come into play.

Considering only man's biological adaptations to cold, we can distinguish at least six mechanisms that can come into play when heat requirements are greater than those met by basal metabolism. The first is, of course, voluntary muscular activity during waking hours. The second mechanism is shivering, which can produce two to three times the resting heat production for short periods. Another mechanism is vasoconstriction, the constriction of superficial veins that reduces blood flow to the skin surface and conserves deep body heat (the "core" temperature) while dropping the skin temperature. Another effect of vasoconstriction is that more venous blood returns through the deep veins and is therefore warmed before reaching the viscera.

With acclimatization to cold, some groups exhibit an increase in the basal metabolic rate (BMR). This has therefore been called *metabolic acclimatization.* European test groups have been found to increase their BMR a maximum of about eight percent with cold acclimatization. A fifth mechanism that can be observed, but which seems to have little or no importance to heat conservation in man, is the formation of "goose bumps" from the contraction of pilomotor muscles attached to each hair follicle. In animals with a furry coat the erection of the hair shafts creates a layer of air turbulence and a lower layer of dead air. This lowers the rate of convective cooling. In man this appears to have little significance.

A final mechanism that can be recognized is the occurrence of postural changes in response to cold. These are responses that reduce the surface area exposed to cold. They occur in all animals, although they have been given little attention. In man they include hunching the shoulders, curling or clenching the hands, holding the arms close to the body, sleeping in a curled position, and so forth.

In addition to these biological responses man copes with his environment through technological means. In dealing with cold this involves devising clothing, housing, and heating that will help to avoid cold stress as much as possible. Indeed, the success of Eskimo cold-climate technology has led one investigator to suggest that Eskimos have been able to "cheat evolution" by "surrounding themselves successfully with a little piece of the same tropical microclimate upon which we also depend" (Scholander, 1955, p. 23). But as Newman pointed out (1956) the arctic cultural adaptation of the Eskimo did not spring fullblown from the minds of a tropical people moving into the arctic. It was the result of long-term experimentation as these people pushed further and further into an inhospitable climate. Even today there are limits to the protection provided by technology. To live, to hunt, and to fish outdoors in the arctic requires considerable exposure of the hands and face. In addition to these normal cold exposures, we must consider the very common sense idea that it is the unusual cold exposure, always a possibility in the arctic, that is critical. The more severe types of cold exposure that may occur only once in five or ten years or more are amply documented by anthropologists and arctic travelers.

We now ask, are there differences between groups of man in their response to cold, and what is the nature of these differences? Clearly there are irreversible growth responses to cold among human groups. Growth differences in terms of overall body size is shown in the study of Newman and Munro cited earlier in this section. Newman (1960) showed that 50 percent of the variance in mean weight between Amerindian groups of the New World can be accounted for by the mean coldest-month temperature of the region. This may of course be due to genetic variation between groups as well as to growth plasticity.

All groups of men show some ability to acclimatize to cold. As mentioned earlier, scientists who normally worked outdoors in

Antarctica were found to be able to expose their hands and faces longer without frostbite and to better predict the onset of frostbite than those who worked indoors. Europeans in the Arctic develop dietary preferences favoring fats, and their basal metabolism rate rises by as much as a mean of eight percent. Other kinds of acclimatization are easily documented. An important question from the standpoint of adaptation is, are there differences between groups of men in their ability to acclimatize to cold?

There is certainly no unanimity on the answer to this question but the bulk of the evidence seems to indicate that there are such differences. One kind of evidence comes from the experience of American army units that have found great differences in the rate of cold injuries suffered by White and Black troops. Of the American combat troops in Korea in the winter of 1951–1952, 9 to 10 percent were Black but they suffered 41 percent of the total frostbite injuries. This means that the rate of frostbite incidents per individual was about 7 times as high ir Black as in White troops. Approximately the same was found when the data were controlled for region of origin, northern or southern United States. Furthermore, among soldiers who suffered frostbite of the feet the Black soldiers not only showed the higher incidence but showed greater severity of effect. Again this remains true when the data are divided by region of origin. The same tended to be true of frostbite of the hand (Schuman, 1954).

There has been some question as to whether or not differences in equipment use could not account for the marked differences in frostbite reported from Korea. But similar figures have been reported from Army maneuvers in Canada and Alaska, where equipment and clothing use was better controlled. Meehan (1955) mentions army winter field exercises in Alaska in 1954 in which 10 percent of the troops were Black and suffered 50 percent of the cold injuries.

Studies of experimentally induced cold stress, although less dramatic, have provided evidence with greater control over extraneous variables. The two common kinds of cold stress experiments involve either whole-body cooling in air or the cooling of extremities, the hand or fingers, by immersion in an ice water bath. Brown and Page (1953) carried out one of the first comparative studies of this kind. They compared the responses of Eskimos of Southhampton Island (Canada) with those of medical students in Kingston, Ontario. One set of observations was made with the subjects lightly clothed and at rest in the room at an air temperature of 20°C. The Eskimos were comfortable at this temperature but the Canadian Whites felt cool. The Eskimos maintained a higher average skin temperature throughout and a higher rate of blood flow in the hand (Figure 7.7).

When the hands and forearms of the subjects were immersed in cold water the surface temperature of the hand dropped more rapidly in the Whites than in the Eskimos. The surface temperature of the Whites dropped to its equilibrium temperature within 10 minutes in baths of 5°, 10°, and 20°C. The Eskimos took 15 minutes to reach equilibrium at 5°C and 30 to 40 minutes to reach equilibrium in the 10° and 20°C waterbaths. Brown and Page noted that the ability to maintain higher temperatures in cold water for a greater length of time may be related to with the fact observed by others that the Eskimos are able to maintain tactile sensitivity and manipulative ability in the cold longer than many other peoples.

If one is asking, as we are, if there is any evidence for adaptive (genetic) differences in cold response, it is important to compare subjects who represent groups most likely

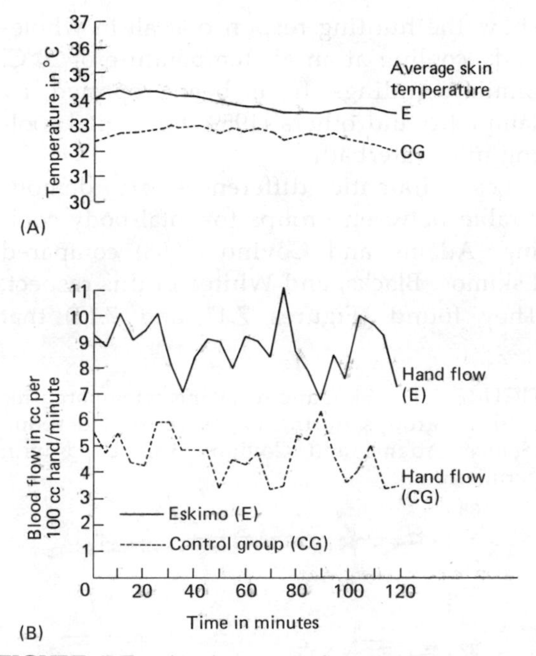

FIGURE 7.7 (A) Average skin temperature of the hand and (B) blood flow in the hand in Eskimo and in White "control groups." (*Source:* Brown and Page, 1952. Used with permission.)

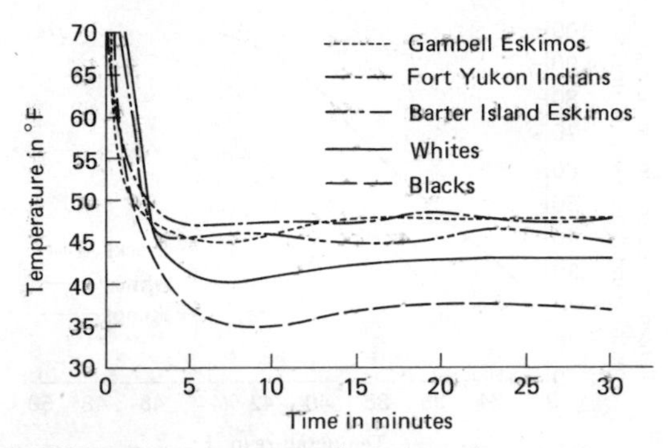

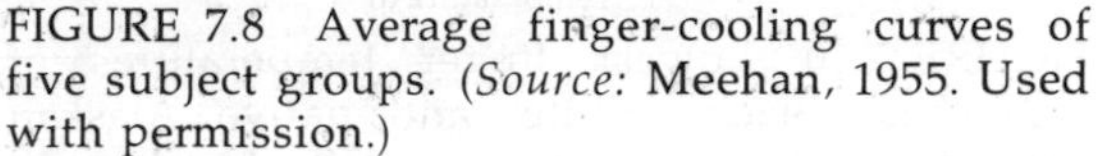

FIGURE 7.8 Average finger-cooling curves of five subject groups. (*Source:* Meehan, 1955. Used with permission.)

to have diverged in this respect. The subjects used by Brown and Page are not ideal by this criterion. The group designated White, or Caucasian, includes peoples who range from those long resident in subarctic areas to people of semitropical regions. Somewhat more suitable in this respect is a study reported by Meehan (1955).

Meehan studied cold response of the fingers of five groups. Three groups were native Alaskans: Barter Island Eskimo, St. Lawrence Island Eskimo, and Alaskan Indians from Fort Yukon. In most comparisons these three groups were considered a single group since their responses did not differ significantly. The remaining two groups were American Black servicemen and American White servicemen stationed at Ladd Air Force Base, Alaska. Most of the individuals of the latter two groups had been born in the southeastern part of the United States.

Thermocouples were attached to the index fingers of subjects and the fingers were immersed in ice water baths for 30 minutes. Figure 7.8 shows the mean rate of cooling registered for the various groups. Even more important as responses to cold are the mean temperature and lowest temperature reached by each subject in the different groups. These are shown in Figures 7.9 and 7.10, respectively. Only 4 percent of the native

FIGURE 7.9 Mean finger temperature of individual Black, White, and native Alaskan subjects. (*Source:* Meehan, 1955. Used with permission.)

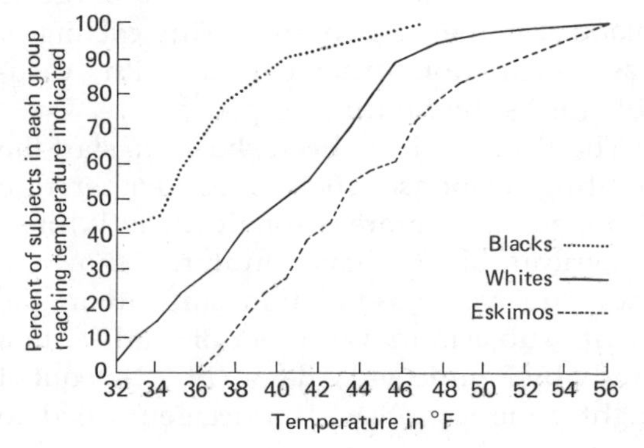

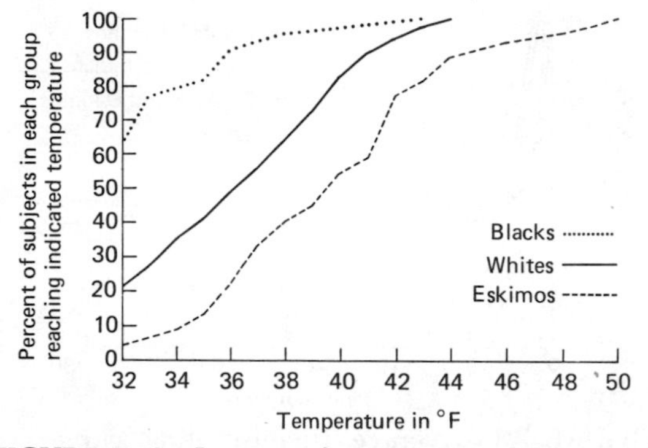

FIGURE 7.10 Lowest finger temperatures of individual Black, White, and native Alaskan subjects. (*Source:* Meehan, 1955. Used with permission.)

Alaskan subjects reached a low of 0°C, whereas 22 percent of the Whites and 63 percent of the Blacks did so.

Much of the difference in response of the groups reported in these and similar studies is related to the time of the onset of what has been called the "hunting" response. As noted earlier, one response to cold is vasoconstriction, which conserves heat by reducing peripheral blood flow. This seems to be a quite efficient adaptation to moderate cold where there is no danger of frostbite. If there is danger of damage to the tissues due to freezing this is obviously a poor response. But in most people we can observe a spasmodic vasodilation that permits a surge of blood to superficial tissues. This cycling of vasoconstriction interspersed with vasodilation is the hunting response.

The Eskimos have been shown to start the hunting response sooner, as temperature dropped, and more strongly than Whites. American Blacks have, in turn, shown a more delayed vasodilation surge than the White subjects in a number of studies. In a study by Rennie and Adams (1957), six out of eight American Black servicemen failed to show the hunting response at all in whole-body cooling at an air temperature of 20°C. Similar findings have been reported by Iampietro and others (1959) for digital cooling in a waterbath.

Less dramatic differences are demonstrable between groups for total-body cooling. Adams and Covino (1958) compared Eskimos, Blacks, and Whites in this respect. They found (Figures 7.11 and 7.12) that

FIGURE 7.11 Skin and rectal mean temperatures of three groups during exposure to cold room. (*Source:* Adams and Covino, 1958. Used with permission.)

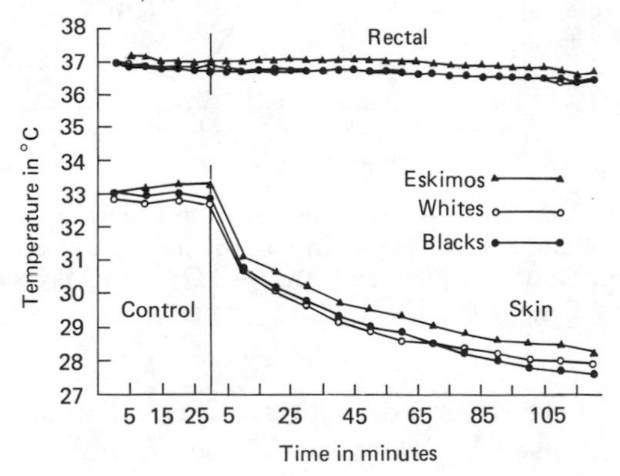

FIGURE 7.12 Mean metabolic rates of three groups during exposure to cold room. (*Source:* Adams and Covino, 1958. Used with permission.)

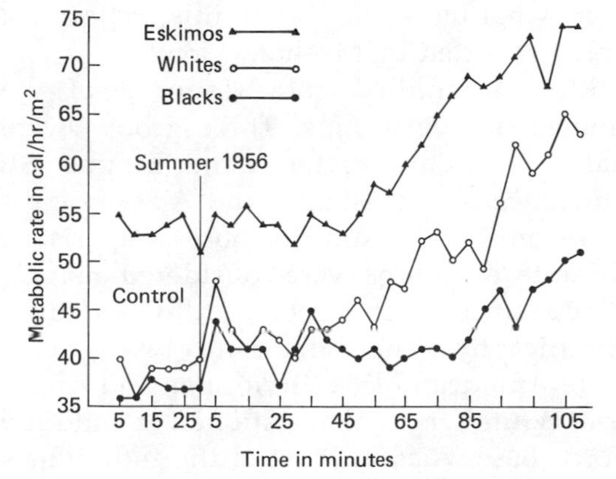

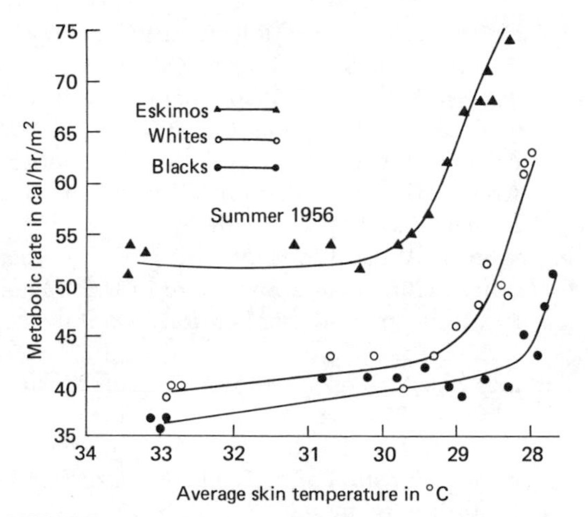

FIGURE 7.13 Mean metabolic rate versus mean skin temperatures of three groups during exposure to cold room. (*Source:* Adams and Covino, 1958. Used with permission.)

when the test subjects were resting nude in an air temperature of 17°C, the Eskimo mean skin temperature and mean rectal temperature remained higher than did the temperature of the Black and White samples. The Black and White means did not differ significantly. The Eskimo initial metabolic rate was also higher than the Black or White. In this respect Black and White were again alike. But as skin temperature began to drop the onset of shivering occurred at the same skin temperature in Eskimo and White subjects, 29.5°C; in Black subjects, however, it did not begin until the skin temperature reached 28°C. The time at which shivering began accounts for differences in the time at which the metabolic rate begins to increase as shown in Figure 7.13.

These and other studies indicate that not only can individuals acclimatize to a certain extent to cold stress but, also, that there may be between-group mean differences in their ability to alter responses with acclimatization. These differences are in the direction that would be expected under a hypothesis of genetic adaptation. They are small differences, and more studies are needed to confirm their reality. Also needed are direct studies that can tell us how much of the variation in response to cold is due to genetic variation between similarly acclimatized individuals. No such "heritability" studies have been carried out.

Comparable kinds of data as have been shown here for cold adaptation can be shown for heat adaptation in man. Other physiological responses and anatomical features of man show gradients (clines) that can be, in some cases, crudely correlated with ecological variables. Similarly there are large numbers of studies on disease stress and nutritional stress among human groups. This literature is too large to survey here. Again there are differences between regions and between groups of peoples in response to these, sometimes rather dramatic, environmental forces. Clearly there are differences in adaptation and accommodation of different populations to these factors. The problems and prospects are intriguing. But as yet there are no pat answers to any hypotheses on microevolutionary differences between human groups. But there are many tantalizing bits and pieces of information suggesting that some answers are not far off.

REFERENCES CITED

Adams, T. and B. G. Covino 1958 Racial Variations to a Standardized Cold Stress. *Journal of Applied Physiology* 12:9–12.

Blum, H. F. 1945 The Physiological Effects of Sunlight on Man. *Physiological Reviews* 25:483–530. 1959 *Carcinogenesis by Ultraviolet Light.* Princeton, N.J.: Princeton University Press. 1961 Does the Melanin Pigment of Human Skin Have Adaptive Value? *Quarterly Review of Biology* 36:50–63.

Brown, G. M. and J. Page 1952 The Effect of Chronic Exposure to Cold on Temperature and Blood Flow in the Hand. *Journal of Applied Physiology* 5:221–227.

Commoner, B., J. Townsend, and G. E. Pake 1954 Free Radicals in Biological Materials. *Nature* 174:689–691.

Cowles, R. B. 1959 Some Ecological Factors Bearing on the Origin and Evolution of Pigment in the Human Skin. *American Naturalist* 93:283–293.

Daniels, F. J. 1964 Man and Radiant Energy: Solar Radiation. *In* D. B. Dill, E. F. Adolph, and C. G. Wilbur (eds.), *Handbook of Physiology, Section IV, Adaptation to the Environment.* Washington, D.C.: American Physiological Society.

Darwin, C. 1839 *Journal of Researches into the Geology and Natural History of the Various Countries Visited by H.M.S. Beagle.* London: Henry Colburn.

Eveleth, P. B. 1966 Effects of Climate on Growth. *Annals of the New York Academy of Sciences* 134:750–759.

Gloger, C. L. 1833 *Das Abändern der Vogel durch Einfluss des Klimas.* Breslau: A. Schulz.

Iampietro, P. F., R. F. Goldman, and D. E. Baas 1959 Response of Negro and White Males to Cold. *Journal of Applied Physiology* 14:798–800.

Kaplan, B. 1954 Environment and Human Plasticity. *American Anthropologist* 56:780–800.

Loomis, W. F. 1967 Skin-Pigment Regulation of Vitamin-D Biosynthesis in Man. *Science* 157:501–506. 1970 Rickets. *Scientific American* 223:76–91.

Mason, H. S., D. J. E. Ingram, and B. Allen 1960 The Free Radical Property of Melanin. *Archives of Biochemistry and Biophysics* 86: 225–230.

Meehan, J. P. 1955 Individual and Racial Variations in a Vascular Response to a Cold Stimulus. *Military Medicine* 116:330–334.

Murray, F. G. 1934 Pigmentation, Sunlight, and Nutritional Disease. *American Anthropologist* 36:438–445.

Newman, M. T. 1953 The Application of Ecological Rules to the Racial Anthropology of the Aboriginal New World. *American Anthropologist* 55:311–327. 1956 Adaptation of Man to Cold Climates. *Evolution* 10:101–105. 1960 Adaptations in the Physique of American Aborigines to Nutritional Factors. *Human Biology* 32:288–313.

Newman, R. W. and E. H. Munro 1955 The Relation of Climate and Body Size in U.S. Males. *American Journal of Physical Anthropology* 13:1–17.

Pathak, M. A. 1967 Photobiology of Melanogenesis: Biophysical Aspects. *In* M. Montagna and R. L. Dobson (eds.), *Advances in Biology of Skin, Vol. 8, The Pigmentary System.* Oxford: Pergamon.

Rennie, D. W. and T. Adams 1957 Comparative Thermoregulatory Responses of Negroes and Whites to Acute Cold Stress. *Journal of Applied Physiology* 11:201.

Roberts, D. F. 1953 Body Weight, Race and Climate. *American Journal of Physical Anthropology* 11:533–558.

Ronge, H. E. 1948 Ultraviolet Irradiation with Artificial Illumination. A Technical, Physiological and Hygienic Study. *Acta Physiol. Scand.* 15 (Suppl. 49):1–191.

Scholander, P. F. 1955 Evolution of Climatic Adaptation in Homeotherms. *Evolution* 9: 15–26.

Schuman, L. M. 1952 Epidemiology of Cold Injury in Man. *In* M. Irené Ferrer (ed.), *Cold Injury,* pp. 11–84. New York: J. Macy.

Thompson, M. L. 1951 Dyshidrosis Produced by General and Regional Ultraviolet Radiation in Man. *Journal of Physiology* 112:22–30. 1955 Relative Efficiency of Pigment and Horny Layer Thickness in Protecting the Skin of Europeans and Africans Against Solar Ultraviolet Radiation. *Journal of Physiology* 127: 236–246.

Urbach, F., R. E. Davis, and P. D. Forbes 1966 Ultraviolet Radiation and Skin Cancer in Man. *In* W. Montagna and R. L. Dobson (eds.), *Advances in Biology of Skin. Vol. 7, Carcinogenesis.* pp. 195–214. Oxford: Pergamon.

8 Evolutionary Changes Today

We know that man continues to evolve. Unfortunately the contrary assumption was made by some authors in the past. Their argument was that natural selection occurs among animals living in a "state of nature" and that man, enjoying the benefits of a skilled medical profession as well as other cultural amenities, is shielded from the effects of natural selection. This view—as the reader should now be able to recognize—is a pregenetic, nineteenth-century one of the nature of evolutionary change.

We know of many specific genetic defects that reduce the fertility of their bearers. Further, we know that there are great reproductive inequalities between different societies as well as between different classes of individuals within a society. Since in many cases there are genetic differences among such groups, the result is evolutionary change.

That evolutionary change continues to occur is a simple deduction, given the principles of population genetics. Identifying the trends of change is a more difficult task. Even more difficult is identifying the causes of change. In looking at some of the evidence for ongoing change, it is again useful to distinguish between change identifiable at the gene frequency level and change in anthropometric traits measured quantitatively.

CHANGES AT THE GENE FREQUENCY LEVEL

Causes of change in the form of differential susceptibility to diseases were discussed extensively in Chapter 5. It is appropriate to point out again, however, that associations between given diseases and given alleles do not inevitably lead to gene frequency change. Even if the disease state leads to differential reproduction this may be balanced by selection in the opposite direction due to other causes. The net result may be a balance with no change occurring.

In Chapter 5 it was suggested that the frequency of hemoglobin S is lower in the American Negro today than can be explained by an admixture of European genes alone. In a similar manner it can be shown that in Central American Indian populations the frequency of the blood group allele L^N has risen faster than can be explained by the admixture of European genes. These represent examples of ongoing evolutionary change.

An even more striking example of change in the frequency of hemoglobin S occurs in selected populations of the American Negro. One group on James Island, South Carolina, has a frequency of $\bar{S} = 0.20$. Another group in Evans and Bullock counties, Georgia, is similar in having little European admixture and roughly the same gene pool of the founders, but this latter population has a frequency of only $\bar{S} = 0.09$. The major difference seems to be that James Island was one of the most malarious areas in the United States. This maintained the high frequency of hemoglobin S while it decreased elsewhere.

In relatively well-studied genetic systems, such as hemoglobin S it is even possible to predict future frequencies of the trait. Again using as an example the Gullah (Negro) population of James Island, South Carolina, we can predict a decline in the frequency of hemoglobin S with malaria no longer a health problem. Based on past experience, we expect the fitness of *AA* and *AS* individuals to be equal, in the absence of malaria, and the fitness of *SS* to be nearly zero. In such a case the frequency of the allele, *S*, decreases over *t* generations as

$$t = \frac{1}{q_t} - \frac{1}{q_o}$$

where q_o is the frequency of the allele at the beginning of the period and q_t the frequency of the allele after t generations. Therefore we can predict that the frequency of S will be reduced to one-half its present frequency ($q_t = 0.10$) in five generations.

This kind of prediction assumes that the present environmental conditions affecting the hemoglobin phenotypes will remain constant for five or more generations. This, undoubtedly, will not be the case, because man is altering his environment, including health practices, at an ever increasing rate. Some of the consequences of preserving deleterious genes will be discussed later in this chapter.

While some environmental changes occurring today may be beneficial to man, other changes undoubtedly are not. One environmental change that clearly is not beneficial is the increase in level of air pollution. It has recently been discovered that specific genotypes may be severely affected by air pollution, which causes pulmonary emphysema, a serious disease that impairs breathing and leads eventually to impaired heart action.

Serum alpha$_1$-antitrypsin deficiency is a genetic polymorphism discovered in 1963. This antitrypsin deficiency is inherited as a simple codominant allele. Alpha$_1$-antitrypsin, as its name suggests, inhibits the activity of a proteolytic (protein splitting) enzyme, trypsin. It has recently been discovered that antitrypsin-deficient persons are far more susceptible to pulmonary emphysema than are nondeficients. The homozygotes are affected earlier in life and more severely than are the heterozygotes. At the same time we find that the prevalence of pulmonary emphysema rises with increased levels of air pollution and is more common among smokers. This suggests that the increasing smog of our urban areas may have a specific genetic effect operating against carriers of the antitrypsin-deficiency gene. Large-scale studies have not been carried out but it seems, based on small samples, that as much as 10 percent of the American population is heterozygous or homozygous for this condition.

CHANGES IN POLYGENIC CHARACTERS

We know that most populations of man are changing slowly with respect to many anthropometric variables. The average European male today cannot be fit into the armor of the fourteenth-century knight. The American high school student cannot sit comfortably at the desk where his grandfather sat in high school early in this century.

We can document such changes with accuracy, but do they represent evolutionary changes? We cannot give a complete answer to this question but we can begin on an answer. We begin by establishing whether or not a trait is susceptible to evolutionary change in a given population. What I mean by this can be illustrated in the following way. If a population is genetically homogeneous for a given trait then all variation in this trait must be environmental in origin. Continued selection among variants in this population will result in no evolutionary change since there is no genetic variation underlying the phenotypic variation. Also, if no genetic variation is introduced into this population any change over the years must be due to environmental changes.

Therefore, in lieu of identifiable genes it is important to know what proportion of the phenotypic variation in a trait is due to genetic causes (genotypic variation) and what proportion is due to environmental causes (environmental variation). The simplest model of genetic and environmental components of phenotypic variation is

$$V_P = V_G + V_E$$

where

V_P = phenotypic variance
V_G = genetic variance
V_E = environmental variance

Variance is a number which is a measure of variation in a trait. The first step taken in the genetic analysis of a polygenic trait is to estimate the two variance components, V_G and V_E. These may be stated directly or may be given as a *heritability* estimate. The heritability of a trait is the proportion of total phenotypic variance ascribable to genetic variance. Representing heritability as H^2 we can say $H^2 = V_G/V_P$.

In experimental animals where environments and matings can be controlled, it is possible to estimate V_E and V_G directly. It is even possible to estimate different subcomponents or environmental or genetic variance. In human populations this is not so easily done. Because of this anthropologists and others have relied heavily on twin studies for heritability estimates in human traits. We can get a rough estimate of heritability by equating

$$H^2 = \frac{V_{DZ} - V_{MZ}}{V_{DZ}}$$

where

V_{DZ} = within-pair variance for dizygous (fraternal or two-egg) twins

and

V_{MZ} = within-pair variance for monozygous (identical or single-egg) twins

Dizygous twins are genetically no more alike than siblings. They have, on the average, half of their genes in common. Therefore there will be both a genetic and an environmental component of V_{DZ}. Monozygous twin pairs are genetically identical. Therefore there is only an environmental basis of differences within the monozygous pair. And for a given population and environment V_{DZ} will always be larger than V_{MZ}. If there is no environmental component to the within-pair variances $V_{MZ} = 0$ and

$$H^2 = \frac{V_{DZ}}{V_{DZ}} = 1$$

In other words a value of $H^2 = 1$ indicates that all of the variation in that trait is due to genetic differences. At the other pole, if all variation is environmental and affects equally the monozygous and dizygous twin pairs, $V_{DZ} = V_{MZ}$ and $H^2 = 0$. In summary, H^2 is an index ranging from 0 to 1 and the higher its value the greater the degree to which phenotypic variation is determined by genotypic variation.

Table 8.1 gives heritability estimates for some of the measurements of man. One word of caution is in order. The two members of a twin pair are raised in an environment more similar than that shared by two individuals picked at random from the population. This restriction of within-pair environ-

TABLE 8.1
The Heritability of Body Dimensions

MEASUREMENT	HERITABILITY, H^2
Weight	0.69[a]
Stature	0.88[a]
Span	0.85[a]
Sitting height	0.72[a]
Biacromial diameter	0.31
Bi-iliac diameter	0.59[a]
Forearm length	0.81[a]
Hand length	0.82[a]
Middle finger length	0.88[a]
Hand breadth	0.80[a]
Foot length	0.81[a]
Minimum waist circumference	0.25
Minimum neck circumference	0.67[a]
Maximum hip circumference	0.63[a]
Maximum midarm circumference	0.62[a]
Maximum forearm circumference	0.53[a]
Minimum wrist circumference	0.65[a]
Maximum calf circumference	0.75[a]
Minimum ankle circumference	0.54[a]
Bicondylar diameter, elbow	0.71[a]
Bicondylar diameter, knee	0.61[a]
Head length	0.54[a]
Head breadth	0.72[a]
Minimum frontal breadth	0.61[a]
Bizygomatic breadth	0.60[a]
Bigonial breadth	0.71[a]
Interpalpebral breadth	0.60[a]
Bipalpebral breadth	0.41[b]
Interpupillary breadth	0.70[a]
Nose breadth	0.66[a]
Head height	0.69[a]
Total face height	0.74[a]
Upper face height	0.72[a]
Nose height	0.76[a]
Ear length	0.75[a]

[a] Significant at the one percent level.
[b] Significant at the five percent level.
Source: Adapted from J. N. Spuhler, "Empirical Studies on Quantitative Human Genetics," in *The Use of Vital and Health Statistics for Genetic and Radiation Studies* (New York: United Nations' World Health Organization, 1962).

mental variance means that H^2 tends to overestimate "true" heritability, or the heritability estimates found by animal breeders where environments are controlled. A number of other cautions concerning the use of heritability estimates have been pointed out by Clark (1956).

The figures given in Table 8.1 are comparable to those found in a number of twin studies. Linear measurements tend to be more highly heritable than circumferences. A different kind of evidence for the relative magnitude of heritability in different body dimensions has been provided by Hiernaux (1963). Hiernaux, in a study of the Hutu of Rwanda, divided his sample into those Hutu living above 1900 meters and those living below that altitude. Both subsamples were culturally identical and, as the higher-altitude residents represent fairly recent expansion, are presumed to be genetically very similar. The primary environmental differences between the two subsamples were that the higher-altitude residents enjoyed better nutrition and a lower prevalence of malaria.

Hiernaux reasoned that those traits which, in the studies of others, showed the highest heritability should show the least "ecosensitivity" if heritability did, in fact, measure the degree to which variation in the trait was genetically controlled. He compared the Hutu subsamples to see which traits showed the greatest differences. One set of results is shown in Table 8.2. In this figure the *F ratio* is used as a measure of heritability and is defined as

$$F = \frac{V_{DZ}}{V_{MZ}}$$

The measures F and H^2 are related simply as

$$F = \frac{1}{1 - H^2}$$

It can be seen from this figure that, as Hiernaux states it, "a more favorable environment (with respect to diet and malaria) tends to increase all physical measurements in a

TABLE 8.2
Significance of the *F* Ratios of the DZ and MZ Within-Pair Variances in a Prior Heritability Study by Osborne and De George Compared to the Significance of the Difference Between the Two Hutu Subgroups

CHARACTER	SIGNIFICANCE OF THE F RATIO OF DZ AND MZ VARIANCES	SIGNIFICANCE OF THE DIFFERENCE BETWEEN HUTU SUBSAMPLES
Cephalic index	**	
Total arm length	**	*
Stature	**	
Total leg length	**	
Nose height	**	*
Upper face height	**	
Bigonial breadth	**	*
Forearm length	*	**
Upper arm length	*	
Biocular width		**
Biacromial breadth	*	**
Chest breadth		**
Mouth width		**
Chest depth		**
Total face height		*
Interocular width		*
Max. calf girth		**
Bi-iliac breadth		**
Nose breadth		**
Upper arm girth		**
Weight		**
Bizygomatic breadth		*
Thigh girth		**
Head height		
Head length		**

Note: Two stars indicate a probability of less than 0.01 that such a difference could be due to chance. One star indicates a probability of less than 0.05 that such a difference could be due to chance.
Source: J. Hiernaux, "Heredity and Environment: Their Influence on Human Morphology. A Comparison of Two Independent Lines of Study," *American Journal of Physical Anthropology*, 21 (1963), 575–590.

proportion inverse to their heritability'' (1963, p. 579).

This we can take as an independent confirmation that heritability scores do provide a relative measure of genetic control of variability.

Evolutionary change requires genetic variability. The high heritability of a number of anthropometric traits indicates that the requisite genetic variability is available for what could be rather rapid change. Let us now look at some of the known and supposed kinds of change that have been going on in human populations in the past few generations.

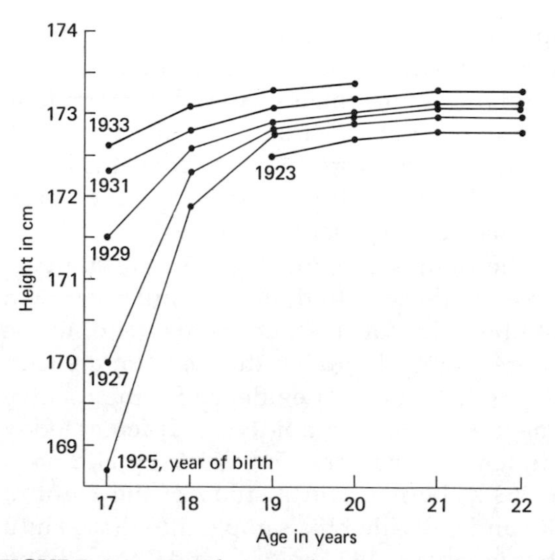

FIGURE 8.1 Trend toward more rapid growth and earlier maturation among French University students. Those who were 17 in 1933 had almost reached adult height. (*Source:* Data of Auberque from Tanner, 1962. Used with permission.)

SECULAR TRENDS

Secular trends in growth, as the phrase is used in anthropology, refer to changes in the pattern of growth or the rate of growth observed to be occurring from generation to generation. Two of the best-documented cases of secular change in Western societies are change in stature and change in age of menarche, or beginning menstruation.

Norwegian data on the stature of adults goes back as far as 1741. They show little change in adult height prior to 1830. During the period 1830 to 1875 adult height increased 0.3 cm per decade, and from 1875 to 1935 the increase averaged almost 0.4 cm per decade. But even greater, during this period, was the change in height of the subadult population. This is due to accelerated rates of growth associated with earlier maturity. Data that show this trend among French university students are shown in Figure 8.1. Those born later in the century are much taller at age 17 than those born earlier. Those born later not only grow faster but also reach their adult height at an earlier age. This then means that growth ceases at an earlier age today but not so early that adult height is not increased.

The trend toward earlier maturation expresses itself strikingly in the decreasing age of menarche. Where records are available the age of menarche has dropped in 100 years time from over 16 years of age to the present age of less than 13. Figure 8.2 shows graphically the rate of change. The average rate of drop in menarcheal age from these data is roughly 3.5 months per decade and there is no indication that the trend is decelerating. And, as in the case of stature, while there is good evidence that nutrition and other environmental variables affect menarcheal age, there is also evidence for a heritable component in the variation found both within and between groups.

Can evolutionary change be going on in these cases in addition to the undoubted effects of environmental improvements? The answer is yes, it can be, but there is no conclusive evidence. The two most obvious pos

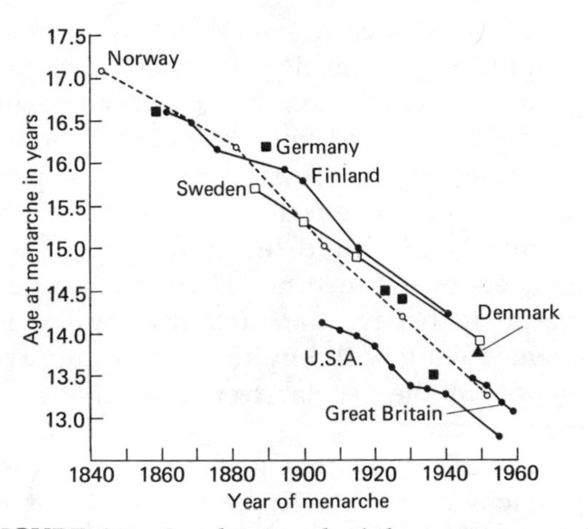

FIGURE 8.2 Secular trend of decreasing age at menarche from 1830 to 1960. (*Source:* Tanner, 1962. Used with permission.)

sibilities that suggest themselves are selection and heterosis effects. Most people have rejected the selection hypothesis because there is no evidence that tall individuals have a larger mean number of offspring than do short individuals. But this conclusion may need reconsideration, as there are many other opportunities for selection to occur. One point at which such selection could occur is in the perinatal period. Studies among well-nourished populations have shown that the optimal survival weight for newborns is greater than the mean birth weight for the population. For example, Karn and Penrose (1951) found that in 13,730 English births the weight of optimal survival of males was 300 grams greater than the mean birth weight of males. And similarly the optimal survival weight of female offspring was 431 grams more than the mean birth weight of females. This may represent selection for larger size in the perinatal period. Similarly, in a study of 179,327 births in Hawaii, Morton, Chung, and Mi found that "perinatal and infant mortality is at a minimum for birth weight about 390 grams greater than the mean" (Morton, Chung, and Mi, 1967, p. 129). Further, although the data on the fertility of tall versus short males is equivocal, there is consistent evidence from England and the United States that shorter women have a higher frequency of stillbirths and generally poorer reproductive performance than do taller women. These findings suggest that it is premature to dismiss natural selection as a contributing factor in secular trends toward increased size.

The heterosis hypothesis has been given more serious consideration by anthropologists. Heterosis can occur as a result of outcrossing, or mating between breeding isolates. The effect of outcrossing is just the opposite of the effects of inbreeding in that, if there are gene frequency differences between the two breeding isolates, outcrossing will produce an increase in the frequency of heterozygotes. A simple numerical illustration will show how this operates. Assume two populations with gene frequencies as follows:

	Frequency	
	$p(A)$	$q(a)$
Population I	0.6	0.4
Population II	0.4	0.6

then genotype frequencies in these parental populations are expected to be

	p^2	$2pq$	q^2
Population I	0.36	0.48	0.16
Population II	0.16	0.48	0.36

If matings occur with one parent from each population, the expected genotype frequencies in the F_1 generation are

$$\overline{AA} = (0.6)(0.4) = 0.24$$
$$\overline{Aa} = (0.6)(0.6) + (0.4)(0.4) = 0.52$$
$$\overline{aa} = (0.4)(0.6) = 0.24$$

The gene frequencies in the F_1 are $p = q = 0.5$

in this case. Therefore, with one subsequent generation of random mating, the F_2 will have

$$\overline{AA} = 0.25$$
$$\overline{Aa} = 0.50$$
$$\overline{aa} = 0.25$$

This shows that the F_1 "hybrids" have a greater degree of heterozygosity than expected under equilibrium conditions, and the F_2 and subsequent generations, although in equilibrium, maintain a higher degree of heterozygosity than the parental populations.

Continuing with our single-locus model let us now ask, what will a change in the degree of heterozygosity in a population do in changing the mean value of a trait? If the heterozygote is exactly immediate in size between the two homozygotes, there will be no change in the population mean as a result of changing the degree of heterozygosity alone. If the heterozygote is not exactly intermediate then a change in degree of heterozygosity will change the population mean. In particular, when the heterozygote is larger than the mean of the two homozygotes, termed *positive dominance effects*, an increase in heterozygosity will increase the mean value of the trait in the population. This is the basis of metric heterosis. The same trait that exhibits heterosis can be expected to be depressed in value due to inbreeding, which decreased heterozygosity.

The secular trend of increased stature does correspond to the period of breeding-isolate breakdown as a consequence of the industrial revolution, increased communication, and so forth. We would then like to know if stature and other anthropometric traits exhibit positive dominance effects. The best evidence on this point comes from studies of inbreeding effects in man. One of the largest and most carefully done of such studies was carried out on Japanese children (Schull and Neel, 1965). Measurements taken included weight, height, head girth, chest girth, calf girth, head length, head breadth, head height, sitting height, and knee height. All showed an inbreeding depression confirming positive dominance effects. In a related study Morton (1958) found that birth weight was reduced by inbreeding. Therefore we have reason to believe that heterosis can be involved, along with environmental improvements, in the secular trends of change in man.

A small study that gets at the heterosis problem more directly was carried out by Hulse (1957). He studied a number of traits among the residents of the Italian-speaking Canton of Ticino in Switzerland as well as their relatives living in California. In Canton Ticino, village endogamy (marriage within the group) has been the normal marriage practice of the past. This is now changing both in Switzerland and among the Californian Ticinesi, so Hulse was able to measure adults who were the offspring of village exogamous marriages as well as those who were the offspring of the village endogamous marriages that were the earlier norm. He was also able to compare those who did not migrate but were born and raised in Ticino ("sedents"); those who were born in Ticino but raised in California ("immigrants"), and those Ticino descendants born and raised in California ("Californians"). The effects on stature of both environmental change and exogamy can be seen in Table 8.3. It appears that environmental differences account for about 4 cm difference in stature, and exogamy (marriage out of the group) accounts for about 2 cm difference. A study of Italian-Americans in Boston similarly revealed a difference of 1.65 cm that was ascribable to exogamy (Damon, 1965).

Saldanha, in a study of Brazilian-Italian "hybrids" in Sao Paulo, Brazil, found that the

TABLE 8.3
Mean Stature (cm) of Ticinesi Showing Both Environmental and Endogamy/Exogamy Effects.

	ENDOGAMOUS	EXOGAMOUS
Sedents	166.21	168.51
Immigrants	166.90	168.71
Californians	170.50	172.33

adult offspring of such marriages were closer in mean height to the mean of the tallest parental group, the Italians, and that the variance in stature was reduced in the F_1 hybrids. Both of these conditions are expected where there are dominance effects that lead to metric heterosis. The data are shown in Table 8.4 (Saldanha, 1962).

These and other studies reveal a great deal of plasticity in human growth responses. They also suggest that a part of this plasticity may be a response to changing breeding structures in human societies. This in turn can be involved in secular trends of change.

TABLE 8.4
Means and Variances for the Distribution of Stature, (cm) Among Brazilians, Italian Immigrants, and the Offspring of Brazilian and Italian Matings

	MEAN	VARIANCE
Brazilian	162.94	61.81
Italian	166.46	61.14
Combined parental Mean	164.70	
Brazilian-Italian	166.08	49.02

Note: The combined parental mean is the unweighted mean of the two parental populations.
Source: Data from P. H. Saldanha, "The Genetic Effects of Immigration in a Rural Community of São Paulo, Brazil," *Acta Geneticae Medicae et Gemellogogiae* 11 (1962), 158–224.

The Supposed Trend of Declining Intelligence

A secular trend of change that has often been postulated but never demonstrated is that of declining intelligence. This concern dates to the old eugenics movement of the first three decades of this century. But the concern is still expressed by a number of knowledgeable investigators today. Therefore it is of some interest to look at the basis of this prediction.

In such a problem the phenotype of interest is performance on an intelligence test. These tests are usually standardized so that test results for a population show a normal, continuous distribution of scores. These can be treated, in quantitative genetics, much the same way as stature, weight, or many other metrics. And as with any other metric trait we can ask if some part of the variation in test results can be ascribed to genetic variation. Twin studies yield a relatively high heritability score for generalized tests of intelligence such as the Stanford-Binet. Different studies have given H^2 values ranging from 0.50 to 0.80. Tests that are designed to examine different "mental abilities" that have been factored out show a high heritability for some abilities and a very low or nil heritability for others. Some of these results are shown in Table 8.5.

The same cautions of interpretation apply to these heritability scores as to those on anthropometric traits. Strictly speaking they apply to the population, and the environment, where the study was conducted. But there are no more limits on their interpretation than in the case of anthropometrics, and these figures are adequate to show that there is a genetic component to between-individual variation in a number of intelligence test results.

Those who predict a decline in intelligence argue that since there is a genetic component

TABLE 8.5
Summary of the Results of Four Independent Studies of Heritability, in Terms of F Ratios, on the Chicago Primary Mental Abilities Test

NAME OF SCORE	BLEWETT (1954) $N_{DZ}26 N_{MZ}26$	THURSTONE ET AL. (1955) $N_{DZ}53 N_{MZ}45$	VANDENBERG (1962a) $N_{DZ}37 N_{MZ}45$	VANDENBERG (1964) $N_{DZ}36 N_{MZ}76$
Verbal	3.13**	2.81**	2.65**	1.74**
Space	2.04*	4.19**	1.77*	3.51**
Number	1.07	1.52	2.58**	2.26**
Reasoning	2.78**	1.35	1.40	1.10
Word Fluency	2.78**	2.47**	2.57**	2.24**
Memory	not used	1.62*	1.26	not used

* $p < .05$.
** $p < .01$.

Source: Vanderberg. Reprinted from J. N. Spuhler (ed.), *Genetic Diversity and Human Behavior* (Chicago, Aldine Publishing Company, 1967); Copyright © by the Wenner-Gren Foundation for Anthropological Research, Inc.

to variation in tested intelligence, if those of lower IQ out-reproduce those of higher IQ, the mean IQ of the population must, in all probability, decline. They have argued that this is happening. The following is a clear, if somewhat dramatic statement of this position.

> Even where unchecked fertility has not impelled human beings to scramble for survival, what is happening today has other sinister effects. In the United States, England, and every Western country, misplaced and badly distributed human fertility is leaching away the inborn qualities of tomorrow's children. This biological "erosion" is insidious in its action: No barren, gullied hillsides meet the eye, but in the foreseeable end such erosion of the biological quality of the people may, after no great lapse of time, result in disaster.
>
> In England, a Royal Commission of experts in sociology and population has been studying this part of the fertility problem. Its 1949 report concluded that the average intelligence quotient of the British people was declining about 2 points every generation. The same pattern exists in the United States, where the experts consider a similar decline to be a "moral certainty." If this trend continues for less than a century, England and America will be well on the way to becoming nations of near half-wits. (Cook, 1951, pp. 5–6)

There are two major kinds of associations between fertility and IQ upon which such predictions are based. First, there is a well-established negative correlation, $r = -0.20$ to -0.30, between family (sibship) size and IQ of offspring in the sibship. Second, class or occupational differentials in IQ are generally the inverse of class or occupational differentials in fertility. In other words those classes with the lower mean IQ have the higher fertility. Both these kinds of evidence and their frequent confirmation have led to the conclusion that the unintelligent are contributing more than their share to future generations.

Those who do not agree with the declining-intelligence hypothesis point out that, in the face of a predicted decline, IQ scores have actually increased in tests conducted over long time periods. Proponents of the declining-intelligence hypothesis argue that the improved performance on IQ tests in the past 40 years is due to an improved environment, including great test-taking sophistication of children, and may mask a decline in genetic potential.

But let us now question the frequently confirmed association between fertility and IQ. That this might be a fallacious association due to errors in method was first suggested in a small report in 1940 but not noticed by many investigators concerned with the problem. Willoughby and Coogan (1940) did a *prospective* study; it was prospective in that they started with a group of high school graduates of tested intelligence, traced them to their present residences, and counted the number of living offspring. They found no evidence of an overall fertility differential favoring those of lower tested intelligence.

The difference between results of a *prospective* versus a *retrospective* study was brought home forcefully in a report of a large study by Higgins, Reed, and Reed (1962). The classical studies of intelligence and family size were retrospective in that offspring were tested. For example, each child of a given school class might be given an IQ test and asked how many brothers and sisters he had. It was then a simple procedure to calculate a correlation coefficient between IQ and size of the sibship. A negative correlation was always found.

One of the major unrecognized faults of this retrospective procedure (although not the only fault) was that it gave no information on people who had no children. Were they bright or were they dull?

Some of the results from the Minnesota study are shown in Tables 8.6 and 8.7. It can be seen from Table 8.6 that, as in the retrospective studies, there is a strong tendency for IQ to decline with increasing sibship size. But, as in the older studies, this says nothing about those who produce no children. These people were accounted for in the Minnesota study and it was found that of those with IQ 55 and below only 38 percent married. This figure rose to 100 percent for the highest IQ bracket. A picture emerged in

TABLE 8.6
Comparison of Sibship size ("Family Size") and Mean IQ of Children per Sibship Size

FAMILY SIZE	IQ OF CHILDREN	STANDARD DEVIATION	NUMBER
1	106.37 ± 1.39	16.46 ± 0.98	141
2	109.56 ± 0.53	12.74 ± 0.37	583
3	106.75 ± 0.58	14.20 ± 0.41	606
4	108.95 ± 0.73	12.95 ± 0.51	320
5	105.72 ± 1.15	15.86 ± 0.81	191
6	99.16 ± 2.17	19.57 ± 1.54	82
7	93.00 ± 3.34	20.58 ± 2.36	39
8	83.80 ± 4.13	20.24 ± 2.92	25
9	89.89 ± 2.94	17.61 ± 2.07	37
10	62.00 ± 7.55	28.23 ± 5.34	15

Source: J. V. Higgins, E. W. Reed, and S. C. Reed, "Intelligence and Family Size: A Paradox Resolved," *Eugenics Quarterly*, 9 (1962), 87, © The University of Chicago Press.

TABLE 8.7
Comparison of IQ of Parents and Their Reproductive Rate When Never-married Males and Females of the Parental Generation are Included

IQ RANGE	AVERAGE NO. OF CHILDREN	STANDARD DEVIATION	NUMBER
0–55	1.38 ± 0.54	2.87 ± 0.38	29
56–70	2.46 ± 0.31	2.69 ± 0.22	74
71–85	2.39 ± 0.13	1.83 ± 0.09	208
86–100	2.16 ± 0.06	1.46 ± 0.04	583
101–115	2.26 ± 0.05	1.47 ± 0.04	778
116–130	2.45 ± 0.09	1.53 ± 0.07	269
131 and above	2.96 ± 0.34	1.65 ± 0.24	25
	Average 2.27	Total 1,966	

Source: J. V. Higgins, E. W. Reed, and S. C. Reed, "Intelligence and Family Size: A Paradox Resolved," *Eugenics Quarterly*, 9 (1962), 89, © The University of Chicago Press.

which those in the lowest IQ range had the largest number of children when married but only a minority of them married or reproduced.

Table 8.7 shows the result of taking this factor into account in arriving at the fertility of the various IQ classes. As can be seen, the distribution of "average number of children" is bimodal, but the highest peak is in the highest IQ category. When these averages are weighted by the number of families in each category is turns out that the persons of IQ 100 or above produced an average of 2.3 children per family. The persons below IQ 100 produced an average of 2.2 children per family.

We come to the rather reassuring conclusion that the earlier studies reached a spurious conclusion due to faulty methodology and there is no evidence for an IQ decline due to differential fertility. A subsequent study in Michigan that also took generation length into account has shown essentially the same results, and suggests that if it has any effect, differential fertility could be slightly increasing the mean IQ of the population (Bajema, 1963).

THE JENSEN HYPOTHESIS

In 1969 an article was published (Jensen, 1969) which launched an acrimonious debate that continues to the present. In the early 1960s a great many programs were initiated in the United States to reduce the educational inequality between the average Black and the average White. This was a political program initiated quickly and many of the educational intervention programs, as a result, were ill-conceived. Almost all of these educational programs seemed to assume that if only cultural differences were involved these could be altered easily with the application of money. In other words, the premise was that cultural differences are, in some sense, trivial differences. Almost all of the programs failed to some extent.

In a period of reassessment of these programs, Arthur Jensen, an educational psychologist, argued that since these programs had failed, the difference in educational attainment between Blacks and Whites was probably due to genetic differences. Jensen reviewed the existing data which had long indicated that IQ scores have a high heritability within Whites. He reviewed the data which had long indicated that Blacks, on the average, scored much lower on IQ tests than did Whites. Jensen concluded that "it [is] a not unreasonable hypothesis that genetic factors are strongly implicated in the average Negro-White intelligence difference" (Jensen, 1962:82). This, although hardly new, has come to be called the Jensen hypothesis.

In these few paragraphs we will be concerned primarily with the logic of the genetic argument. We will be less concerned with questions about the worth of IQ testing. The *validity* of IQ tests is established in terms of how well they predict future scholastic performance. In these terms, IQ tests in use today have greater validity than most psychological tests and measurements. On the other hand, there is, for example, little correlation between childhood IQ score and economic success as an adult, once family socioeconomic differences have been controlled. Similarly, there are many other achievement measures which correlate very little with IQ scores although this is contrary to the mystique that many people seem to associate with IQ testing. But whatever else they are, or are not, IQ scores are numbers which can be treated as phenotypes and analyzed in exactly the same way as stature, weight, and so on.

In considering the genetic argument involved it is important to first note that heritability scores are *within-population* measures. They are specific to a given gene pool and a given environment. The genetic variance, V_G, may differ in different populations. This is best illustrated with a single locus trait such as the ABO blood group system. An American Indian group, having only the *O* allele, would have no genetic variation with respect to this trait, $V_G = 0$, and no phenotypic variation; therefore $H^2 = 0$. A European population would have a high phenotypic variance, all of which would be due to genetic differences between individuals in the population, and therefore have an $H^2 = 1$. The same point can be made for environmental variance when we are discussing polygenic traits. This does not mean that heritability estimates on polygenic traits will vary greatly between populations. On a number of anthropometric traits for which we have studies, the traits with high heritability in one population tend to have a high heritability in other populations.

We are now in a position to ask and answer the following question: Given that a trait has a high heritability in each of two populations, can we assume from this that differences between these populations are, in part, genetic? The answer is no. A hypothetical example will help to clarify this.

As a strictly hypothetical example, let us assume we can randomly assign individuals from a single population to two different islands. We wish to measure the heritability of stature on each island. Let us further assume that we can completely control the environment so that everyone on island A is raised in a uniform environment and everyone on island B is raised in a uniform environment. Due to random assignment of individuals each island should have approximately the same array of genotypes. There is genetic variability within each island but no environmental variability within each island. So the heritability of stature in each case is $H^2 = 1$.

Can there be important differences in stature between island A and island B? The answer is yes. Suppose that the residents of island A received normal amounts of ultraviolet radiation from sunlight. But residents of island B got no ultraviolet radiation and no nutritional sources of vitamin D. It is still the case that $H^2 = 1$ on each island. But the residents of island B are markedly shorter because they suffer from rickets. In other words, we have a hypothetical example of a trait with a heritability $H^2 = 1$ but in which large between-population differences are due entirely to environmental differences. We could design a similar hypothetical situation in which $H^2 = 0$ in each population and yet have large genetic differences underlying phenotypic differences between two populations. Heritability measures, as now conceived, measure within-population genetic differences and, alone, say nothing about between-population genetic differences.

Although the preceding experiment is hypothetical, something similar occurred repeatedly in the industrialization of the West as rural migrants came into the dark and sunless slums of the early industrial cities. This was not hypothetical. It is also a fortuitous example in that it illustrates the difficulty of identifying relevant between-population environmental variables. Differences in ultraviolet radiation are not directly visible. Prior to the 1920s this variable was totally unrecognizable as an environmental variable affecting growth.

From the preceding discussion it should be clear that heritability estimates alone, no matter what their magnitude, provide no information about the genetic or nongenetic nature of between-group differences. What is required is either equalization of environments or estimation of the magnitude of environmental differences between groups. The latter is often referred to as the "statistical control" of environmental effects.

Many studies of Black-White IQ score differences have attempted to statistically control environmental differences. Unfortunately the environmental variables used are almost always picked for convenience. They are usually items readily available in census data. They seldom index many of the environmental variables known to affect IQ test scores.

Recently Jane Mercer and her associates published a study (Mercer and Brown, 1973) which investigated the effects of nine environmental variables on between-group differences in IQ scores. The three groups tested were the Anglo, Black, and Mexican-American populations of Riverside, California. The environmental variables included such things as socioeconomic status, home ownership, parents from urban or rural setting, English spoken in the home, and so forth.

IQ tests produced the usual average differences, with Mexican-American and Black children scoring an average of about 15 points below Anglo children. Then *within* each group the effect of each environmental variable on IQ was estimated and this effect removed from scores. This reduced individual differences (the within-population variance) only a little, less than 20 percent, in each population. If these nine environmental variables index most of relevant environmental variation, then the remaining variance should be largely genetic. This is consistent with genetic studies showing $H^2 = 0.80$ for IQ scores. Again, this is *within* populations.

Mercer and associates then estimated the effect of each environmental variable *between* populations. When the population average IQ score was adjusted for these effects the between-population differences remaining were effectively zero. In other words, when environmental differences were statistically controlled between groups there was no remaining between-group difference to be explained by genetic differences. To date the Mercer study is by far the most carefully executed of these kinds of studies.

Another kind of study, a genetic approach, leads to the same conclusions as in the Mercer study. The reasoning is as follows. The gene pool of American Blacks represents about 70 to 80 percent genes of African origin and 20 to 30 percent genes of European origin. But the African or European contribution to the genome of individuals varies greatly among individuals self-identified and sociologically identified as American Black. If the Jensen hypothesis is correct, there should be a relationship between the average proportion of European-derived genes per individual and average IQ score. Using a number of single locus genetic markers, individuals can be categorized in terms of European versus African derived genes. Only two small studies of this kind have been done. In each case there was no effect of different proportions of European- or African-derived genes. Again we have to conclude that, insofar as we have evidence, our evidence is that it is the environmental differences associated with the ethnic categories, Black or White, which are associated with IQ differences and not the genetic differences between these groups.

At this point it would be well to be precise about what can be demonstrated and what has been demonstrated. We have not argued that different ethnic groups are absolutely identical in anything. This is a logical impossibility. For example, we may have two measuring rods of equal length. In all applications we can use them interchangeably. But we cannot argue that no one will come along in the future who, by finer technique, will show that our two measuring rods differ in length. This is one of the reasons that science proceeds by the use of the *null hypothesis.* The null hypothesis is the hypothesis that there is no difference. If we wish to demonstrate a difference in any characteristic we first set up the null hypothesis, the hypothesis of no difference, and see if we can reject it. Only if we can reject the null hypothesis do we accept the opposite, that real differences exist. With respect to Black-White IQ difference, we have shown that we cannot reject the null hypothesis, the hypothesis of no genetic basis for Black-White IQ differences. Therefore we reject the opposite hypothesis, the Jensen hypothesis.

RELAXED SELECTION

Another set of statements about ongoing trends of change can be found under the name of relaxed selection. As in the case of the declining-intelligence hypothesis, the earliest statements on relaxed selection date to the pregenetic eugenics movement that, up until the 1930s, had a strong element of unsavory political or economic views disguised as biological theory. These statements were usually directed against various forms of social welfare because—it was alleged—such programs preserved less fit types by interfering with the process of natural selection. The reader can now recognize this as a pregenetic use of "natural selection," and this kind of relaxation of selection has now been dismissed by thoughtful persons.

So we must now ask, in what context can we speak meaningfully of relaxed selection? We might speak of a general relaxation of selection in the sense that individuals of all genotypes leave more offspring than they did before. This means that the population grows at a more rapid rate. If a population is not controlling fertility, the rate at which it increases is certainly some indication of the stringency of the conditions of life. But such a view of relaxed selection assumes no change in relative fitness of different genotypes and consequently no genetic change within the population.

If we speak of relaxed selection in terms of relative fitness as in the models of gene frequency change developed in Chapter 3, there can be no such thing as a general relaxation of selection. Increasing the fitness of one genotype decreases, by definition, the relative fitness of alternative genotypes. This means that we can speak of change due to relaxed selection only in terms of specific genotypes and not as an overall phenomenon. But this means that relaxed selection is no more and no less than a change in fitness values of specific genotypes due to changes in the environment.

A recently discussed case of this kind of relaxed selection is the apparent rise in frequency in red-green color blindness cited in Chapter 4. Some of the subsequent studies on red-green color blindness have supported Post's contention that the frequency of red-green color blindness is proportionate to the length of time since a society has depended on hunting as a way of life. Others have been critical of this conclusion.

Post has (1966; 1969) applied a similar hypothesis to deformed nasal septa. The bony septum that divides the posterior chamber of the nose into right and left halves is, ideally, smooth and perpendicular. In a certain proportion of individuals examined, the nasal septum will be found to be buckled, warped, or otherwise deformed in such a way that there is partial occlusion of the nasal air passage. This can lead to a variety of respiratory and other health problems as well as limiting the person's ability to maintain strenuous activity.

Although not a simple genetic trait, as in

the case of the different forms of color blindness, deformed nasal septa have been shown, in twin studies, to have a heritability of 0.4–0.6. In arguing a relaxed-selection hypothesis the nasal septum could have the advantage of being recoverable in archeological populations, thus making it potentially possible to compare true ancestral and descendant populations. Thus far Post has made comparisons similar to those for color blindness. And, as in the case of color blindness, he has found a positive association between the length of time a group has been "civilized" and the frequency of septal defects.

There is now the possibility that a number of rare genetic defects may increase in frequency due to a form of relaxed selection with improved medical care. As an example we can use a condition known as phenylketonuria. Phenylketonuria, or PKU, is inherited as a recessive defect in the metabolism of a substance known as phenylalanine. In homozygous normal or heterozygous persons, an enzyme, phenylalanine hydroxylase, is present to convert phenylalanine to tyrosine. But in the homozygote for PKU this enzyme is not produced. In such cases the body builds up byproducts that eventually affect the central nervous system to produce severe mental deficiency and early death.

Since PKUs seldom or never reproduced in the past, the condition appears to have been maintained by mutation alone. In most populations the frequency of PKU is one in 10,000 or less. This would be a gene frequency of 0.01 or less. Tests have been developed that permit the PKU infant to be identified in two to five days after birth. An infant so identified can be placed on a low phenylalanine diet, and if this is begun within the first few weeks of life he can escape the more severe consequences of PKU.

A number of states in the United States now have laws that make screening for PKU mandatory among the newborn. This should enable those persons homozygous for the recessive trait to lead nearly normal lives and to reproduce if they so desire. This, like some other genetic conditions that can now be treated, raises the possibility that the gene for this condition will now begin to rise in frequency in such a way that future generations will have ever-increasing problems with the care and treatment of their load of genetic diseases.

IS THE GENETIC LOAD INCREASING?

Such considerations as those detailed in preceding paragraphs impelled the Nobel Prize–winning geneticist Herman J. Muller to deliver an address entitled "Our Load of Mutations" (Muller, 1950). Muller started from the simple point that, at equilibrium, the mutation rate and elimination rate of a gene are equal. He then reasoned that relaxed selection would permit deleterious mutants to increase in frequency to a higher level at which the elimination rate would again equal the mutation rate. But the result would be an increase in the load of deleterious genes in the population and in the individual. By making a number of assumptions and utilizing what data were then available, Muller was able to estimate the average number of deleterious mutant alleles in the individual. He concluded from this that "the average man must be . . . at least 20 percent below par of the fictitious all-normal man" (Muller, 1950, p. 142). The fictitious all-normal person was the person who would have the proper alleles at all loci to maximize Darwinian fitness. Muller's major concern was not the present state of affairs but which way things were going.

Since this time the theory of genetic load has become greatly elaborated. At present it is assumed that any genetic variability imposes a certain "load," in terms of selection, on the population. But it was very soon pointed out that a large part of the genetic load might not be a mutational load – that is, the classic model of deleterious new mutants being eliminated by selection – but might be due to the selection associated with a balanced polymorphism due to heterozygote advantage. This is the so-called "segregational load." If most of the genetic load is segregational load, then there is little that can be done to change the situation, as sexual reproduction guarantees the continued production of the less fit segregants.

The relative importance of these and other components of the genetic load are being investigated and debated at length. One of the arguments put forward is simply that there is too much polymorphism in man for it to be maintained by heterozygote advantage. We can illustrate the argument by the following model. Assume the importance of heterozygote advantage and assume that the fittest possible couple (most heterozygous) can have twenty offspring. But in a random-mating population the presence of persons homozygous at these loci reduces the mean fitness of the population. Let us assume that the mean fitness of the population at each locus is 95 percent that of the most fit possible couple. What this means in terms of reproductive success is that maintenance of a polymorphism at one locus reduces average reproduction to 95 percent of the maximum possible for the idealized couple. The maintenance of polymorphisms at a second locus reduces the average possible reproduction to 95 percent of 95 percent, that is, (0.95)(0.95)(20). Maintaining three such loci reduces this to $20(0.95)^3$ and so forth. How many such loci can be maintained before each couple has less than two offspring and the population begins to decline? In this hypothetical situation the answer is given by solving for N where $20(0.95)^N = 2$. We find that in this case if $N > 44$ the population can no longer replace itself from generation to generation. Even if the cost of such balancing selection reduces fitness to only 99 percent of that of the most fit genotype there can only be 229 polymorphic loci.

At the rate at which new genetic polymorphisms are being discovered in man, it is assumed now that the number of polymorphic loci in any population will far exceed these figures. What then maintains this variability? Some have suggested that much of this is residual variability from an earlier period and that at any one period in time only a small part of the total genetic variability is being maintained by selection. Others have suggested that the model of how selection acts in the segregational load is incorrect.

Whatever the solution to this problem, it remains true that the mutational load is real and constitutes a medical problem that could be increasing. On the other hand it is not true that the accumulation of deleterious genes to new levels renders the population less fit. This may sound like a paradox and requires some further explanation.

If we remember to stick to our definition of fitness as Darwinian fitness, it becomes apparent that advances in medical genetics have improved the fitness of those who suffer from PKU. The same is true for a host of other conditions. This means that the environment has changed in such a way that these conditions are not as deleterious as they once were. In effect, the overall population fitness has improved. But with a constant mutation rate the formerly more deleterious mutations accumulate over a period of many generations, at which time an equilibrium again exists between mutation and elimination of the mutants. At this point the population is no more and no less fit than it was when the process began. What has changed is that a part of the population is now being maintained in a new and much more expensive environment, a point to which we will return.

One way in which the mutational load can be increased with a decrease in fitness of the population is by increasing the mutation rate. It is likely that some changes in our industrial society of today are increasing the density of mutagens in our environment. As the atmosphere is enriched with complex hydrocarbon polymers, mutagenic effects on somatic cells increase. Hopefully these seldom reach a site in the body where they affect the sex cells.

The same is not true of radiation effects. The two major sources of high-energy radiation having increasing mutagenic potential are atmospheric contamination by radioactive products and the diagnostic use of x-rays. Nuclear weapon testing has been the major source of atmospheric contamination by radioactive isotopes. The diagnostic use of x-rays has been recognized as a potential source of danger and current practices limit the amount of exposure of the individual much more than in the recent past. But with the increasing complexity of our petrochemical-dependent society it is certain that the problem of controlling mutational sources will increase.

BIOLOGY AND CULTURE

An obvious conclusion that emerges from the preceding pages is that ongoing evolutionary change, including the cases discussed and the many cases not discussed, is as much a social concern as a biological one. This must be the case when, with sufficient depth of time for our evolutionary perspective, we can see that social and biological phenomena are inseparable. As man's own activities loom larger and larger in determining his ecological relationships with the animate and inanimate world, it becomes a more crucial matter to be able to foresee the possible effects on man himself.

We know that the ever accelerating rate of cultural change will make it impossible simply to extrapolate from the past to understand man's future. The rate of increase of PKU and other genetic diseases will not be as predicted because genetic counseling services, unknown a few years ago, are becoming widespread. Not only will rates of transmission of deleterious genes be affected but it is only a matter of time before more direct "genetic engineering" will be feasible.

Biological and cultural phenomena are interrelated in yet another way. The increasing genetic load of man, to the extent that it may be increasing, is not a genetic load in the sense of a true reduction in population fitness but is, in fact, an economic load. In other words an environment can be produced in which many individuals who formerly would not live or would not reproduce can now do so. This means that the population is living in a new environment where, if anything, mean fitness has increased. But this is achieved at some economic cost. The question of how much society is willing to pay for such an environment is faced each year in decisions that concern, for example, the cost of medical insurance, whether or not voters will support bond issues to develop more medical facilities, or how much money is to be spent in the training of research workers in biomedical and related fields.

At an even more general level, human biology and culture are intertwined in the assumptions made about human nature that are currently influential in social or political thought. Each age has its own assumptions about the nature of man. We are all familiar with children's stories of "the little lost prince" who, for any of a variety of reasons, is raised in mean and anonymous circumstances. Always, it turns out, his innate princely qualities shine through and he is recognized as belonging to a higher station. These stories are relics from the age of aristocracy. They contain, except when altered for present-day purposes, an assumption about innate differences between the nobleman and the commoner. This assumption about human nature supported, of course, the social order of a certain period.

The *tabula rasa* assumption is characteristic of an upwardly mobile middle class. It has been a dominant theme in American life. *Tabula rasa* is the concept that each individual is born as a blank page, ready to receive the impressions of his environment. The environment, it follows, is the only determinant of the individual's behavior. This we know to be just as fallacious as the earlier assumption about innate differences between ruler and ruled. Behavior is a phenotype that is always, as with any phenotype, a function of genotype and environment. But this assumption is today enshrined in much of our educational philosophy and is influential in behavioral science (Dobzhansky, 1967; Hirsch, 1967).

An inchoate assumption on the nature of man has recently appeared and may fit all too well the temper of the times. This assumption is based on the legitimate recognition that man, as a primate, has behavior common to many primates. The assumption as it is emerging is that man is an innately vicious, territorial primate. Since he can really do nothing about his nature, he can at least try to be successful in terms of this nature. The last phrase suggests the social and political implications of this assumption. At a personal level this suggests that all forms of interindividual competition are legitimate. At a political level it suggests that nations able to prevail in territorial competition are doing nothing but fulfilling their primate heritage. The assumption about human nature on which this is based is, of course, as fallacious as the previous assumptions in that it ignores the behavior-as-phenotype orientation. Our best evidence now suggests that man has indeed performed as a highly aggressive species throughout much of his past. Undoubtedly he has the genetic and physiological capacity to perform in this way. But since behavior is also a function of environment, we cannot conclude that man will behave aggressively in all possible environments.

We can see that man's future, whether we can foresee it or not, contains inseparably intertwined features that are simultaneously biological and social phenomena. For this reason, we can expect that the individuals who will be able to deal intelligently with such a future will be those with the sophistication in human biology and culture to be able to recognize the possible consequences of change in either area.

REFERENCES CITED

Bajema, C. J. 1963 Estimation of the Direction and Intensity of Natural Selection in Relation to Human Intelligence by Means of the Intrinsic Rate of Natural Increase. *Eugenics Quarterly* 10: 175–187.

Clark, P. J. 1956 The Heritability of Certain Anthropometric Characters Ascertained from Measurements of Twins. *American Journal of Human Genetics* 8: 49–54.

Cook, R. C. 1951 *Human Fertility: The Modern Dilemma*. Chicago: W. Sloan.

Damon, A. 1965 Stature Increase Among Italian-Americans: Environmental, Genetic, or Both? *American Journal of Physical Anthropology* 23: 401–408.

Dobzhansky, T. 1967 On Types, Genotypes, and

the Genetic Diversity in Populations. *In* J. N. Spuhler (ed.), *Genetic Diversity and Human Behavior*. Chicago: Aldine.

Hiernaux, J. 1963 Heredity and Environment: Their Influence on Human Morphology. A Comparison of Two Independent Lines of Study, *American Journal of Physical Anthropology* 21:575–590.

Higgins, J. V., E. W. Reed, and S. C. Reed 1962 Intelligence and Family Size: A Paradox Resolved. *Eugenics Quarterly* 9:84–90.

Hirsch, J. 1967 Intellectual Functioning and the Dimensions of Human Variation. *In* J. N. Spuhler (ed.), *Genetic Diversity and Human Behavior*, pp. 19–31. Chicago: Aldine.

Hulse, F. S. 1957 Exogamie et Hétérosis. *Archives Suisses d'Anthropologie Générale* 22:103–125.

Jensen, Arthur R. 1969 How much can we boost IQ and scholastic achievement. *Harvard Educational Review* 39(1):1–123.

Karn, M. N. and L. S. Penrose 1951 Birth Weight and Gestation Time in Relation to Maternal Age, Parity and Infant Survival. *Annals of Eugenics* 16:147–164.

Mercer, Jane R. and W. C. Brown 1973 Racial differences in IQ: fact or artifact? *In* Carl Senna (ed.), *The Fallacy of I.Q.* New York: The Third Press.

Morton, N. E. 1958 Empirical Risks in Consanguineous Marriages: Birth Weight, Gestation Time, and Measurements of Infants. *American Journal of Human Genetics* 10:344–349.

Morton, N. E., C. S. Chung, and M. P. Mi 1967 *Genetics of Interracial Crosses in Hawaii. Monographs in Human Genetics*, vol. 3. New York: S. Karger.

Muller, H. J. 1950 Our Load of Mutations. *American Journal of Human Genetics* 2:111–176.

Post, R. H. 1966 Deformed Nasal Septa and Related Selection, *Eugenics Quarterly* 13:101–112.

Post, R. H. 1969 Deformed Nasal Septa and Relaxed Selection: II. *Social Biology* 16:179–196.

Saldanha, P. H. 1962 The Genetic Effects of Immigration in a Rural Community of São Paulo, Brazil. *Acta Geneticae Medicae et Gemellogogiae* 11:158–224.

Schull, W. J. and J. V. Neel 1965 *The Effects of Inbreeding on Japanese Children.* New York: Harper & Row.

Spuhler, J. N. 1962 *Empirical Studies on Quantitative Human Genetics. The Use of Vital and Health Statistics for Genetic and Radiation Studies.* New York: United Nations World Health Organization.

Tanner, J. M. 1962 *Growth at Adolescence*, 2d ed. Oxford: Blackwell.

Vandenberg, S. G. 1962 The Hereditary Abilities Study: Hereditary Components in a Psychological Test Battery. *American Journal of Human Genetics* 14:220–237.

Vandenberg, S. G. 1967 Hereditary Factors in Psychological Variables in Man, with a Special Emphasis on Cognition. *In* J. N. Spuhler (ed.), *Genetic Diversity and Human Behavior*. pp. 99–133. Chicago: Aldine.

Willoughby, R. R. and M. Coogan 1940 The Correlation Between Intelligence and Fertility. *Human Biology* 12:114–119.

PART THREE

THE EVOLUTIONARY PAST

In Part One of this book, we surveyed the mechanisms underlying organic evolution. We considered some of the results of the evolutionary process as expressed in living populations. These are studies of human diversity today. These are largely what we earlier termed microevolutionary changes.

Now we turn to macroevolutionary changes: changes which have taken place slowly but which, over long periods, have resulted in profound changes in our species. The record of such change is the archeological and paleontological record; the record of human activities and of animal and plant forms buried in ancient soils. Interpreting this record is not an easy task. It requires the collaboration of scientists from many fields.

The history of human evolution revealed by these scientists is subject to continuing modification as new finds are made. But as more fossil ancestors have come to light our picture of human evolution becomes clearer, even though each new find seems to stir additional debate. The changes through time in our lineage that are revealed in these studies are a result of the operation of the same evolutionary processes as were described in Part One of this book. In almost no case, however, can we identify genes or talk of changed gene frequencies in the fossil record. But, at the least, our hypothesis on changes in the fossil record must be consistent with evolutionary theory as outlined in previous chapters.

Our species is a member of the order *Primates* and the evolution of our species and of immediate ancestral species is the evolution of primates. By looking only at the material brought back from excavations we can learn quite a lot about how these earlier creatures looked, the habitat in which they lived, and perhaps how they moved about and what they ate. But we would also like to infer something about their behavior. To have some insight into primate behavior we begin Part Three with a survey of behavior in living primates.

9

Primate Behavior and Evolution

One of the most important and interesting aspects of human evolution which we wish to understand is the evolution of human behavior. This is also one of the most difficult areas in which to make real progress.

The more complex the behavior the less directly is it related to skeletal anatomy. Locomotor behavior, whether bipedal walking or quadrupedal leaping, or otherwise, is far more easily inferred from fossil remains than whether or not the animal defended a mating territory, lived in monogamous pairs, or shared food. Inferences about complex behavior is, in a sense, more indirect, and must be made on the basis of as wide a variety of evidences and approaches as is available.

One consequence of this situation is that it is often difficult to disprove or reject a false hypothesis. In other words, there may be several hypotheses for the origin of a given behavior, not all of which can be correct, but none of which can be clearly disproven. This, plus the high intrinsic interest of the subject matter, has recently induced more speculative popular writing, bordering on science fiction, in this area than in any other area of evolutionary thought.

Most of the speculative writing which has been so popular promotes one or another simple view of, or bias about, the nature of man: man the sexy primate; man the nasty, territorial and warlike primate; man the happy fruit and shoot eater corrupted by culture; and so forth.

Without the wilder speculative works, the solid findings and research being done on the roots of human behavior are fascinating enough. Serious study of the evolution of behavior began only recently. Scientists began to understand the evolution of morphology through comparative anatomy. Similarly, studies of behavioral evolution have begun with comparative studies.

The first systematic studies of primate behavior in the wild began with the work of C. R. Carpenter in the 1930s. Before this time what we knew of nonhuman primate behavior was based on travelers' reports and observations of captive animals. The behavior of primates in capativity, often derogatorily called "zoo behavior," is useful in understanding important aspects of primate behavior but gives no picture, or at best a very distorted picture, of primate behavior in a normal habitat. The behavior of nonhuman primates in cages changes in many of the same ways as that of humans caged in penal institutions.

Field studies of primates were few in number until the late 1950s and 1960s when S. L. Washburn and his students, and many others, began research which today provides dozens of studies of primates in the wild. Many others, using controlled conditions in the laboratory, explore development and variation in nonhuman primate behavior.

There is nothing clearer than the fact that the behavior of an animal is important to its survival. This does not mean that each act of an individual contributes to that individual's survival. For example, certain acts such as defense of a group or giving a warning call in the presence of a predator may contribute to the survival of a group while endangering the individual giving the warning or defending the group. These are the so-called altruistic behaviors.

Most, if not all, patterned behavior affects the survival of a group. This group may be a flock of birds, a troop of baboons, or a band of paleolithic hunters. It is in this sense that we say that behavior is adaptive. That behaviors are adaptive is the assumption made by Charles Darwin in his studies of animal behavior and emotions. It is the assumption we make today and is the key assumption in attempts to understand the evolution of behavior.

Behavior as an adaptive system is an important concept. It means that to understand primate behavior, we have to view it in an evolutionary context. This has been emphasized by Hans Kummer in studies of baboon social behavior.

Baboons are fairly large, ground-feeding, Old World monkeys. Some baboons of East Africa live in rather open areas making them easy subjects to observe. Baboons live in large, multi-male troops. Troop size ranges from 12 to 100 individuals. Within the troop the strongest social bond is between mother and infant. The infant is the object of attention of males as well as other females, but the mother is quite protective and a strong bond is maintained between her and the infant until it approaches maturity.

A particular male may be associated with a particular female during the estrus period of the female. Then consort relationships are formed between an adult male and the estrus female. The consort male remains close beside the female, mounting her occasionally. The consort male need not be the most dominant male in the group. But the dominant male usually accedes to the position of consort male at least by the height of estrus swelling of the female. After the estrus period the male and female may go their own ways within the troop. But copulatory relationships are not the only social bonds between males and females. There are affiliative bonds which may develop or operate during pe-

FIGURE 9.1 Male baboon permitting infant of female companion to take meat from a kill he is eating (T. W. Ransom).

riods when the female is not sexually receptive. Pairs in such a relationship we can simply refer to as companions (Figure 9.1).

There is pronounced sexual dimorphism among baboons. This means that the males and females differ greatly. The males are about twice as large as females, have heavily hair-mantled forequarters, and have longer, daggerlike canines. This sexual dimorphism is associated with the male's role in defense against predators.

Baboons protect themselves from predators, primarily the large cats and cheetahs, by fleeing to the trees or by standing together as a group in a "mob" defense. Adult males position themselves to confront and threaten the predator. The advantage of large troop size and cohesiveness is obvious with such a defense in open savanna areas.

Kummer has shown that the behavior of the hamadryas baboon differs from the pattern just described for most baboon species. And hamadryas behavior differs in ways which are clearly adaptive in the habitat in which it is found.

The hamadryas baboon of Ethiopia lives in a more arid environment than do other baboon species. Suitable sleeping groves are few in this environment so the hamadryas return each evening to sleeping cliffs. These cliffs provide them the same protection from predators at night as do the sleeping groves of other baboons.

The hamadryas male, unlike other baboons, maintains a "harem" of females throughout the year. This consists of the adult male, one or more adult females and young, an average of about five individuals. Subadult males may attach themselves to one or more of these one-male groups, but if they attempt openly to copulate with the females, will be driven off by the adult male unless he is quite old. The male herds this group, keeping it closely together at all times. A young female learns to follow a male when he begins to herd her and punish her, with bites to the nape of the neck, for wandering away.

Unlike the savanna dwelling baboons, which remain together as a troop during the day, the hamadryas troop breaks up into the separate harem groups during the day. These forage independently and reunite at the sleeping cliffs in the evening.

The adaptive advantage of this behavior seems related to the food supply. In this area, the Danakil Desert, the food resources (primarily vegetation) are sparse and scattered. The troop which has slept together for defense during the night scatters during the day to exploit the scattered food resources. The harem system places at least one adult male with each small group of females and subadults, an evolutionary compromise which distributes predator defense among the widely dispersed components of the larger troop.

This apparently represents genetic adaptations in the behavior of the hamadryas male. In zones where the hamadryas baboons overlap the range of anubis baboons, the hamadryas males will occasionally capture and herd anubis females. The anubis females eventually learn to conform to harem behavior as do the hamadryas females. Anubis males, who do not form harems within their species, do not herd hamadryas females. In an area described by Kummer, occupied by troops which were hybrids between anubis and hamadryas, the one-male groups were unstable and herding behavior was variable and imperfect. Clearly gene flow from the anubis groups disrupted the gene complex associated with distinctive hamadryas behavior.

LEARNING

The comparison of two kinds of baboon behavior also provides some interesting illustrations of what is learned and what is not. The females of both species will herd if they are taught to do so. On their part, this is learned behavior. But, on the part of the male, he herds if he is hamadryas and does not herd if he is anubis. This variation, then, seems to be genetically determined. Anubis troops normally sleep in trees. When they are in areas with no trees they too will sleep on cliffs. Hamadryas, in turn, will occasionally sleep in trees when they are available. We are not sure they do this as readily, however, as do the savanna dwelling species.

Learning is a process of behavioral adaptation without the long process of genetic adaptation. All animal species learn to some extent but the order Primates is outstanding in its elaboration of learned behavior.

Learning, as opposed to genetic adaptation, is not an unmixed blessing. Learning implies more maleable behavior. This means there will be more error or more misdirected responses in the formation of learned behavior. Also the more prominent learning is to a species, the slower the maturation of individuals. This means there will be a longer period of infant and childhood dependency (Figure

FIGURE 9.2 Male baboon grooming subadult companion (T. W. Ransom).

9.2). This is a period in which the young must be fed and protected and an appropriate social structure maintained. The longer period of infant dependency also means the female reproductive potential is lowered. Increases in learning ability must have a high pay-off in fitness to overcome these biological disadvantages.

The period in which much learning occurs in primates, as in all mammals, corresponds to the period of playful behavior. Play behavior (Figure 9.3) is more pronounced among the primates than in any other order. This period of playfulness, and rapid learning, ends or decreases markedly with sexual maturation. And in primates sexual maturation is delayed, relative to other phases of the life cycle, as compared to other vertebrate species.

One way to illustrate this is to compare the duration of fetal development (the gestation period) to the age at sexual maturity. In females the ratio of age at menarche (first menstruation) to gestation period for females is approximately 4–5 in cercopithecoid monkeys, 13–14 in pongids, and 18–19 in humans. In a number of primate species the males mature even later than the females. These figures indicate the relative length of time involved in the maturation of the nervous

FIGURE 9.3 Subadult baboon play group (S. Strum).

system of individuals prior to birth and after birth. As such, they are a developmental index of learning ability.

Among the nonhuman primates the chimpanzee is best known for its ability to learn in some ways similar to man. This was recognized early in laboratory studies. Wolfgang Köhler carried out a remarkable series of studies on chimpanzee learning and tool using abilities in 1913–1917, almost 50 years before successful field studies of the chimpanzee were carried out.

Köhler's usual technique was to place a desired object, usually a banana, just out of the reach of a chimpanzee under various conditions which tested the ingenuity of the animal in obtaining the object. When a string attached to the banana was within reach of the chimpanzee it soon realized the banana could be drawn within reach by pulling on the string. That seems simple enough but it involves the perception of relationships which are, apparently, beyond the capacity of most animals.

When sticks were left on the floor of their enclosure, Köhler found the chimps would soon learn to use them as implements to pull in the banana. The chimps would also use rags, food pans, and anything else around to attempt to drag the banana to themselves. One chimpanzee, Sultan, discovered he could join a small diameter and large diameter bamboo piece to form a compound tool which would reach a banana beyond reach with either piece alone. He later learned to join three pieces to reach bananas placed at a greater distance. And he appeared to enjoy such achievement.

Chimpanzees and other primates exhibit quite a bit of learning in their native habitats. This is most easily seen in primate troops which are provisioned, that is, provided food occasionally so they will stay in the area of the observer. The chimpanzees studied for many years by Jane Goodall practice "termiting" by opening a hole in the tall termite hills and inserting a twig prepared for the purpose (Figure 9.4). Termites attack the twig, which is then withdrawn, and the termites, a chimpanzee delicacy, eaten. The young apparently learn this technique by watching adults. Infants pick up the abandoned tools used by adult chimpanzees and attempt to follow suit in termiting.

Learning can lead to the rise and transmission of new behavioral traditions among nonhumans primates. This has been best documented in long-term studies of provisioned troops of the Japanese macaque. In one case Japanese investigators began to put sweet potatoes on the beach for a small troop of macaques on Koshima Island. Monkeys do not like gritty food and would brush the sand off the potatoes with their hands before eating. In the next year an immature female dipped her potato in the water, rinsing it, before eating (Figure 9.5). Over the years this practice spread through the troop.

Fortunately the mode of spread was well documented. The mother of the immature female was the next to learn to wash sweet potatoes in a brook or, later, in the sea. The other offspring of this female later took up the practice. It spread to play group members. It spread through most of the troop by transfer between pairs in which the monkey which learned potato washing had some preexisting social relationship which caused it to be especially attentive to the monkey from which it learned. Where such relationships did not exist this learning did not occur. Because of this, the high ranking adult males were the last who were not washing their potatoes. A strongly developed dominance relationship is, in many ways, a one-way relationship. The dominant animal is attended to but pays little attention to others lower in the dominance hierarchy.

FIGURE 9.4 Chimpanzee "fishing" for termites (G. Teleki).

Other learned traditions have developed and spread in the Koshima monkeys in the time they have been under observation. Wheat washing is one such tradition. Wheat was scattered on the sand as a normal part of provisioning. The monkeys would laboriously pick the wheat grains out of the sand and eat them. Eventually the same female who began the potato washing tried dropping handfuls of wheat and sand into the ocean; the wheat floated and could be picked from the surface. Again this practice spread in the troop.

These monkeys usually avoid the water and avoid getting wet, if possible. But potato washing and wheat floating resulted in a

FIGURE 9.5 Potato washing by adult female Japanese macaque (M. Kawai).

greater tolerance of getting wet. This plus other provisioning practices of the investigators led to swimming and bathing. Now many of the Koshima monkeys bathe to escape the heat on summer days. They swim and dive as playful behavior. But there appear to be genetic factors in the monkeys' behavior which are inappropriate to their new bathing practices. Mothers are quite solicitous of their infants wherever they are. They permit the infant to attach themselves in a firm grip before moving and are careful that they are not dislodged by quick movements. They are equally solicitous in the sea, but they seldom notice if the infant is above or below water as long as it is clinging well to the mother.

There are genetic factors between species which condition what is easily learned or what is learned with difficulty or not at all. Chicks learn to recognize the hen which is then treated as "mother hen" during a very specific period of life, hours after hatching. Male zebra finches do not tolerate the presence of other male zebra finches, but only if they have been chased by an adult male between the thirty-fifth and thirty-eighth days of life, the normal situation in the wild. When learning is specific, has a critical period for development, is stereotyped and essentially irreversible, it is called imprinting. But even more generalized kinds of learning are influenced by species-specific genetic factors. Dogs, using the sense of smell, readily learn the identity of individuals. Primates, in contrast, learn very little through a sense of smell. How much this difference is due to differences in primary sensors and how much to cognitive differences we do not know.

Primates are outstanding in their ability to learn to manipulate material objects. This was the case in the illustrations of learning in the Japanese macaque given earlier; it is even more true of the chimpanzee.

Chimpanzees seem to have a particular proclivity for manipulating sticks. They use sticks as tools in a variety of circumstances. They also use other objects as tools and, running a poor second to humans, are the next most innovative tool users around.

Termiting, or fishing for termites, with a twig has been mentioned. A chimpanzee will select a flexible twig or grass stalk, break it to the proper length, and remove any leaves or side branches. He selects a spot on one of the tall hard clay termite mounds where a tunnel will be near the surface. Human investigators have been unable to discern the cues used in locating such a spot (Teleki, 1974). He inserts the twig carefully so that it follows the contours of the tunnel; rotates it so the soldier termites attack the twig; then carefully withdraws the twig with termites attached. In West Africa where chimpanzees eat ants which have loose dirt mounds the chimps select longer and stronger twigs more appropriate to stirring the loose soil.

Other tool use by chimpanzees in the wild includes using sticks and stones to pound hard-shelled fruits; crumpling up leaves to sponge water from a crevice in the hole of a tree; and generalized using of sticks in poking, probing, and prying to get to a variety of foods. Chimpanzees have also been observed throwing objects at baboons who were competing for a favorite food source. In test situations chimps have also picked up heavy wooden sticks and threatened a stuffed leopard suddenly presented to them. Whether this is tool use or not is not clear since chimps also grab and wave branches and sticks in general excitement displays.

DOMINANCE

The concepts of dominance and of dominance hierarchy have been very important in comparative studies of animal behavior. Traditionally dominance is judged by who defers to whom or who gives way to whom. A simple dominance hierarchy is best seen, and was first studied, in birds. In a flock of domestic chickens, each bird can be ranked on a single scale, lowest to highest, in terms of who pecks whom. Such a *pecking order* is a simple and easily observed dominance hierarchy.

In some primate species an analogous dominance hierarchy exists. The alpha male has first access to food and others have access in rather regular order. This order will also be reflected in who gives way to whom on a trail. It will be reflected in who initiates movements of the troop. But for many primate species a simple unilineal dominance ranking does not adequately describe the social order.

In long-term, intensive studies it is often found that the animal that is the more powerful of the two in paired interactions is not dominant within a troop. One reason for this is that among some species, such as the baboon, coalitions are important. In such cases two or three adult animals may form a coalition helping one another in conflicts and enabling them to perhaps dominate an individually more powerful animal.

Encounters which establish dominance between male baboons are rather dramatic, and for years it was assumed that the dominant males controlled the movements of the troop. Recent long-term studies of the olive baboon have shown that the group of reproductively mature females of the troop strongly influence group movements. This is independent of dominance relationships among males. It is this group of females who are the permanent core of the troop.

In earlier studies, by assuming a simple unilineal dominance hierarchy, the true complexity of the situation was missed. For this reason many students of primate behavior have deemphasized the concept of dominance and utilized the concept of role in much the same way it is used in human sociology. Roles are distinctive sets of behaviors characteristic of categories of individuals within the troop. "Nursing mother" is a role found in all primate societies; "subadult male" is another; and so forth.

Rather than use outcomes of different kinds of behavioral interactions as indicators of dominance, it is possible to define dominance in terms of the direction of submission gestures. Within each species of group-living primates, there is a set of gestures indicating submission or subordinance. Many of these are common across primate species. Some of these gestures develop, ontogenetically, out of gestures used by infants to placate adults. A dominance hierarchy defined in this way does not have a one-to-one relationship to dominance defined in the more traditional manner as outcomes of certain behavioral interactions. It is a structure of dominance-subordinance signals which affect behavioral interactions.

Defined as a system of dominance-subordinance signals, dominance relationships are fairly clear-cut in most primate societies, and dominance behavior has been important in the evolution of human behavior. Some primate species, baboons, macaques, chimpanzees, and humans for example, have more than one sexually mature male in the social group throughout the year. Evolutionarily this is a good trick. It is rare. Very few species of any animal order have more than one sexually mature male together when females are sexually receptive. Altruism is apparently not easily achieved in this area. The ability to learn dominant and subordinate roles seems to be critical in keeping adult males together under such conditions.

TERRITORIALITY

One rationale for studying comparative primate behavior is simply to establish to what extent there is a common primate heritage in specific behaviors. As a hypothetical example, if we establish that staring is a behavior common to all primates in establishing dominance, and that it occurs in this context in humans also, we assume it is present in humans as a common primate heritage. The burden of proof would then be on those who claim that staring behavior in humans is only a cultural tradition or learned behavior.

Some popularizers of science have claimed that territoriality is a primate heritage present in humans. For these writers it then follows that territorial defense, including wars between nations, are "natural" and inevitable. This position confuses at least two major issues: first, whether or not genes make any kind of behavior "necessary" and, second, the nature of territoriality.

The concept of territoriality was, unfortunately, defined on the basis of the behavior of song birds and does not fit primate behavior too well in any case. The early focus of animal behaviorists on song bird behavior led them to define territoriality as the defense of an area, by its occupants, against competing members of the same species. There are primate species which are territorial in this sense. These are usually forest species which live in small troops occupying a relatively small area.

The lar gibbon (*Hylobates lar*) is an example of a territorial "family group." These arboreal acrobats of Southeast Asia live in troops consisting of only one adult male and female and their young. The territory they occupy may be less than one square mile. Each group sleeps near the center of its territory but moves out toward the boundaries each morning. Calling as they go, they are met by members of the adjacent troop. The adult male, in particular, goes through a highly stylized, acrobatic, and noisy threat display toward the opposite male at the boundary of the territories. Actual contact is rare, but on those rare occasions a nasty bite will be given.

The gibbon territorial display will occupy the whole troop for a short time and the males continue for one to two hours. There is little overlap of troop areas with this kind of territoriality. Gibbon behavior fits the common stereotype of what territorial behavior is like in primates, but it is not the common condition in primates.

The spatial relationships between troops of any given primate species represents an ecological adaptation and, consequently, varies quite a bit between species. In most, but not all, species there are advantages to aggregation. These aggregates, which in nonhuman primates we call troops or groups, tend to occupy different areas. This reduces, or at least regulates, the competition for resources such as food, sleeping groves, and so forth. In many cases behavioral mechanisms have evolved to facilitate and maintain this between-troop dispersal. In species in which the troop occupies the year around an area which is very small relative to the animals' ability to travel across it, that species usually shows highly specific between-groups spacing mechanisms including boundary marking, display, and occasional conflict. This is the classic conception of territorial behavior.

Where the area utilized by a troop is large relative to daily movement, the behavioral mechanisms which contribute to spacing between troops are, themselves, less specific. As in the case of the gorilla, they may be little more than a display of nervous excitability on the part of the adult silverback males of opposite troops. This is a small but real pressure causing troops to move apart. In such cases the area utilized by adjacent troops may overlap extensively.

The latter situation, with no overt defense of ground, no marked boundaries, and extensive overlap between troops, is usually designated a *home range* rather than a territory. A home range has been defined as a region which an animal habitually occupies without excluding others of the same species. In all primate species studied, the spacing of animals is a little more regular than is implied in this definition of home range. So neither territoriality or home range accurately classify the behavior of most primates. These two concepts represent extremes in a continuum of group spacing adaptations.

GROUND-DWELLING HIGHER PRIMATES

The ground-dwelling higher primates are of particular interest to us because it is in the ground-dwelling situation that important aspects of human social behavior evolved. Do such primates have common features of social behavior? One of the major general findings of recent nonhuman primate studies is that there is great diversity in social behavior among primates. Still, we can discern some more or less common features of organization and behavior.

If we look at group spacing mechanisms

among the ground-dwelling anthropoids, we find these largely dependent upon the behavior of males toward males. Males, as single animals, do transfer between groups. But this is a slow process where the immigrant male, with some difficulty, manages to get accepted in a new group. On the other hand, when groups of males of one group meet groups of males of another group there are a variety of agonistic behaviors that will be displayed. These agonistic or aggressive behaviors seldom involve fighting although this can occur. In effect, the males of one group keep their females from the males in the other group. Since this occurs on the part of both groups the groups move apart.

Any specific case of an encounter between groups may not take the simple form just described. Primates have long memories; they are capable of assuming different roles. For example, baboon troops which have fissioned into two independent troops may be able to comingle to some extent when they encounter one another for some time after the two have separated.

Chimpanzee social structure is complex enough to still not be clearly understood. For many years it was assumed that there was no defined group membership among chimpanzees. Varying kinds of groups were observed: some consisted of a few adult males only, some contained only mothers with young and other females, some included all sexes and ages except mothers with dependent young, some contained all ages and sexes. Membership in these kinds of chimpanzee groups was fluid and designated "open groups" by the Reynolds as opposed to the "closed groups" found among most primates (Reynolds and Reynolds, 1965).

But it appears now that these are open groups within a larger group which is bounded. The larger group may be, typically, 50 to 60 individuals who do not all come together at one time. This group does range over a common area, in contrast to other groups. And it is within this larger unit that individuals rearrange themselves into small groups of different composition and carrying on different activities.

The social structure of ground-dwelling anthropoids shows some features in common with that of the ungulate herds (that is, herds of hoofed mammals). This is related to similarities in the ecological niche occupied by each. Both subsist primarily on vegetation and are, in turn, an item of subsistence of a number of carnivores. The social structure of both provides mechanisms for reducing predation pressure while exploiting their food resources.

As in an ungulate herd, the core of each group is a group of females and their young. The major continuity of group membership is within a line of females. As they approach sexual maturity young males are "peripheralized" both socially and spatially, by the actions of the mature males. Many eventually leave the group of their birth to attempt to join another group. In this process they may live singly or in a small all-male group for a short time. Older males also get displaced or, for reasons not well understood, leave to join a different group. The net result is that, over years, the male membership of a group changes. The female membership changes little if at all through the transfer of individuals. The female membership depends, primarily, on birth and death processes.

Many social roles within a group are predictable in terms of two variables, age and sex of individuals. There are characteristic roles for males and for females. There are characteristic roles depending on age when animals are classified as infant, juvenile, subadult, and reproductive adult. But there are more limited roles determined by other factors. Some animals are larger and stronger than others. Some are better at forming coalitions than others. Among rhesus monkeys (*M. mulatta*) a male has a better chance of becoming dominant if his mother was a dominant female. This is probably true in other ground-dwelling cercopithecoids also, but it requires long-term studies to confirm.

In the studies of tool using and manipulative abilities, not all individuals of a troop will be equally skilled. Some chimpanzees, as adults, are not as good at termite fishing as others. Some, within the same troop, never fish for termites. The same is true of other tool using. As more detailed studies are carried out, behavioral variability within, as well as between, troops becomes more and more apparent.

HUNTING

One of the surprises of the primate field studies of the 1960s was the discovery that some nonhuman primates catch, kill and eat other animals, at least occasionally (Figure 9.6). Rural residents of southern Africa have told of baboon and chimpanzee predation for many years but no such incidents were described in specific field studies until the 1960s.

The best documented examples of nonhuman primate predation on other mammals comes from baboons and chimpanzees. Baboons do not often eat meat and they usually ignore dead animals. But they do catch and kill birds and small mammals. It appears that adult males engage in the chase and kill most frequently, but females and the young do also, and seek a share of the meat at other kills. Baboons of East Africa, where studies have been most extensive, kill and eat birds, hares, smaller monkeys, and the young of a number of antelope species from Thomsons gazelle to impala.

Long-term studies of baboon troops near Gilgil in Kenya have recently shown that the nature of predation, both capture and consumption, may change quite a bit over a period of a year or two within a single troop. One troop increased the systematic pursuing of prey during the time it was under observation. Strum (1975) observed the males of this troop increase hunting until pursuits covered two miles and lasted as long as two hours. This is true hunting behavior. Although this has not been observed in other baboon troops, these primates are obviously not only capable of hunting but of increasing the frequency of hunting under appropriate

FIGURE 9.6 Male chimpanzee eating a young baboon (G. Teleki).

circumstances. The ecological changes contributing to this change are not clear. But in the area of this study the traditional predators have been eliminated, and the pressure of grazers has been increased both from native species and from domestic cattle. The baboons are also safer from predation themselves.

Chimpanzees engage in no less frequent predatory behavior than do baboons, and they occasionally prey on baboons. The catching and eating of other mammals, primarily other primates, by chimpanzees is extensively documented by Teleki (1973). The chimpanzees of the Gombe Stream Research Centre, eastern Tanzania, occasionally catch and eat individuals of almost all the mammalian species which are potential prey within the region. The studies of predation in this area indicate that the success rate of chimpanzees in making a kill is about the same as in other mammalian predators. But since pursuits are unusually silent, an unsuccessful attempt may go unnoticed more often than a successful pursuit, which is crowned with a great deal of excitement and vocalization among the chimpanzees.

FIGURE 9.7 Supplication, or begging, gesture in a group of chimpanzees eating meat (G. Teleki).

Some of the catches made by the chimpanzees are opportunistic. But predation, overall, is fairly regular and some episodes have quite an organized character. *Stalking* involves the most cooperation among males. In stalking, one chimpanzee may move quietly up a tree occupied by a prey monkey. Other males station themselves near adjacent trees. As the monkey attempts to flee the closest chimp takes up the stalk blocking escape while the entire group shifts its position to again block the new set of possible escape routes.

Chimpanzees, by any criterion, like the meat obtained from a kill. They eat slowly and with concentration. Other chimpanzees attempt to obtain a share of the kill by taking some or by begging gestures (Figure 9.7). In this manner meat is shared. There may be some conflict involved in the distribution process. But, overall, it is a rather calm and relaxed process compared to the excitement of the kill. A single kill may get distributed quite widely. Larger portions of the carcass that have been torn off and yielded to another chimp become the focus of another small group of individuals seeking portions.

Although chimpanzees are the outstanding nonhuman tool users, there is only one case in which a chimpanzee has been observed using a tool in the pursuit of prey. One kind of tool use has been observed in the consumption of prey. In eating the brain a male was observed to crumple up a leaf sponge or "wadge" which was used to wipe and sponge out all bits of the brain tissue.

Hunting was, not long ago, considered a distinctive feature of *Homo* among the primates. This supposed qualitative distinction has now been dropped. This is the case for a number of traits once considered distinctly human. This is important in considering what is to be explained in the evolution of human behavior. Evolutionary explanations must deal with how hominid behavior came to differ from that of their nearest primate relatives. Since the end products of this process, man and the African apes, differ less today than we once thought, we have a bit less to explain than we once thought necessary.

HUNTERS TODAY

Small societies of hunters living today or in the recent past are economically and demographically more like early human societies than any other human societies today. These small groups who hunt and gather wild foods will be considered briefly for insights they can give us on human preagricultural behavioral adaptations.

Only a few thousand years ago, hunter-gatherer populations occupied most of the habitable parts of the earth from the tropics to the arctic. They lived in a greater variety of habitats than any other mammal. Of course, there was quite a bit of diversity in cultures associated with this ecological diversity. But there were common features among those who were described before their destruction as cultures. And there were some social behaviors which, although not present in all hunting-gathering societies, were common enough throughout the range of our species that we can take these features, too, as typical of hunting-gathering society.

The usual residence group of hunters is small. It consists of perhaps five to seven households totaling 25 to 35 individuals. This residence unit is called the band. It corresponds to the nonhuman primate troop. The band occupies a home range or territory sufficient in size to provision the group the year around except in exceptionally bad years.

Bands within a region maintain close ties. These ties are almost always based on intermarriage. Marriage and kinship ties are reinforced by visiting and trade along kinship lines. In times of food shortage or other distress, these kinship ties can be utilized to seek aid or refuge. Visits can occasionally lengthen into permanent residence in the band of a kinsman. This means that territorial boundaries are quite permeable. But this permeability applies only to well-known kin sharing reciprocal obligations to one another. It does not apply to strangers.

Households are based on a more or less permanent marriage between a man and a woman. Most marriages in hunting societies are monogamous although few or no hunting societies have only monogamous marriage. A good hunter or otherwise high-ranking male might have more than one wife. But this seems to have been, numerically, a minor variant everywhere except in Australian cultures. Australia was unusual in its degree of polygyny.

In very few instances do households remain at the same place the year around. One to several moves are made each year within the band area. With seasonal changes in food resources households might disperse for a part of the year to come together later. Where resources permit, or where a large resource windfall occurs, a number of related bands might come together for a brief season of visiting, ceremonies, perhaps common hunts, and marriage arranging.

Marriage normally does not occur within the band. Age mates are considered to be

FIGURE 9.8 Young Birhor men, sons of three brothers, who live together in a single hunting band in east-central India (B. J. Williams).

like brothers and sisters. Therefore bands are exogamous; they seek mates in other bands. These marriages maintain the kinship alliances between bands. In most, although not all, known hunters and gatherers, a greater effort is made to keep sons within the band than to keep daughters. So the nucleus of a typical band is a group of brothers (Figure 9.8) with wives from adjacent bands (Figure 9.9), where in all likelihood their father's sisters now lives. "Brother" or "sister" in this sense is a social category and does not imply the same genetic relationship among all "brothers" or all "sisters."

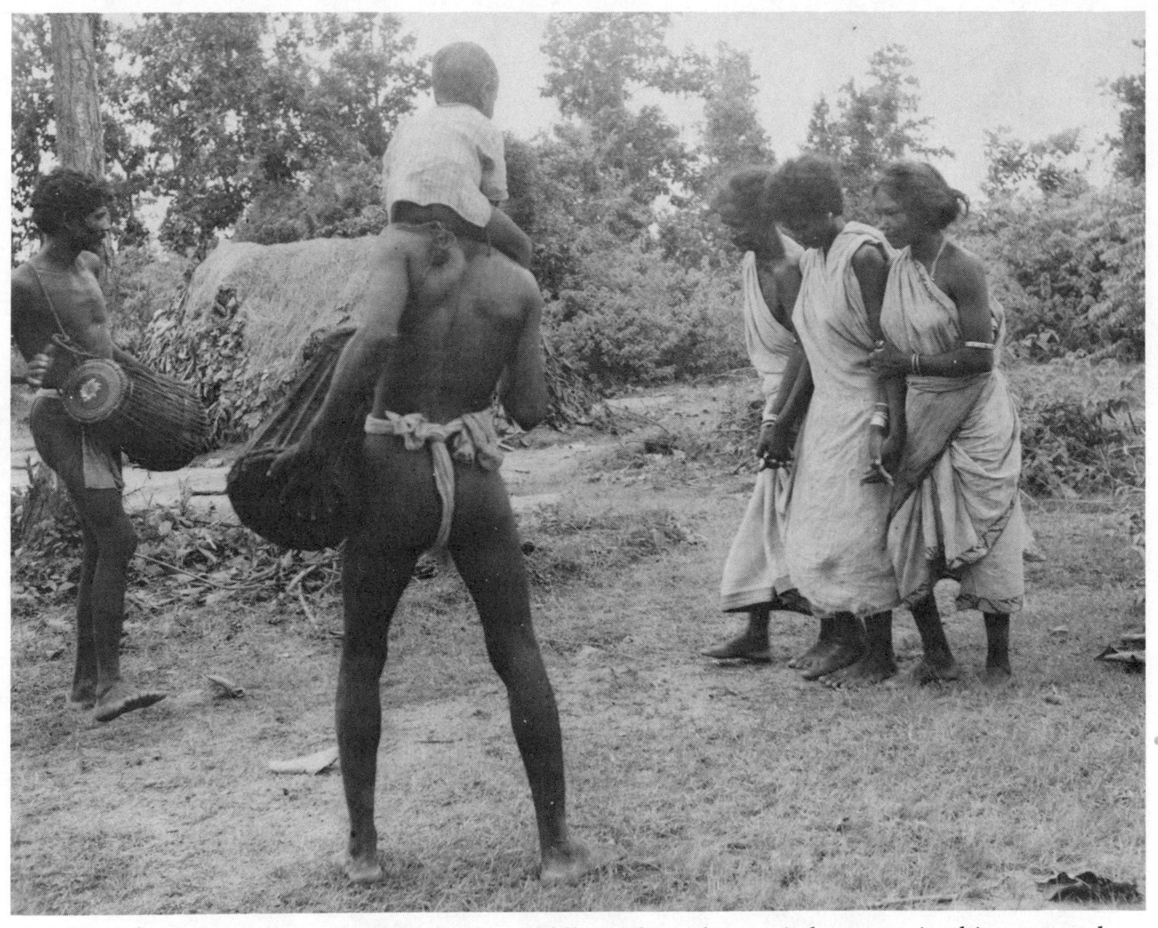

FIGURE 9.9 Dance among hunters. In the middle is the only married woman in this group whose husband has chosen to join her rather than have her move into his group (B. J. Williams).

The cooperating group of males within the band ranges from two to eight individuals, depending primarily on the technology of hunting or collecting. The cooperating group of females is smaller, two to three individuals. Human hunter-gatherers always exploit their environment with tools. Tropical hunters tend to have the simplest technology: perhaps little more than a digging stick, bow and arrow or spear, axe and carrying bag. Many tropical and temperate region hunters make nets and snares (Figures 9.10 and 9.11). The Eskimo, in pushing into an inhospitable arctic region, developed the most elaborate technology of any hunting culture.

The extent to which hunter-gatherers rely upon meat in their diet varies greatly from region to region. In general, tropical groups rely less on meat than groups in higher latitudes. Some tropical hunters might have as little as 20 percent of their total caloric intake provided by meat; the remainder provided

FIGURE 9.10 Young Birhor male making a small hunting net used for snaring tree squirrels (B. J. Williams).

by gathered vegetal items. The Coastal Eskimo, on the other hand, rely almost entirely on meat. But all hunter-gatherers seek meat. In tropical areas where the bulk of the food comes from gathered items, the men will hunt even when it appears that there should be greater caloric pay off in gathering for the same length of time.

Certainly there were ethnographically known hunting groups which did not fit the pattern described in the preceding paragraphs. But many hunter-gatherers did, and

FIGURE 9.11 Late adolescent Birhor male making rope, from the bark of a vine, for a large hunting net (B. J. Williams).

some still do, fit this pattern, and they occur in all latitudes on all continents. It is a very general pattern. We do not know how far back in time this pattern extended. We do know, from archaeological evidence, that fishing and the exploitation of aquatic resources is recent, within the last 20,000 years. But the archaeology of recent hunters is very similar, except for types of tools, to the archaeology of hunters through the Upper Paleolithic, the past 30,000 or more years.

FROM APE TO HOMINID

Genetically our two closest relatives among the primates are the gorilla and the chimpanzee. The gorilla has evolved into a large, lethartic, browser. In body size and in ability to exploit a savanna-woodland environment, the chimpanzee remains least changed from what the fossil evidence indicates a common ancestor to have been like. And, in any case, we know more about the chimpanzee than we do the gorilla. Therefore we will consider briefly some differences between chimpanzee and human adaptations.

First let us consider the technology of the food quest. The most obvious difference is that humans utilize tools far more extensively. Indeed humans are continuously dependent on tools. In most cases human hunter-gatherers could not do without tools. Chimpanzees do a limited amount of food collecting with tools, fishing for termites, digging for ants, probing and sponging. Some chimpanzees never use tools and, overall, chimpanzees would suffer very little without them. On a qualitative level we find a major difference in that chimpanzees seem rarely to apply their tool using abilities to hunting. Tools are all-important in human hunting.

Meat provides perhaps one percent of the caloric intake of the Gombe Stream chimpanzees studied by Goodall, Teleki, and others. Human hunter-gatherers in a comparable environment would obtain 25 to 40 percent of their calories in the form of meat.

Nonhuman primates share food very little. Chimpanzees have been observed to share bananas, but ordinarily a monkey or ape with a fruit in its hand does not willingly give it up, even to young, unless he or she loses all interest in the food and discards it. But meat eating in primates is somewhat different. Chimpanzees never offer meat, but they share it in the sense that, when "begged" by certain other chimps, they will release portions of the meat without conflict. Obviously meat eating could provide an early hominid not only with a concentrated source of protein but a basis for the kind of food sharing necessary for a long period of infant nutritional dependency; a period longer than that provided by the mother's milk.

Within a given group human grouping behavior seems to be a little less flexible or fluid than that of the chimpanzee. Although the hunting band rearranges during the year one does not encounter the diversity of groups in a hunting band's yearly cycle as is encountered within a chimpanzee troop in the course of a year. This, again, is clearly related to the demands of child rearing in humans.

A major difference between human and chimpanzee social organization is the existence of a kinship system in human societies. Chimpanzees and other nonhuman primates maintain bonds of affection between mother and offspring, between siblings, and between certain other companionate individuals. Humans do this too, of course, but the kinship system does more. The kinship system depends on a system of names. These names label role relationships within the coresident unit. For each individual these kinship terms come to be associated with affective values and expectations of certain kinds of behavior. This constitutes an ethic of behavior which is extended outside the family or household by extending the terminological system. The effect of this is, first, it makes role relationships more permanent than would be the case where these role relationships depend only on frequent face-to-face interaction for reinforcement. And, second, the kinship system extends these role relationships to a larger number of people than would be the case without a kinship system.

The kinship system obviously depends upon language, a point we will return to. With a kinship system matings become marriages and marriages between bands constitute alliances between kin groups. The function of alliances between hunter-gatherer bands has been described.

Human hunter-gatherer bands differ from non-human primate troops in that females transfer between bands more often than do males. The movement of females between autonomous groups occurs as a normal event only with arranged marriages. This depends, in turn, on the existence of a kinship system. Its effect is not just to maintain alliances but to preserve a group of males who, through having grown up together, are able to cooperate with little friction in economic activities.

There are also motor control, skeletal, and muscular differences which exist today between chimpanzee and man and which contribute to the hominid tool-using hunting adaptation. The hominid hand is better adapted to skilled manipulation of objects. And more of the cortex of the brain is devoted to control of the hand. As a bipedal walker and distance runner, humans have no equal among primates. With his locomotor adaptation and exaggerated ability to cool by sweating, man can pursue prey for long distances. Although he prefers not to, and seldom has to, man can easily pursue prey for 30 miles in a day. In the context of a primate heritage man would be, and obviously was for a long time, a highly efficient savanna predator.

Obviously an important factor in the human adaptation is the existence of language. One definition of language is an oral system of symbols used in communication. All primates communicate vocally at times. But these vocalizations, even in chimpanzees, reflect emotional states and are controlled by the primitive part of the brain, not the cerebral cortex as is the case in quite a bit of human vocalization.

Symbols are things whose meaning has no necessary relationship to their form. What they convey in communication is arbitrary with respect to the physical properties of the symbol. Recently chimpanzees have been taught to communicate with experimentors by manipulating plastic tokens and, in another case, by a standard sign language for the deaf. In both cases the chimpanzees demonstrated their ability to generate novel, but meaningful messages. That is, they could communicate something by recombining elements into new forms they had not been taught. This means the message had true symbolic content. The ability to generate novel messages has been labeled "productivity" in communication systems. It is one of the characteristics of human language. Chimpanzees can not do this orally although there have been a number of attempts to teach them to do so. The difference resides primarily within the structure of the brain and not in the articulatory mechanisms of the mouth and pharynx. This does not mean that chimpanzees could articulate the same sounds as humans if they wished. But articulatory problems do not seem to be the major block to greater oral communication in chimpanzees.

In an important sense linguistic ability, but not specific languages, is innate to *Homo sapiens*. As has long been pointed out, only human infants go through a babbling stage where they play with sounds, they experiment with sounds. A human is the only animal that can associate a sound with a remembered event and store and recall that from all other senses. Actually we do this best with images and worst with smells. Nevertheless only the human brain seems to have the connections for this kind of cross labeling. And this is critical to language as we think of it.

The importance of language to human adaptations today does not mean that this was the factor which set an apelike ancestor along the evolutionary road leading to *Homo*. Some linguists have estimated that it would have taken less than 50,000 years for all surviving languages in the world to have differentiated from a common ancestral language. This means that language, with all the features we associate with human language today, may be very recent. The linguistic abilities of early hominids may have been limited for a long period after other distinctively human features had developed.

It is impossible to determine the linguistic abilities of earlier species from fossil evidence alone. Tool-using abilities are somewhat easier to deal with, at least when stone tools are made. It should be remembered, also, that tool use may have had adaptive significance for early hominids which are not apparent from studies of present-day chimpanzees. Chimpanzees live in forests or forest fringes. Here they suffer little predation except from man. Without man the distribution of chimpanzees and many other Old World primate species would be quite different from what it is today. As will be described in a subsequent chapter, the first fossil species which everyone can agree was ancestral to *Homo* lived in savanna regions. Defense against predation is a much more prominent feature of nonhuman primates on the savanna today. This means that tools as weapons may have been more important in removing predator pressures than in providing increased food supplies to early hominids. At the moment we are not sure how important this was. In the following chapters we will review not only the sequence of fossils leading to our species today but will look at evidence—where there is evidence—of tool use and cultural capabilities generally.

REFERENCES CITED

Köhler, Wolfgang 1926 *The Mentality of Apes*. New York: Harcourt, Brace and Company.

Reynolds, V., and F. Reynolds 1965 Chimpanzees of the Budongo Forest. *In* I. Devore (ed.), *Primate Behavior*, pp. 368–424. New York: Holt, Rinehart and Winston.

Strum, S. C. 1975 Primate Predation: Interim Report in the Development of a Tradition in a Troop of Olive Baboons. *Science* 187:755–757.

Teleki, Geza 1973 *The Predatory Behavior of Wild Chimpanzees*. Lewisburg: Bucknell University Press. 1974 Chimpanzee Subsistence Technology; Materials and Skills. *Journal of Human Evolution* 3:575–594.

10

Background for the Study of Fossil Primates

With the primates we come to the mammalian order which includes man. At this point it is well to focus with some care on the means used to make inferences about the scraps of fossilized bone which represent close relatives of, and in some cases ancestors of, modern *Homo sapiens*. Three kinds of data are essential to the reconstruction of primate evolutionary history: morphological (structural) affinities of the fossil specimen, its age, and the environmental setting in which it existed.

The methods of the paleoanthropologist differ from those of the paleontologist only in that archeological materials are also available for the interpretation of man's evolution during the past two million years. These evidences of man's activities and way of life become important in Pleistocene deposits and indeed have come to be used as a criterion of hominid status (a member of the family Hominidae) in addition to the normal morphological criteria. Prior to the Pleistocene, throughout the Tertiary period of the Cenozoic, we must rely on time, place, and morphology of the fossil itself.

Most primates are forest dwellers, and these forests are usually tropical. Unfortunately tropical forest soils are generally acidic, and bone decays rapidly in the upper soil zone. Where decay is slow it is more likely that mineralization processes can replace bone structure with insoluble salts and thus produce a fossil. Furthermore the kinds of sedimentary deposits that are most fossiliferous are not generally associated with a tropical forest ecology. These are two of the reasons that fewer primate skeletons are preserved than those of animals such as the horse. The parts of the skeleton that are most likely to be fossilized are those of the greatest density. Foremost, this includes the teeth; secondly, parts of the skull and long bones. Since such fragmentary remains are often far separated in time of origin and in space, the conversion of morphological similarity into a valid phylogenetic sequence is no easy task.

EVOLUTIONARY RULES AND THE INTERPRETATION OF FOSSILS

To assist themselves in the construction of fossil family lines, investigators often resort to evolutionary "rules." Before going further we must consider a few such rules and their validity. One such supposed rule, or law, was termed by Haeckel the *biogenetic law*. The biogenetic law is often stated with that clever phrase, "ontogeny recapitulates phylogeny." In other words, embryonic and fetal development (ontogeny) reflects the evolutionary history of the species. This "recapitulation theory" was based on the assumption, going back to the Greek philosophers, that all animal forms can be arranged in a "ladder of nature" from lower to higher forms. Haeckel added to this the idea that higher forms evolved through the addition of further ontogenetic developments onto the sequence of development of the lower form. Therefore the young stages in the development of an animal should reflect the form of adult ancestors. Had Haeckel been correct the biogenetic law would have provided not only the ancestral forms of living animals but the relative sequence in which they evolved.

Better embryology and more complete fossil sequences demonstrated the biogenetic law to be incorrect. The studies of von Baer provided a sounder set of generalizations to replace the biogenetic law. The most important of these was that "the young stages in the development of an animal are not like the adult stages of other animals lower down on the scale, but are like the young stages of those animals" (de Beer, 1951:3). This generalization is much more reasonable in terms of what we now know of the genetic control of development. Mutational changes are not easily incorporated into earlier stages of embryonic development. Small changes early in embryology have serious consequences simply because many subsequent developmental steps are dependent on earlier steps. The earlier the change the more likely it is that the whole machine does not fit together. Changes late in development have a greater chance of being successfully incorporated into the genome and this is where, in the life cycle of individuals, the most rapid evolutionary change is seen. Early embryonic stages, being evolutionarily conservative, maintain a similarity among organisms which, in the adult form, may diverge widely.

This rule does not help us interpret recent evolutionary change. We can be sure that man had ancestors with external tails. They had longer body hair. The infants had a grasping reflex permitting them to cling to their mother's fur as is the case in most primates today. But this tells us nothing about the most recent 30 to 40 million years of man's ancestry. But von Baer's rule has other uses. It suggests that any recently acquired adaptations of a species that inhibit the expansion of that species in a new environment may be eliminated by a simple retardation in certain developmental processes. This process is called *paedomorphism* where fetal features are retained, or *neoteny* where juvenile features are retained.

Another interpretive principle often misused is that of orthogenesis or straight-line evolution. Orthogenesis is invoked when a person, observing a trend of evolutionary change over a given time, assumes that this same trend must have been present up to the time of his observation or continued after that time. The erroneous assumption is that the future direction of change is inherent in the organism itself and not a function of environmental circumstance and natural selection. This error has even led persons to assume that certain fossil lines became extinct because they evolved in directions inappropriate or maladaptive to the environment in which they existed. This simply does not occur.

A closely related idea is the irreversibility of evolution. This does not mean that a trend of change will continue indefinitely but, rather, that evolution will not reverse direction and return to an earlier form. As a rigorous principle this is obviously incorrect. A population having a low frequency of the blood group allele I^A may, over several generations, develop an increased frequency of I^A and then, with a change in selection, return to its original condition. This is a reversal in evolution. In a genetically complex structure, however, the situation is different. If the change in an organ from condition *X* to condition *Z* involved a network of gene frequency changes in many genetic systems, especially sequentially dependent changes, then the likelihood of a reversal of all these changes and a return from *Z* to *X* would become negligibly small. Mammals that have readapted to the sea represent a classic illustration of this point. Fishes propel themselves by a wave-like lateral flexion of the body and tail. Whales, evolving from quadrupedal carnivores, did not again develop fishlike musculature and fins. From the mammalian structure they evolved tail flukes moving up and down, accomplishing functionally the same thing as the side to side motion of the fish.

In later primate evolution irreversibility has been often invoked in questions of simple size change in a structure. For example, can the sequence small molar → large molar → small molar represent a single line of evolution in teeth? The answer is yes. Relative size difference may be genetically quite simple and irreversibility cannot be used to rule out the possibility of sequences such as that illustrated.

INFERENCES BASED ON SKELETAL MORPHOLOGY

Since teeth are most often recovered, and frequently they are all that is recovered, a great deal depends on their interpretation. Primitive mammals had a dental formula represented as $\frac{3.1.4.3}{3.1.4.3}$, which means that the animal had, on each side of the mouth, three incisor teeth, one canine tooth, four premolar teeth, and three molar teeth. The numbers above the line represent maxillary teeth (upper dentition); the numbers below the line represent mandibular teeth (lower dentition). The teeth represented by the preceding dental formula are shown in Figure 10.1. In primate evolution there has been a general shortening of the muzzle and a reduction in total number of teeth. New World monkeys of the family Cebidae have $\frac{2.1.3.3}{2.1.3.3}$ for a total of 36 teeth. Old World monkeys, Cercopithecoidea, and the Hominoidea, man and the great apes, have $\frac{2.1.2.3}{2.1.2.3}$ for a total of 32 permanent teeth.

There are two ways of referring to individual teeth. One way of doing this is shown in Figure 10.2. Superscripts are used to designate maxillary teeth. Subscripts designate

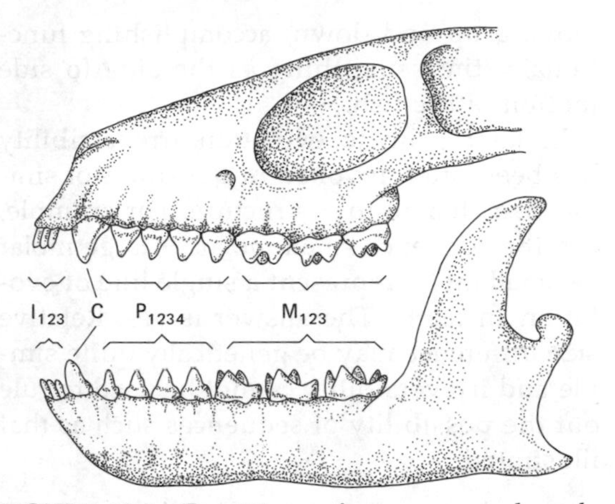

FIGURE 10.1 Dentition of an assumed early placental mammal of the kind that gave rise to the primates. (I) incisors, (C) canines, (P) premolars, and (M) molars. (*Source:* Clark, 1962. Used with permission.)

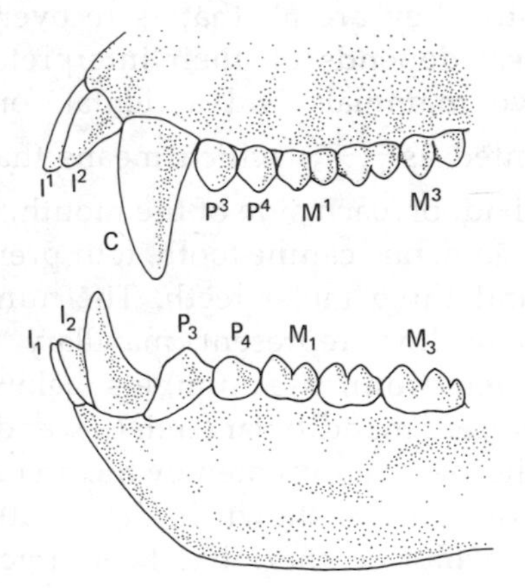

FIGURE 10.2 Designation of individual teeth in the catarrhine monkey *Macaca.* (*Source:* Clark, 1962. Used with permission.)

mandibular teeth. Similarly, lower-case letters may be used to designate deciduous (milk) teeth. This system retains the numbers used for the dentition of primitive placental mammals. Premolars may also be referred to as first and second premolars, ignoring the fine point that these are really the third and fourth premolars of the early mammals.

Some of the changes in tooth morphology from primitive mammal to modern primates (Figures 10.3 to 10.6) are briefly these: Incisors that were simple, pointed teeth become, in most primates, compressed to a spatulate form to function as shearing teeth (Figure 10.4). In some prosimians the lower incisors have become specialized "combs" for cleaning the fur. With some exceptions, notably man, canines have retained their projecting and pointed form for slashing and puncturing. In man (Figure 10.6) and in the lower dentition of some lemurs, the canines have taken on the functions and much of the form of incisors. Primates retaining the projecting lower canine also have a gap, a *diastema,* between the upper lateral incisor and the upper canine into which the lower canine fits when the teeth are in normal occlusion (Figure 10.3). In some, the cercopithecoids, for instance, P_3 shears against the

FIGURE 10.3 Skull of *Colobus* showing the dental occlusion, left lateral view. A diastema between I^2 and the upper canine accommodates the lower canine. The upper canine shears against sectorial P_3 (Figure 10.4) and the surfaces of the upper and lower cheek teeth overlap.

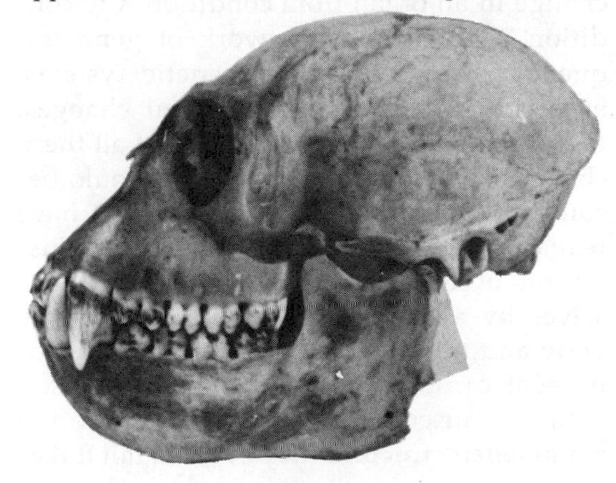

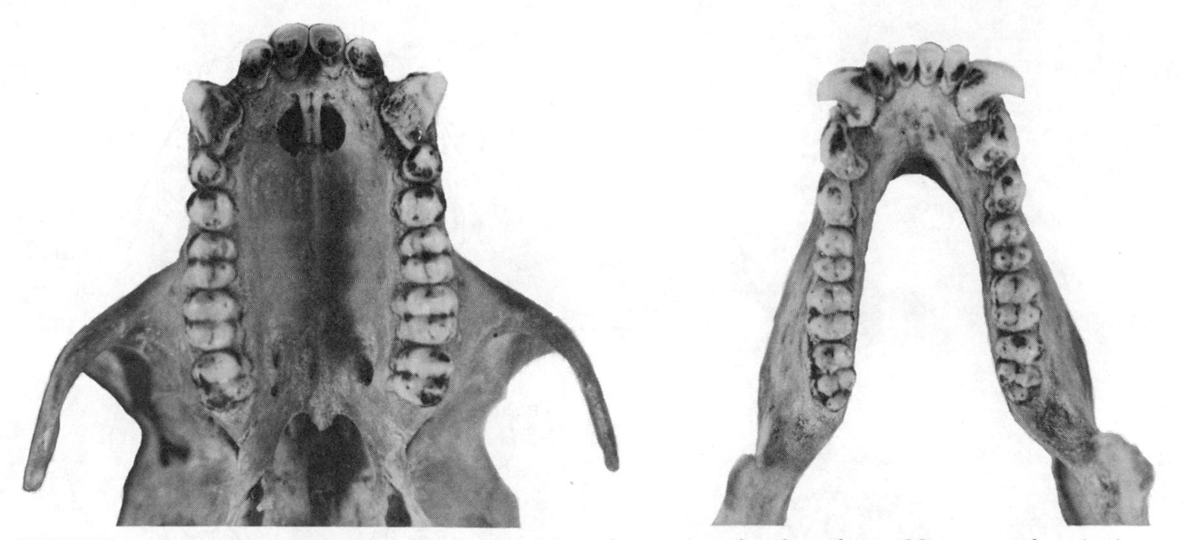

FIGURE 10.4 *Colobus* upper and lower dentition showing occlusal surfaces. Note spatulate incisor shape, sectorial P_3, and bilophodont molars.

upper canine and assumes a specialized cutting shape unlike that of P_4. This is termed a *sectoral* premolar (Figures 10.4 and 10.5).

The premolars change from simple shearing teeth having one prominent cusp to a condition where they are more like molar teeth in form and function. In this change the heavy enamel ring, the *basal cingulum*, around the primitive premolar gave rise to additional cusps that provide the premolars with a grinding or crushing surface. In man all premolars have become bicuspid. The primitive upper molar had three cusps, two lateral and one medial, in fairly sharp relief, the *tritubercular* pattern. Most primates today have developed a medial fourth cusp on the upper molar. The lower molars also evolve, from a slightly more complex lower molar, into a four-cusped form.

The Old World monkeys have developed enamel ridges or *lophs* connecting pairs of cusps transversely across the molar. This *bilophodont* pattern (Figure 10.4) is characteristic of the Cercopithecoidea. All of the Hominoidea are characterized by the occurrence, in different frequencies, of a five-cusped lower molar pattern termed the Y5 pattern. This is frequently referred to as a "dryopithecine" pattern after the fossil ape *Dryopithecus*. In modern man this Y5 pattern is not as common as the simplified +4 pattern. The Y5 pattern in man today is most often seen on the M_1, which has undergone less evolutionary reduction in size than the other molars. These patterns are shown in Figure 10.7.

Many features of the skeleton permit us to draw inferences about changes in posture, locomotion, diet, disease, brain size, and even some features of behavior. A discussion of some of these will be introduced in descriptions of particular fossils.

In the study of primate fossils it is helpful to realize, as will be documented in Chapter 11, that many of the characteristic features of primate anatomy are to be understood as parts of an adaptive complex that developed early in the Cenozoic era. Outstandingly this

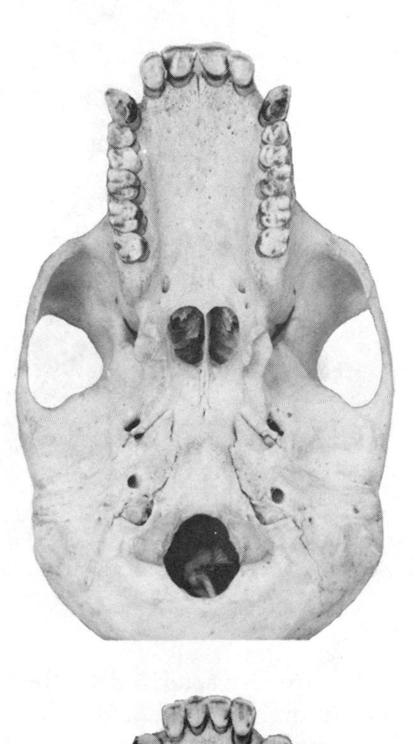

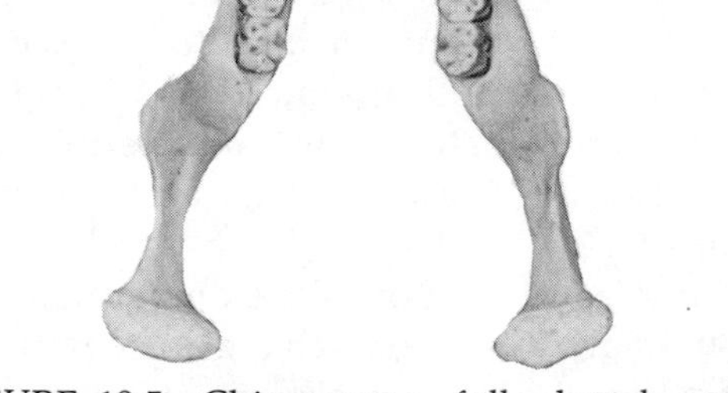

FIGURE 10.5 Chimpanzee full dental arcade upper and lower dentition showing occlusal surfaces.

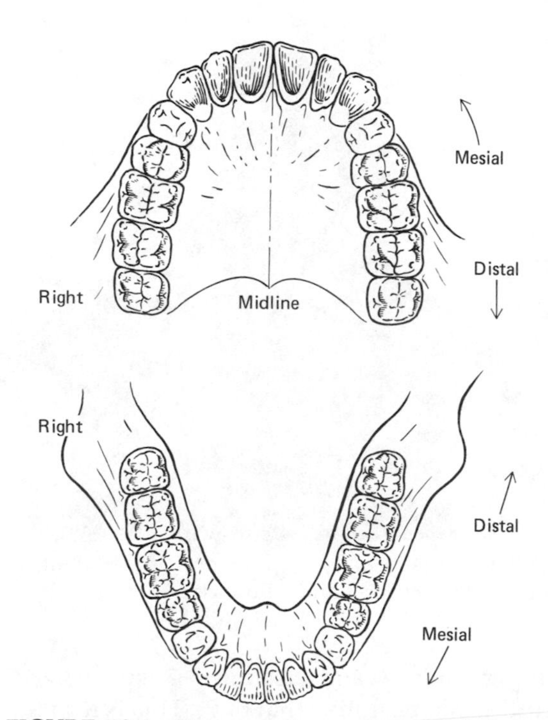

FIGURE 10.6 *Homo sapiens* upper and lower dentition.

was an insectivorous and arboricolous adaptation. But the animal climbed by grasping rather than by the use of claws. In fact, one of the early forms, *Plesiadapis*, had laterally compressed claws rather than horizontally flattened nails. All primates today have flattened nails on one or more digits. The arboricolous adaptation is related to many of the anatomical features characteristic of primates. Flattened nails and dermal ridges (fingerprints and palm prints) are grasping adaptations. Digits are long so that the entire hand or foot is prehensile. It can wrap around a limb. The first digit can be rotated to oppose the palm or other digits in the seizing or grasping motion.

In three-dimensional life in the trees the olfactory sense became less important in orienting the animal. Food seeking was no longer a matter of nosing around in the forest duff. This change, combined with the tendency of the head to be postured in a more upright position, resulted in a general reduction in length of the snout. Along with this there has been some reduction in total number of teeth. Conversely vision and the coordination of vision with movement became very important. Primate eyes evolutionarily rotate to face forward rather than laterally. This becomes stereoscopic vision with extensive overlap of the fields of vision of the two eyes. This is important in fine discrimination of distance, a feature of obvious importance to a creature that hunts its food high above the ground. Accompanying the forward rotation of the eyes, bony structures evolved to protect the eye in this more exposed position. The first occurred as a bone ring and later, among the Anthropoidea, included a complete postorbital plate of bone.

Many features of this arboreal primate life, including hopping, jumping, sitting, and feeding with the forelimbs, predisposed primates to a more orthograde, (upright) posture. This postural change was accompanied by anatomical changes accentuated in brachiators. Brachiation is the form of locomotion in which an animal proceeds hand over hand, hanging underneath a limb or supporting structure. Just which primates should be considered brachiators has been a topic of some debate in anthropological literature. There has been little agreement on this, just as there has been little agreement on the more inclusive classifications of primate locomotion.

FIGURE 10.7 A modern mandible with molars showing +4 and Y5 cusp patterns.

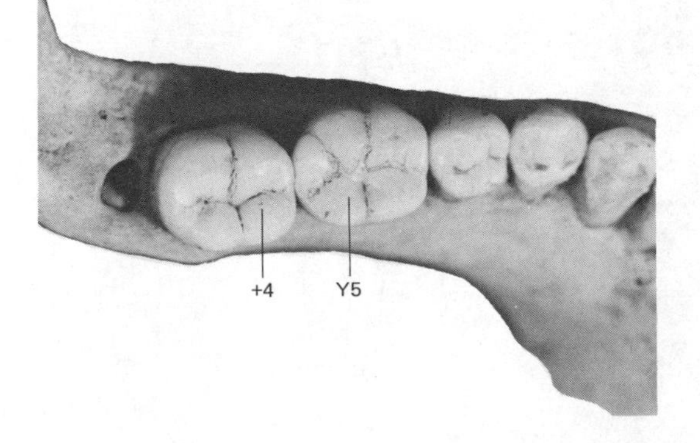

At least part of the problem of discussing locomotor patterns lies in the failure to distinguish whether the objective is to classify locomotor behavior or to classify primate species. This distinction becomes important when we realize that the locomotor pattern shown by a single primate will change depending on whether he is on the ground or in a tree, the slope of the surface, how fast he is traveling, and so forth. If we classify locomotion and then look at one primate species, we find, inevitably, that the species observed engages in more than one kind of locomotor behavior.

As a strictly informal classification we can recognize three locomotor-postural patterns that occur in terrestrial progression among primates. These include

1. plantigrade or quasiplantigrade quadrupedalism of many primates in which the foot is placed flat on the ground or in which the weight rests on the lateral edge with the digits partially flexed,
2. knuckle-walking quadrupedalism with the weight of the forequarters resting on the flexure of the second phalanx of digits of the hand (a stance most common in gorillas),
3. bipedalism, which is the common form of terrestrial locomotion found only in man.

Arboreal locomotor patterns of primates include

1. the hopping leap of the tarsier,
2. quadrupedal walking and leaping among limbs on the part of many monkeys,
3. generalized brachiation, which is best described as "clambering" in which the animal will raise or lower itself by its forelimbs or hang from a branch by its fore-

limbs when moving from branch to branch or when moving out on smaller branches,
4. extreme brachiation, which is the arm-catapulted swinging so characteristic of gibbons and siamangs.

A debate that has been carried on to the present is whether or not man went through a *brachiating stage* in his evolutionary past. This debate is not too meaningful in terms of the locomotor classification given here. Man today is a generalized brachiator if forced into the trees. The anatomical evidence from modern and fossil man indicates that man has long engaged in generalized brachiation but that this form of locomotion was much more common for man's ancestors at an earlier period. This earlier adaptation to generalized brachiation facilitated man's bipedal terrestrial adaptation.

A characteristic of all of the Hominoidea is that they are tailless. The remaining caudal vertebrae of the hominoids do not appear as an external tail but form the coccyx. The coccyx is incorporated as part of the pelvic floor, important in visceral support in these more orthograde primates. The larger hominoids cannot manage very great quadrupedal leaps and a tail, no matter how bushy, would be useless as a balancer. Because of their size, the progression of larger hominoids in a tree is typically through generalized brachiation. They use the forelimbs in letting themselves up and down between levels and when moving or reaching out among small branches. The small hominoids, the Hylobatidae, have

FIGURE 10.8 Skeleton of a quadrupedal monkey showing forces supporting the trunk: → ← compression; ← → tension. (*Source:* Clark, 1962. Used with permission.)

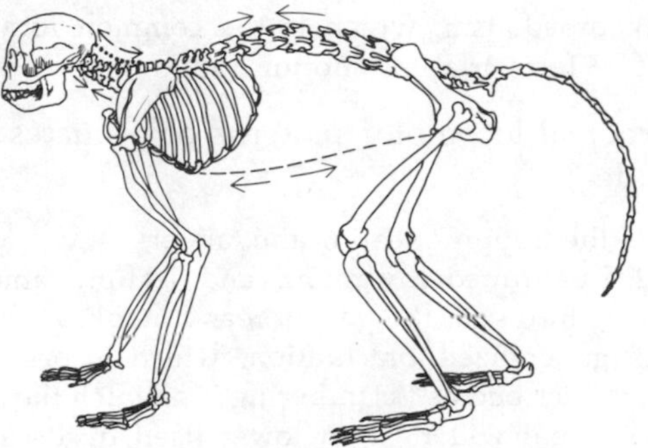

FIGURE 10.9 Curvature of the vertebral column in chimpanzee and man. (*Source:* Adapted from Schultz, 1963. Used with permission.)

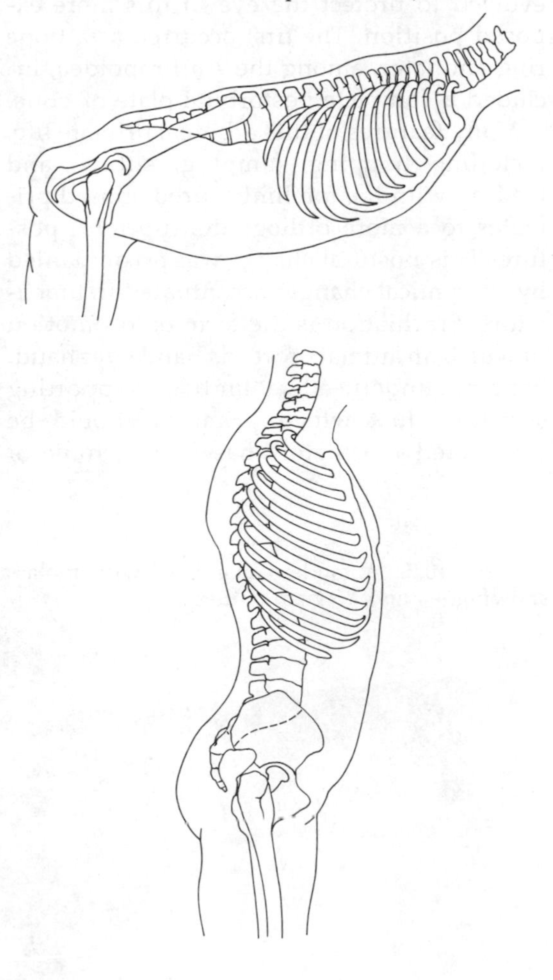

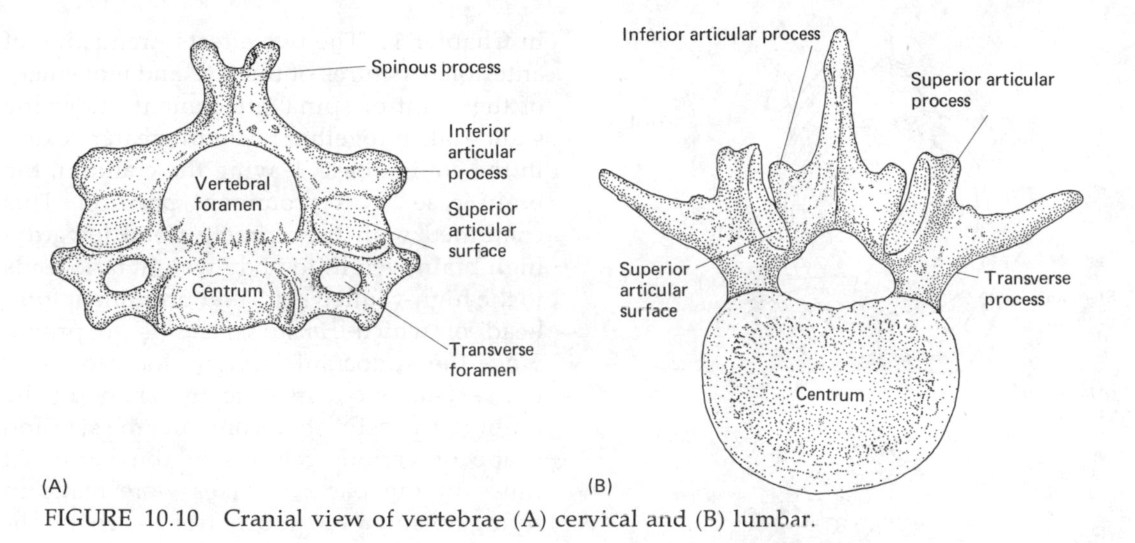

FIGURE 10.10 Cranial view of vertebrae (A) cervical and (B) lumbar.

become great aerial acrobats. The projectile-like propulsion they manage with extreme brachiation is well known to anyone who has watched gibbons in a zoo.

Some of the anatomical changes correlated with these postural and locomotor changes can be seen in the skeleton. The vertebral column of the quadruped forms a simple arch supporting the thoracic and abdominal viscera held rigid by tension of the abdominal muscles and sternum. This situation is illustrated in Figure 10.8. As can be seen in the diagram, the body is held in a horizontal position by tension in the abdominal muscles and sternum of the rib cage and compression in the vertebral column. Nuchal muscles and ligaments, under tension, and cervical vertebrae, under compression, support the head. The vertebral columns of chimpanzee and man are shown in Figure 10.9. Man being completely orthograde has a recurved spine that bends back sharply in the lumbar region. In this case the largest part of trunk weight is carried by compression down the vertebral column and, secondarily, by the basin-shaped pelvis. The centrums of the vertebrae increase in support area down the vertebral column to accommodate this unidirectional load as shown in Figure 10.10.

In the quadrupedal primate the rib cage is deep and narrow. In more orthograde forms the rib cage becomes shallower in the dorsoventral direction and wider laterally. This change is illustrated in Figure 10.11. In the hominoids the scapula has rotated toward the rear and the clavicle is lengthened to form a sturdily buttressed point of leverage for the arms that are at some distance laterally from the midline of the trunk.

In the quadrupedal primate the pelvis, while providing a birth canal, is primarily a point of rigid support for muscles of the trunk, rear legs, and tail. These muscles act primarily in anterior-posterior direction and remain near a plane through the midline of the animal. In more orthograde primates the innominate bones of the pelvis flare out, providing more lateral muscle attachments. In

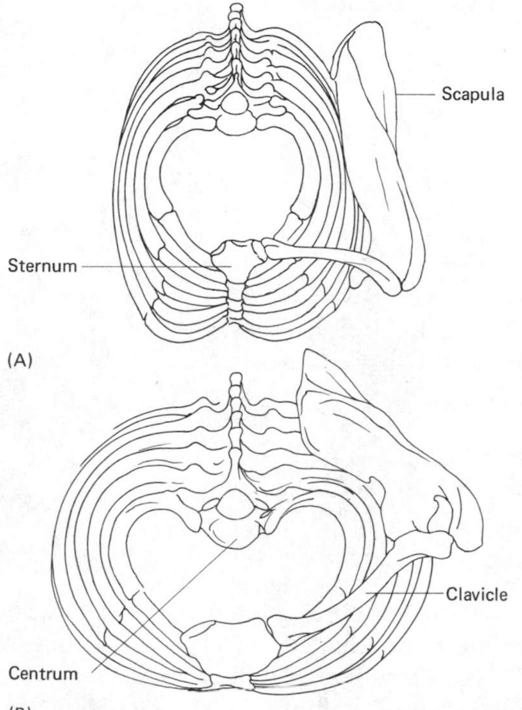

FIGURE 10.11 Thorax and shoulder girdle of (A) *Macaca* and (B) *Homo*, anterior view drawn with same dorsoventral depth. (*Source:* A. H. Schultz, "Man and the Catarrhine Primates" *Cold Spring Harbor Symposium in Quantitative Biology*, 15:35–53, 1950. Copyright © 1950.)

man the superior portion of the innominate, the ilium, curves around to form a more dish-like supporting structure (Figure 10.12). In hominids the increased role of pelvic structures in visceral support has conflicted strongly with the role of the pelvis in providing an open birth canal.

These changes are accompanied by changes in the skull. The jaws recede toward a position more under the orbits. The occipital condyles, the points of articulation of the skull on the vertebral column, move further under the skull. This situation is most extreme in *Homo sapiens* (Figure 10.13) and is a diagnostic of hominid status to be discussed in Chapter 11. The two effects – reduction of anterior structures of the face and movement of the point of spinal attachment under the skull – taken together result in greater flexion in the cranial base, leaving the cranium, the brain case, in a prominent position. This combined with the primate trend toward a high brain weight to body weight ratio leads to the high-vaulted skull and prominent forehead of which *Homo sapiens* is so proud. Since the endocranial cavity does, to a certain extent, mold itself to the shape of the brain it is possible to reconstruct the size and shape of various regions of the brain. At times in the past, mistakes were made in trying to infer too much from such reconstructions. The cranial meninges, three protective sheets of tissue that envelope and cushion the brain, mold themselves over the brain and insure that any reconstruction of brain shape from endocranial surface relief can only reflect fairly gross differences.

When working with skeletal material not too unlike living forms, it is possible to make even finer inferences. The sex of the individual, age at death, certain pathologies, and occasionally postural habits may leave their mark on bone. In primates the bones of the male are more robust than those of the female and points of muscle attachment on the bone are more strongly marked in males than they are in females. In the nonhominid primates the males generally have longer canines than the females. In hominids the sex of the individual is best determined from the bones of the pelvis (Figure 11.18). The acetabulum, the socket for the femoral head, is – relatively – much larger in males than it is in females. The public symphysis is relatively short in females and the inferior border of the pubic bones diverge more rapidly (the subpubic angle) in the female than in the male. The two latter features are adaptations in maintaining an open birth canal in a bipedal pri-

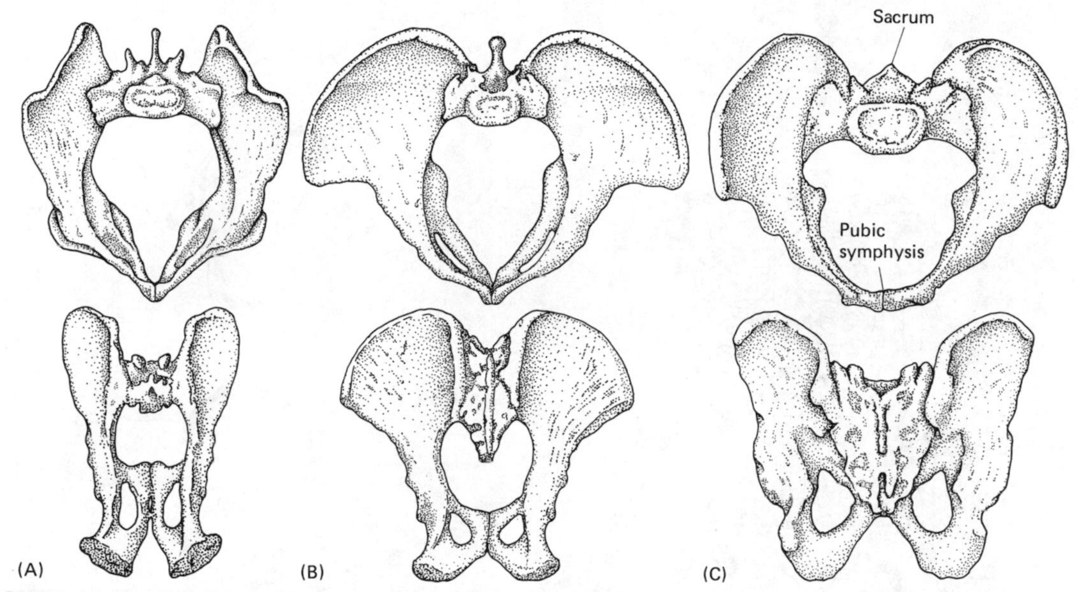

FIGURE 10.12 Pelvis of (A) *Macaca*, (B) *Gorilla*, and (C) *Homo*, viewed from above (top) and behind (bottom). [*Source:* S. L. Washburn (ed.) *Social Life of Early Man* (Chicago: Aldine Publishing Company, 1961.) Copyright © 1961 by Wenner-Gren Foundation for Anthropological Research, Inc. Reprinted by permission of the author and Aldine Atherton, Inc.]

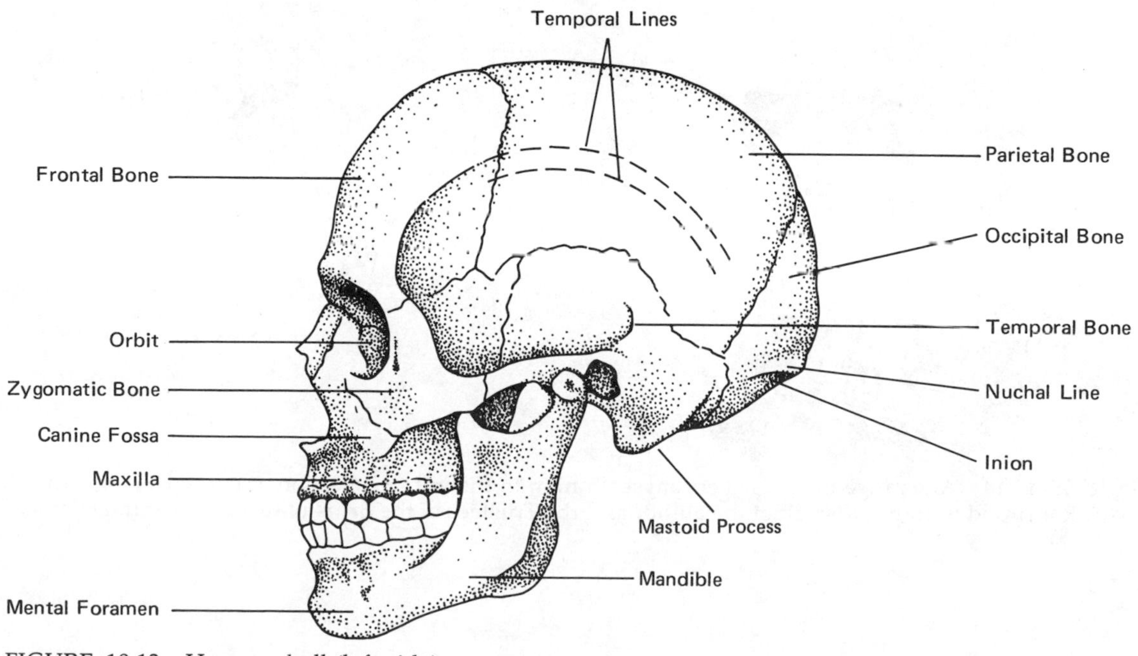

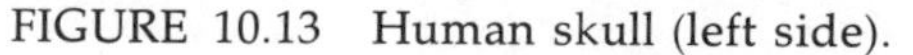
FIGURE 10.13 Human skull (left side).

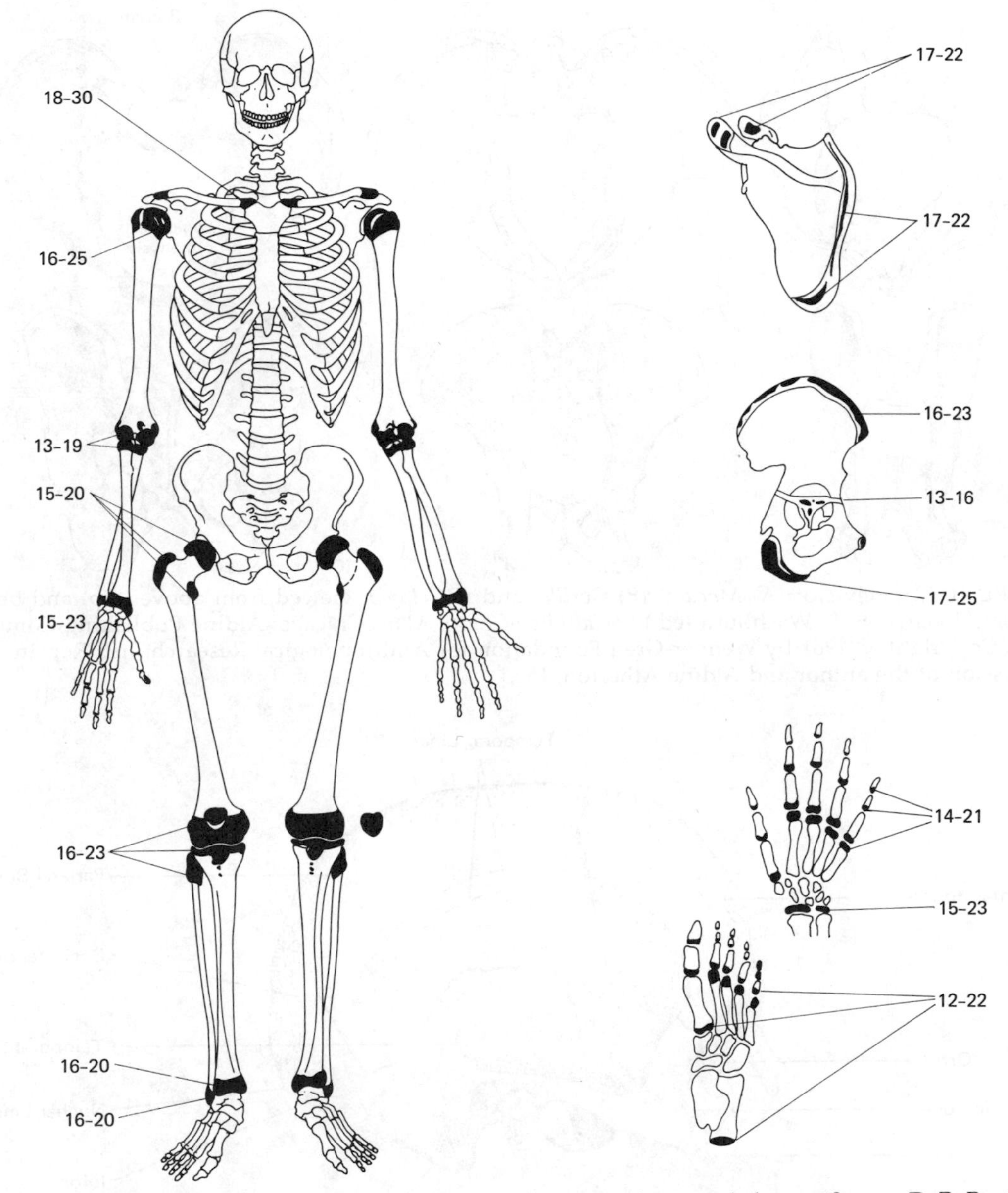

FIGURE 10.14 Age range in years of epiphyseal union in the human skeleton. (*Source:* D. R. Brothwell, 1965. Reprinted with permission of the author and the Trustees of the British Museum, Natural History.)

mate. Also related to this is the relatively short and broad sacrum and a wider and more open sciatic notch in the ilium of the female. In the female the pubis is almost as long as the ischium, but in the male the ischium is much longer than the pubis.

The skull is a bit less reliable than the pelvis in establishing sex but does show a number of relative differences between male and female. Male skulls are generally larger and more rugged than female skulls from the same population. The rugosity of the male skull is related to the heavier muscle attachments of males. Males have more pronounced supraorbital ridges, external occipital protuberances, larger mastoid processes, and broader palates. The male mandible tends to be heavier and has greater symphyseal height and a broader ascending ramus. The male generally has more rectangular-shaped orbits than the female. The upper margin of the female orbit is more sharply angulated than in the male, probably due to greater pneumatization of the bone in males. The female tends to have a higher and more continuously rounded forehead than the male.

Age (age at death) can be judged from a number of features most of which have to do with maturational changes in the skeleton. For the very young up to adolescence, the state of eruption of the dentition, both deciduous and permanent, is a fairly reliable indication of age. Table 10.1 presents data on age at eruption of teeth. In archeologically well-known populations the state of dental attrition can be used to estimate age in adults. The rate of attrition will be influenced by the diet, manner of food preparation, and other uses to which the teeth may be put. The age estimate attached to a particular state of dental wear will therefore vary from one culture to another.

From adolescence into adulthood the long bones of the skeleton cease growing and the epiphyses fuse with the main shaft of the bone. Times of epiphyseal union are given in Figure 10.14.

The state of closure of sutures of the skull has been used extensively in the past to judge age in the 20 to 30 year range. Recent studies have indicated, however, that suture closure is not too reliable as an index of age. Much more reliable in assessing age is the condition of the symphyseal surface of the pubic bone (McKern and Stewart, 1957). At present the characteristic age changes in the pubes are well studied only for male skeletal material.

Other inferences possible from skeletal material include estimates of stature and robusticity and the prevalence of certain nutritional and infectious diseases that leave their mark on bones or teeth. And skeletal populations, just as living populations, can be typified by their measurements. These measures can then be used to assign an individual, with a certain degree of probability, to one population or another. Such identification of individuals has been attempted in the past, but only in terms of racial taxonomies. Differences between populations and ethnic groups include differences in rate of maturational changes in the skeleton, differences in robusticity, and so forth. Because of this the standards used for establishing sex and age of skeletal material should be from populations as similar as can be obtained to the skeletal population being assessed.

It must be remembered that these latter paragraphs are descriptions of what can be done with skeletal material of *Homo sapiens*. With the earliest primates, to be taken up in the coming chapter, the job is simply to identify those forms that are in, or are closely related to, the line of evolutionary development that led to man.

TABLE 10.1
The Sequence of and Age at Gingival Eruption of the Deciduous and Permanent Teeth in Man

DECIDUOUS OR MILK-TEETH

Eruption Order	*Tooth*	*Age (in Months)*
1	lower central incisor (i_1)	6–8
2	upper central incisor (i^1)	9–12
3	upper lateral incisor (i^2)	12–14
4	lower lateral incisor (i_2)	14–15
5	upper and lower first molar (m^1 and m_1)	15–16
6	upper and lower canine (c)	20–24
7	upper and lower second molar (m^2 and m_2)	30–32

PERMANENT TEETH

Eruption Order	*Tooth*	*Average Age (Years, Months)*
1	lower first molar (M_1)	6,2
2	upper first molar (M^1)	6,5
3	lower central incisor (I_1)	6,6
4	upper central incisor (I^1)	7,5
5	lower lateral incisor (I_2)	7,7
6	upper lateral incisor (I^2)	8,7
7	upper first premolar (P^3)	10,3
8	lower canine (LC)	10,8
9	lower first premolar (P_3)	10,10
10	upper second premolar (P^4)	11,0
11	lower second premolar (P_4)	11,6
12	upper canine (UC)	11,7
13	lower second molar (M_2)	11,11
14	upper second molar (M^2)	12,6
15	lower third molar (M_3)	18,10
16	upper third molar (M^3)	19,11

Source: Data taken, in part, from A. H. Schultz, "Eruption and Decay of Teeth in Primates," *American Journal of Physical Anthropology*, 19 (1935), 489–581.

DATING FOSSIL REMAINS

Fossil primate finds may be dated by any of the methods of the geologist or paleontologist that are applicable. In fact, most fossil primates are dated on the basis of faunal associations. The various epochs of the Cenozoic era are characterized by distinctive faunal assemblages. This means that the nonprimate faunal sequence must be known. In some regions such sequences have not been well established or have not been well correlated with the sequences in other areas.

Dating techniques can be classified as relative dating or absolute (or chronometric) dating or some combination of these. A relative date places an event in a sequence of events. You then know that the event is younger or older than the other events. An absolute date states the duration of time that has elapsed since the occurrence of the event. As an example, if we conclude that a fossil comes from Miocene deposits this date that we decide upon is a relative one. But since we have some absolute dates for Miocene events we also have an estimate of the date in years of the fossil. In the past few years the possibilities for obtaining absolute dates has been greatly enhanced by development of isotope dating techniques.

The first isotope dating technique developed was the carbon-14 technique of Libby. The ratio of ^{14}C to ^{12}C is constant in the tissues of a living organism. When the organism dies and cells are no longer being renewed, the amount of ^{12}C, a stable isotope, remains the same but the amount of ^{14}C, an unstable isotope, decays to nitrogen-14 at a constant rate. This decay rate is measured by the half-life of the isotope, which for ^{14}C is 5730 ± 40 years. Therefore the amount of ^{14}C remaining, as a proportion of the total carbon, provides an estimate of age since death of the organism. This method can be used to about 50,000 years BP but not much longer, as the amount of ^{14}C becomes too small in older samples. Similar methods involving radioactive isotopes with longer half-lives can be utilized whenever (a) the rate of decay of the isotope is known, (b) the proportionate or absolute amount of the isotope present when the deposit was formed is known, and (c) the isotope is known to be associated with the fossil or the event in question. Obviously condition (b) presents the greatest problem, but this has been solved for a number of isotopes that are now used to build absolute chronologies.

The interval of time which spans primate evolution is shown in Table 10.2. This is the Cenozoic Era. The various epochs of the Cenozoic will be referred to frequently in the next chapter. The age estimates of these epochs are only approximate and are subject to revision as dating procedures become more refined. The only notable change in Table 10.2 and time charts published earlier is the shortening of time assigned to the Pliocene. Earlier estimates on the beginning of the Pliocene were approximately twice what they are today.

Two relative dating techniques that are applicable directly to skeletal material involve the determination of the amount of fluorine and the amount of nitrogen in the bone. Buried bone slowly absorbs fluorine from ground water. Therefore the longer the bone has been interred the greater the fluorine content. Local conditions, principally the fluorine content of ground water, influ-

TABLE 10.2
Geological Time in Primate Evolution

ERA	PERIOD	EPOCH	BEGAN MILLIONS OF YEARS AGO	EVOLUTIONARY EVENTS
Cenozoic	Quaternary	Recent (Holocene)	0.01	
		Pleistocene	1.8	Evolution of Modern Man
		Pliocene	5	Appearance of Hominids
		Miocene	25	Hominoid Radiation
	Tertiary	Oligocene	40	Appearance of Monkeys and Apes
		Eocene	55	Appearance of Placental
		Paleocene	68	Mammals Including Early Primates

ence the rate at which fluorine is taken up by the bone. Because of this, fluorine absorption cannot be used to establish temporal correlations except in restricted areas. Its primary use has been to establish whether all the faunal material of a site is in fact contemporaneous. This was the critical evidence that showed that the "Piltdown Man" fossil of England was not a contemporary of the approximately 250,000-year-old animals with which it was found. This led to the discovery that this "early man" fossil was a forgery (Weiner, 1955).

Nitrogen determinations from bone are used in much the same way as fluorine determinations. But nitrogen, being a component of the organic matter of bone, *decreases* with age due to the gradual breakdown of the organic bone matrix.

Methods developed earlier for establishing the age of fossils include the rate of sedimentation in the build-up of deposits, the association of the fossil with a particular terrace in a sequence of terraces formed by a river, or a similar association with raised beaches along a lake or seashore. A good general account of these and other methods used to date skeletal and cultural remains has been provided by Hole and Heizer (1973).

Another radiometric dating technique which satisfies the preceding conditions is the decay of potassium-40 to argon-40. Potassium-40 (^{40}K) has a half life of 1.3 billion years which means it can be used to date extremely ancient events. The K-Ar method is limited primarily to volcanic deposits. This is because the K-Ar "clock" is started when rock containing ^{40}K has been heated sufficiently to drive off all the gaseous ^{40}Ar which has accumulated in the material in previous years. From this time the ratio of ^{40}Ar to ^{40}K builds up at a known rate.

REFERENCES CITED

Brothwell, D. R. 1965 *Digging Up Bones.* London: The British Museum (Natural History).

Clark, W. E. Le Gros 1962 *The Antecedents of Man,* 2nd ed. Edinburgh: University Press.

De Beer, G. R. 1958 *Embryos and Ancestors,* 3rd ed. Oxford: Clarendon Press.

Hole, F. and R. F. Heizer 1973 *An Introduction to Prehistoric Archeology,* 3rd ed. New York: Holt, Rinehart and Winston.

McKern, T. W. and T. D. Steward 1957 *Skeletal Age Changes in Young American Males, Analyzed from the Standpoint of Identification, Technical Report E P-45.* Natick, Mass.: Headquarters, Quartermaster Research and Development Command.

Schultz, A. H. 1935 Eruption and Decay of Teeth in Primates. *American Journal of Physical Anthropology* 19:489–581. 1950 Man and the Catarrhine Primates. *Cold Spring Harbor Symposium on Quantitative Biology* 15:35–53. 1961 Some Factors Influencing the Social Life of Primates in General and of Early Man in Particular. *In* S. L. Washburn (ed.), *Social Life of Early Man,* pp. 58–90 New York: Viking Fund Publications in Anthropology, no. 31. Wenner-Gren Foundation. 1963 Age Changes, Sex Differences, and Variability as Factors in the Classification of Primates. *In* S. L. Washburn (ed.), *Classification and Human Evolution,* pp. 85–115. New York: Viking Fund Publications in Anthropology no. 37.

Weiner, J. S. 1955 *The Piltdown Forgery.* London: Oxford University Press.

11

The Earliest Hominids

The earliest fossil mammals that are recognizably members of the order Primates appear in the Paleocene, the first epoch of the Tertiary period. Most Paleocene fossil primates come from North America. Indeed, the Paleocene is not well represented in Europe and only a late Paleocene fauna is found there. The Paleocene and Eocene primates are all small prosimians. They are recognizably different from insectivores by middle Paleocene times. A late Paleocene form, *Plesiadapis,* occurs in both North America and Europe. *Plesiadapis* (Figure 11.1) lacks the postorbital bar of bone behind the eye which in all later primates is present to form a closed ring of bone as the orbit of the eye. On the other hand, this form shows an unusual dental specialization. Premolar reduction gives it a rodentlike appearance in its anterior dentition. This and specializations of the postcranial anatomy suggest that *Plesiadapis* was not ancestral to the higher primates.

The great radiation of the Prosimii took place in early Eocene times. The numerous prosimians of this epoch are differentiated into lemurlike and tarsierlike forms as well as some lineages unlike any living primates. The family Adapidae are lemurlike and may be ancestral to modern lemurs. This family gets its name from the first fossil primate to be recognized as such, *Adapis,* described by Cuvier in 1821. Like modern lemurs, the Adapidae had pelvic (hind) limbs that were longer than the pectoral (front) limbs. The

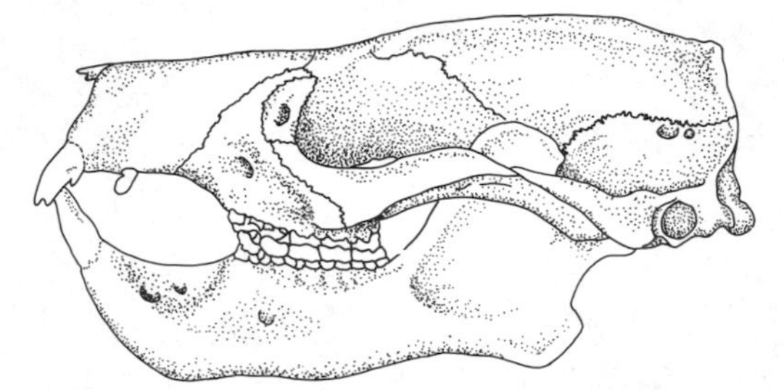

FIGURE 11.1 Skull of *Plesiadapis,* a Paleocene Prosimian, full size. (*Source:* Adapted from Simons, 1964. Used with permission of the author and *American Scientist.*)

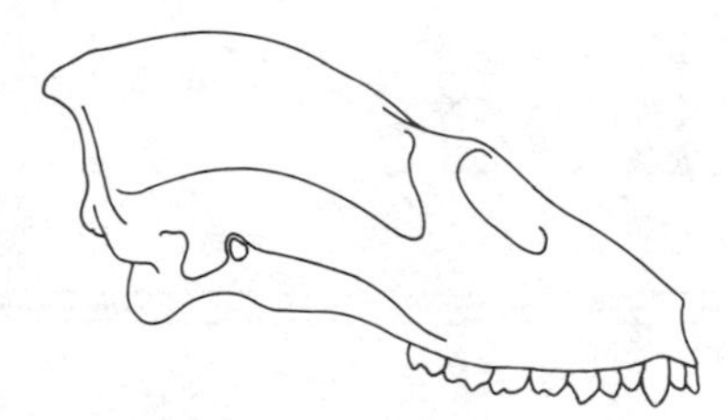

FIGURE 11.2 Skull of *Adapis parisiensis,* full size. (*Source: Annals of New York Academy of Sciences,* vol. 102, article 2, p. 286, Fig. 2. *Source:* E. L. Simons © The New York Academy of Sciences, 1962. Reprinted by permission.)

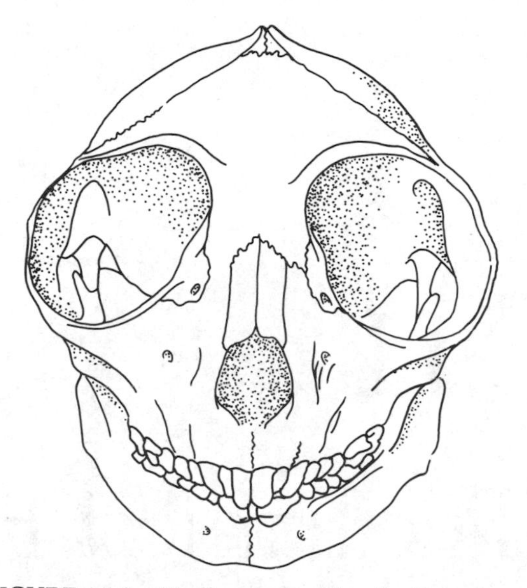

FIGURE 11.3 Skull of the fossil *Necrolemur.* (*Source:* Simons, 1964. Used with permission of the author and *American Scientist.*)

vertebral column appears structured for a horizontal orientation. The first digit, thumb or big toe, was opposable to the other digits. Adapidae were quadrupedal jumpers maintaining an arboreal existence. A postorbital bar was present. The skull of *Adapis* is shown in Figure 11.2.

These early lemuroids had taken, by the Eocene, a line of development that makes them extremely unlikely candidates for the roles of ancestors of higher primates. It has been suggested that the higher primates did go through a tarsioid stage of evolution. As would be expected, tarsierlike primates are not easily distinguished from lemurlike primates in the Paleocene and early Eocene. Perhaps the most convincing member of the Tarsiidae family is the Eocene *Necrolemur. Necrolemur* (Figure 11.3) was a small hopping animal like present day *Tarsius.* This is reflected in changes in the skull, a shortening of the face, and rotation of the foramen magnum to a position slightly further under the skull. *Necrolemur* also had a much larger brain relative to body size than any other animal of the day. But it now seems likely that the ancestor of the Anthropoidea is to be found in another Eocene family, the Omomyidae.

Like the family Plesiadapidae the Omomyidae occurred in both the Old and New Worlds, but unlike the Plesiadapidae the Omomyidae show no specializations that would preclude them from being ancestral to higher primates. Simons has noted that incisors that are large relative to the canines might be considered a specialization in this family (Simons, 1963). But, as noted in the previous chapter, simple relative size difference of this sort is easily altered through time and can be given no great phylogenetic weight. The dental formula of the Omomyidae is $\frac{2.1.3.3.}{2.1.3.3.}$, the same as that of the living Cebidae, the New World monkeys. They had unspecialized molar crown patterns that are reasonable precursors for the catarrhine forms that appear in the Okogocene.

There are over 50 genera of fossil prosimians that have been identified in Eocene deposits. These were small, squirrel-size primates undergoing a radiation into many forms in both the Old and New Worlds. At this time many genera show anatomical specialization in the teeth or in the postcranial skeleton much like that found in modern lemurs or tarsiers. The Omomyidae have no such specializations to rule them out as ancestors of the higher primates. The Omomyidae are predominantly middle Eocene and late Eocene forms. Interestingly enough, there are two late Eocene forms from Burma which have been suggested to be pongids!

Amphipithecus and *Pondaungia* both come from the late Eocene Pondaung formation of Burma. *Amphipithecus* (Figure 11.4) and *Pondaungia* are clearly not prosimians. The molar cusp pattern is much like that of the Hominoidea. There is no bilophodont specialization as there is in modern cercopithecoids (Old World monkeys). But unlike in living Hominoidea, there was a third premolar present. It has been noted, however, that the most anterior of these (lower) premolars had a diminutive root, indicating that the premolar was reduced in size and perhaps vestigial at this time. *Amphipithecus* would serve well as an intermediate form between an early omomyid and both hominoids and cercopithecoids. *Amphipithecus* and *Pondaungia* were not pongids, however, as the superfamily Hominoidea, which contains the Pongidae, does not become clearly differentiated until the following epoch.

The Hominoidea appear to have undergone a radiation in Oligocene times. Most of these fossil hominoids as well as other primate and nonprimate remains have been re-

FIGURE 11.4 *Amphipithecus mogaungenesis,* occlusal and side views. (*Source:* American Museum of Natural History.)

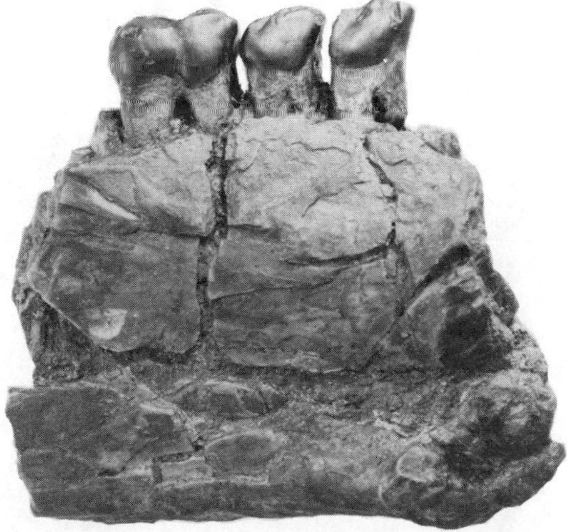

covered from the Fayum depression of Egypt. Within this group of fossils we can identify specimens that indicate that the Cercopithecoidea had differentiated from the Hominoidea and that within the Hominoidea a gibbonlike line had developed. The Oligocene hominoids are small apelike forms. No hominids had differentiated at this time.

The earliest primate from the Fayum has been named *Oligopithecus*. This fossil comes from early Oligocene strata and is much like the Eocene Omomyidae. As such it is probably too late in time to be ancestral to the living hominoids. One small ape from late Oligocene strata appears to be quite gibbonlike, suggesting that this line had differentiated from other pongids as long ago as 27 million or more years. This form has been named *Aeolopithecus*. It was smaller than the modern gibbons, the smallest of living apes. Whether this was in fact a differentiated line leading to modern gibbons or not is still in doubt, as any fossil pongid mandible of such small size (Figure 11.5), unless it has marked dental specializations, can be expected to look more like a gibbon than any of the large apes.

FIGURE 11.5 *Aeolopithecus chirobates*. (*Source:* E. L. Simons.)

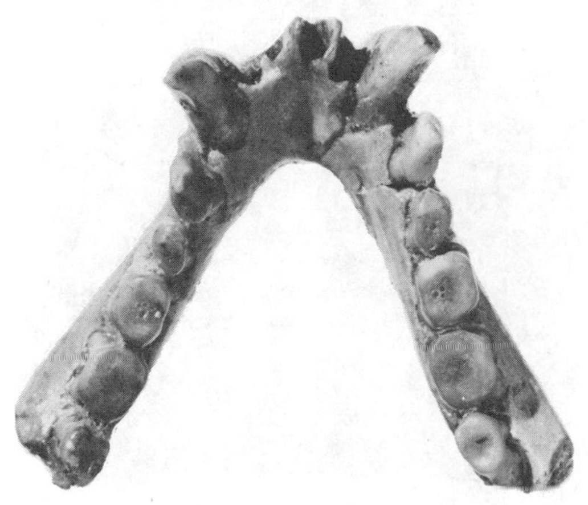

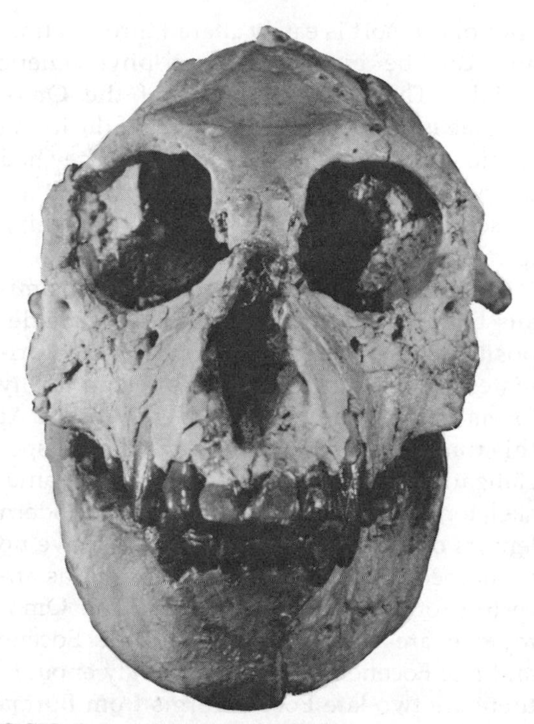

FIGURE 11.6 *Aegyptopithecus zeuxis*. (*Source:* E. L. Simons.)

Another recently discovered late Oligocene form is *Aegyptopithecus*. This animal was about the size of the modern gibbon and about twice as large as its contemporary *Aeolopithecus*. But in a number of relative proportions of the teeth and mandible *Aegyptopithecus* is more like the great apes than like the gibbons (see Figure 11.6). The dental formula is 2.1.2.3 and the molar size sequence is $M_1 < M_2 < M_3$. The canines are large and the upper canines shear on an elongated first premolar that becomes sectorial in shape. The mandible is apelike in form. The snout is more elongated than it is in living apes but the orbits face well forward. Surprisingly enough it seems that *Aegyptopithecus*, unlike modern apes, had a tail.

From the middle Oligocene of the Fayum,

there is a fossil genus *Propliopithecus* that could well be the transitional form between late Eocene *Amphipithecus* and late Oligocene *Aegyptopithecus*. *Propliopithecus* had earlier been proposed as one of the fossil gibbons. *Propliopithecus* did not have the large somewhat everted canines of *Aegyptopithecus*, nor was P_3 elongated as a specialized shearing tooth. Its posterior molars do not show the marked size increase of *Aegyptopithecus*. *Propliopithecus* was slightly smaller overall than *Aegyptopithecus* and, judging from the angle of the mandible, had an even more projecting snout.

At present then it seems that man's phylogeny from Eocene through Oligocene would include the sequence in Figure 11.7. Such a simplified diagram certainly does not include all the change and diversification that was going on during this considerable length of time. Further, because of the kind of reasoning used in its construction it is subject to alteration whenever a more likely form is found to interpose between, or to replace one of, these forms.

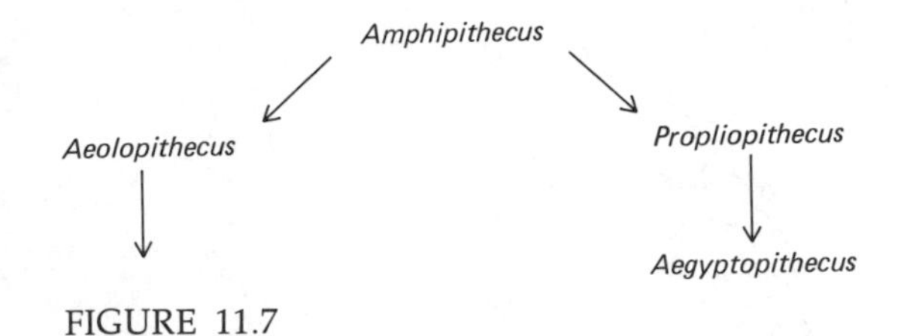

FIGURE 11.7

In the Miocene we find other gibbonlike apes, which now have been put into a single genus *Pliopithecus*. This genus could well be a descendant of *Aeolopithecus*. The great apes of this time, which are critical to an understanding of man's phylogeny, were once placed in a single genus, *Dryopithecus* ("oak apes"). This genus represents a radiation of large, tailless, at least partially tree-dwelling primates who became sufficiently diversified to be recognized now as a subfamily, the Dryopithecinae, containing three or more genera. At present it appears that the first representative of the family of man, the Hominidae, also developed and diverged early in this same radiation of tailless apes. This probably occurred in early Miocene times.

Today a large number of fossil finds are included in the fossil genus *Dryopithecus*. A smaller number are in the fossil genus *Gigantopithecus*. These genera are members of the Dryopithecinae, a subfamily of the Pongidae. The genus *Dryopithecus* is ancestral to the present-day African pongids, the gorilla and chimpanzee. *Dryopithecus* ranged over all of the tropical Old World, its remains occurring from Europe to China to Africa. I include the African "*Proconsul*" forms as members of the genus *Dryopithecus*.

Most *Dryopithecus* finds are teeth. Indeed it is for the genus that the "dryopithecine" Y5 tooth pattern is named. This pattern still occurs in larger lower molar teeth in man. The most complete *Dryopithecus* remains are the "*Proconsul*" species from Rusinga Island, Lake Victoria. These and many other important East African fossils were recovered by L. S. B. Leakey and his coworkers. The long bones available from the medium-sized dryopithecines suggest an animal having linear dimensions that are close to those of the modern chimpanzee, but more slender. It appears that the dryopithecines were generalized brachiators but had not developed extreme brachiation as seen in the Hylobatidae today (Napier and Davis, 1959).

The dryopithecine radiation represents the evolution of relatively large anthropoids who, in response to larger size, were developing new modes of locomotion. From arboreal quadrupedal ancestors they had become arboreal generalized brachiators. Probably the larger species were becoming partially or

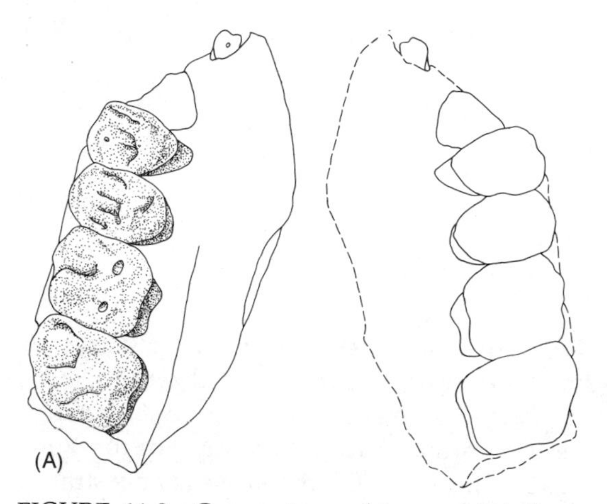

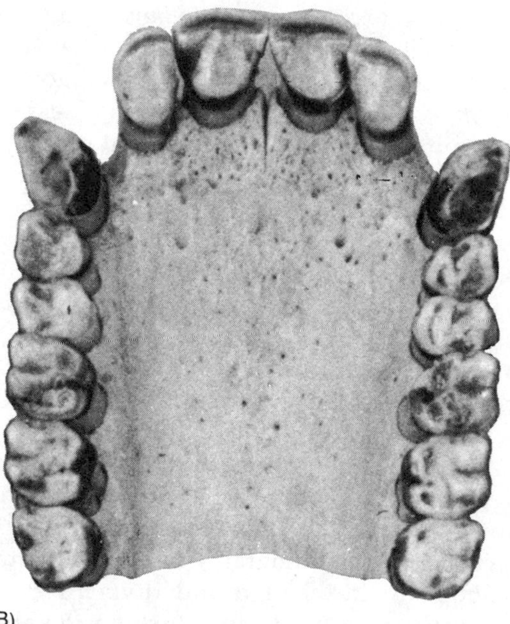

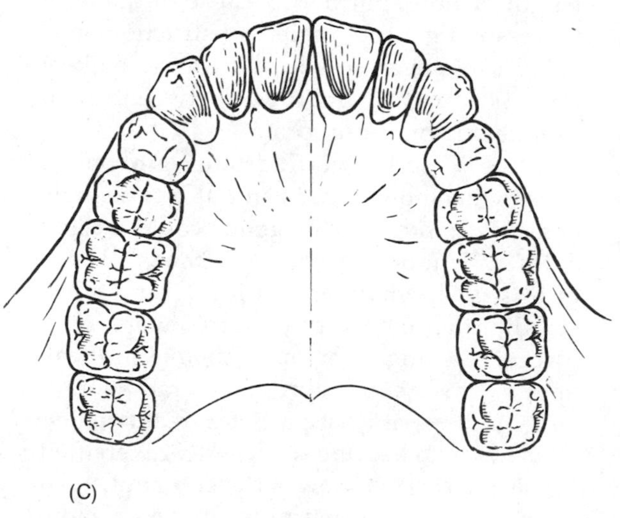

FIGURE 11.8 Comparison of jaws of (A) *Ramapithecus,* full size, upper, (B) chimpanzee, and (C) human, both reduced to comparable size. (*Source:* E. L. Simons, 1964. Used with permission.)

fully terrestrial. The genus *Dryopithecus* itself appears to be directly ancestral to the African great apes of today. Another member of the subfamily, the genus *Ramapithecus,* appears to be the ancestor of man.

Another pongidlike genus—*Oreopithecus*—lived at the same time as *Ramapithecus* and was evolving in the same direction. The pelvis and limbs of *Oreopithecus* indicate an adaptation to upright posture. The face, because it protrudes much less than does the face of other pongids, may also reflect upright posture. The molar teeth of *Oreopithecus* show a cusp pattern unlike that of known hominids. *Oreopithecus* lived in Europe and Africa in late Miocene times as did *Ramapithecus.* Indeed, fossils of both occur in the same deposit at Fort Ternan, East Africa.

The taxon *Ramapithecus punjabicus* includes material described by G. E. Pilgrim in 1910 and other material described by G. E. Lewis in 1934. These fossils, as well as some recently discovered ones, came from the Siwalik Hills in Northern India. The genus *Ramapithecus* has been restudied and reordered by Simons (1961, 1964; Simons and Pilbeam,

FIGURE 11.9 *Ramapithecus,* maxilla; occlusal (above) and lateral (below) views. (*Source:* Simons, 1961. Used with permission.)

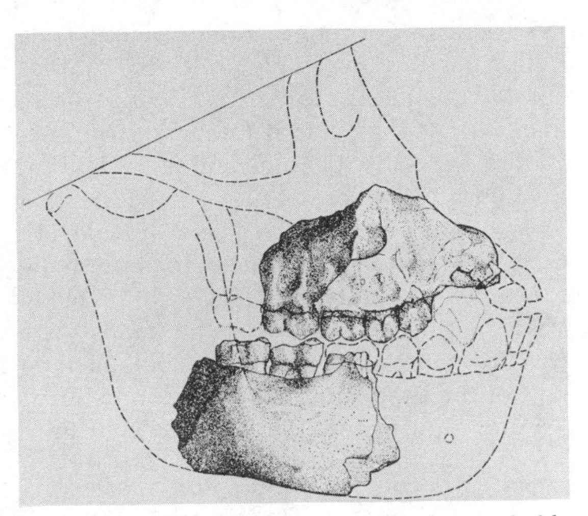

FIGURE 11.10 Reconstruction of the probable appearance of the *Ramapithecus* face based on the maxillary and mandibular fossils. (*Source:* Simons, 1964. Used with permission.)

1965). In 1934 Lewis suggested that *Ramapithecus* was a hominid, and after years of controversy it is now generally agreed that *Ramapithecus* is a member of the family Hominidae or is the immediate ancestor of this family.

All of the *Ramapithecus* fossils now published are mandibular and maxillary fragments (Figures 11.8, 11.9, and 11.10). *Ramapithecus* is smaller than most species of the fossil pongid genus *Dryopithecus.* The incisors and canines are relatively small. The canines projected beyond the level of the tooth row but not as much as in pongids, and accordingly, the lower first premolar was a semisectorial premolar. The face projected more than in modern man, but was much more orthognathous than the face of pongids. The shape of the face may have differed quite a bit between African and Asian *Ramapithecus.* The reconstruction of the Asian form *Ramapithecus punjabicus,* Figure 11.8, presents a rounded arclike dental arcade. The African specimen, now referred to as *Ramapithecus wickeri,* has been reconstructed by Walker and Andrews. They find the tooth row to be more elongated and straight although diverging toward the rear. Also the African specimen, the only specimen preserving the symphyseal region (the region at the front where the two halves of the mandible have fused to form a single bone), shows a simian shelf, a pongid characteristic.

Even more recently there have been *Ramapithecus* finds from Greece and Hungary. There are some differences in tooth morphology between these finds and both the African and Asian finds.

The more recent finds have not changed the overall view of *Ramapithecus.* The cusp pattern of the teeth, the arched palate, and the transitional condition of the premolars all suggest a relationship between *Ramapithecus* and the hominid *Australopithecus,* which lived in late Pliocene and early Pleistocene times. Pongids that have projecting canines employ these teeth in defense, in aggressive displays, and as a tool for bark stripping. With a less salient canine and an overall reduction in size of the face, it seems that *Ramapithecus* may have been employing tools, to at least a small extent, to perform the functions no longer possible with the teeth. This, in turn, implies that the hands were free to manipulate and carry tools at least part of the time. Unfortunately we do not as yet have postcranial remains to confirm a hypothesis of bipedalism.

The *Ramapithecus* fossils from Northern India are not well dated but on the basis of faunal associations they appear to be Miocene in age and to extend into early Pliocene. In East Africa L. S. B. Leakey found *Ramapithecus* remains that can be dated by potassium-argon decay at 14 million years and late Miocene.

At this point, it is appropriate to say something about what constitutes a tool and what constitutes tool use. For years anthropologists spoke of tools as any objects utilized to perform a task. Recognizing an object, a stone for example, as a tool, simply meant finding evidence of use. Over time, more and more examples of this kind of tool use were described among other animal species. Therefore the suggestion was advanced that true tool use was to be considered to occur only if it is clearly learned behavior, not an innate sequence evoked by a specific stimulus.

Such a convention rules out the species of Galapagos finch, known as the woodpecker finch, which has evolved a behavior pattern in which it employs a cactus spine held in the beak to probe for grubs in the crevices of tree bark, thereby occupying the ecological niche similar to that occupied by woodpeckers with their specialized beak in other regions of the world. But such work as that of Köhler (1927) showed clearly that chimpanzees can learn to employ sticks and other objects as tools of this kind. The suggestion was then made that "tools" should not refer to any casually used item but to manufactured items, items that had been modified with some forethought as to how they were to be used (Oakley, 1951). This led to definitions that stressed "systematic" tool making or the existence of a "set and regular pattern" among tools.

The remarkable field work of Jane Goodall has shown that chimpanzees in the wild learn by watching adult chimpanzees make and use rudimentary tools. For example, termite probes are made from twigs or vines that are trimmed to twelve inches or less in length and stripped to be inserted into the termite nests. Goodall has observed a chimpanzee carrying such a tool for half a mile while looking for a suitable termite hill (Goodall, 1965).

This illustration of tool-using behavior also invalidates another supposed distinction between human and nonhuman primate behavior. This was the distinction made in response to Köhler's studies. The latter stated that nonhuman primates could engage in tool-using behavior only if the "tool-object" and the "goal-object" were both in view at the same time. In other words, this distinction does not allow "foresight" in chimpanzee behavior. This distinction is related to the distinction made by some psychologists between perceptual and conceptual abilities. By all objective criteria that can be marshaled, the chimpanzee has a concept of what the tool should look like and of the situation in which he will use this tool.

An even more extreme step in this same tradition, which might be called the "keep the apes their distance" tradition, is the statement that the tool using of *Homo* differs fundamentally from the tool using of apes in that *Homo* uses tools to manufacture other tools. This is somewhat parallel to the caveat that while other primates can learn to make tools only man *teaches* tool making.

This digression on tool using sounds rather polemical but it has a purpose. All sciences in their formative stages rely heavily on qualitative distinctions. Quantitative distinctions are disdained as when it is said, for example, that a difference is "a matter of degree only." These outdated expressions imply that qualitative "fundamental" distinctions are of the greatest significance. But qualitative differences are a special case of quantitative differences. As a science matures many problems are resolved by converting qualitative statements to more sophisticated quantitative statements.

The evolutionary process can give rise to *quantitative* differences of enough magnitude that they will be recognized as *qualitative* differences between species. In the case of tool using it is true, as Bartholomew and Birdsell have said, "that man is the only mammal which is continuously dependent on tools for survival" (Bartholomew and Birdsell, 1953). But this is a distinction describing *Homo* as he differs from his pongid collateral relatives today. This difference originated through a long period of slow, quantitative change. The earliest tool use of man cannot be expected to look like his tool use today. The search for an absolute, qualitative difference between "tool" and "non-tool" becomes irrelevant and even impedes our understanding of human evolution. If *Ramapithecus* was using implements to break open bone 14 million years ago (Leakey, 1968), and if his reduced canine size is evidence of some dependence on tools, then he had taken the adaptive turn that must have been taken by the ancestor of *Homo*.

Ramapithecus is an exciting find in that, to the limited extent we now know this genus, it is the evolutionary step to be expected in the differentiation of hominids from pongids out of the dryopithecine radiation. It would be nice to be able to trace a sequence of fossil forms from *Ramapithecus* through the Pliocene epoch. Unfortunately little has been done with Pliocene deposits and our record of man is blank for several million years until we pick up the hominid line with the late Pliocene *Australopithecus africanus*.

Ramapithecus dwelt in a lush, forested environment that probably shaded into woodland-savanna in the tracts more distant from river courses. To what extent *Ramapithecus* may have been still arboreal we cannot know until postcranial remains are discovered. The Pliocene epoch was characterized by a general cooling of the earth's atmosphere and the spread of grasslands at the expense of forests. *Australopithecus*, from late Pliocene and early Pleistocene, is a savanna-dwelling form. We know then that early and middle Pliocene times include an important ecological transition for the ancestors of man.

THE PLEISTOCENE

Before discussing the finds of *Australopithecus*, we must consider some problems involved in dating the Plio-Pleistocene boundary, the chronology of the Pleistocene epoch, and the major climatic changes of this time. Much of the knowledge that we have of Pleistocene chronology is dependent on the fact that the Pleistocene is defined roughly as the "Ice Ages" of the Cenozoic. As mentioned earlier, a general cooling of the earth's atmosphere characterized the Pliocene. More dramatic climatic oscillations mark the beginning of the Pleistocene. Great climatic fluctuations produced an alternating spread and retreat of glacial ice masses. These glaciations left their marks, directly or indirectly, on a global scale and provided a relative chronology for arraying the major events of man's evolution during the Pleistocene.

The occurrence of a cool glacial period or a warm *interglacial* can be inferred from several kinds of evidence. Major glaciations involved the spread of mountain glaciers in all the high mountain ranges of the world as well as the spread of polar ice masses. The south polar ice mass did not spread over continental areas outside of Antarctica, but the edges of this polar ice cap broke off and, carrying soil and rock detritus, drifted into warmer seas where they melted and dropped the rock to form a layer on the sea floor that can now be dated.

The north polar ice cap became a continental glacier that spread over the northern regions of both the Old and New Worlds. This grinding mass of ice as well as the mountain glaciers scoured deep valleys and both carried and pushed large amounts of soil and rock that were left behind as *moraines*. The outwash from these moraines, as well as from the glacier itself, carried gravel and silt deposits down watercourses to be deposited far from the periphery of the glacier. The periglacial area was inhabited by a cold-adapted flora and fauna.

The weight of the ice sheet can deflect the crust of the earth resulting in altered drainage patterns in the region. A glacier advances, of course, when the annual precipitation in the form of snow exceeds the amount of snow that melts. A great deal of moisture is taken up in the glacial ice. This results in a worldwide lowering of sea levels. This in turn alters patterns of deposition and sedimentation of coastal streams.

The names of glaciations or cold and warm periods are so extensive in the literature that they cannot be dropped abruptly even though we now have many absolute dates on Pliocene and Pleistocene fossils. The first glacial name sequence to be widely adopted was the Alpine sequence set up by Penck and Bruckner (1909). Listing the latest at the top as though it were a geological column these glaciations and their assumed age is shown in column 3 of Figure 11.11.

A recent set of time markers which are world-wide and for which we have absolute dates are shown in column 5 of Figure 11.11. The names represent periods in which geomagnetic polarity was reversed from the previous period. With some regularity in past ages, the earth's magnetic field has collapsed and reversed direction (north to south) within a 1000–5000 year time span. Certain rocks, when heated and recooled in a magnetic field, become magnetized in a direction parallel to this magnetic field. Rocks heated by volcanic activity and then cooled provide a record of the direction of magnetic polarity by remnant magnetism in the rock.

In the terminology of paleomagnetism the long-period reversals separate *epochs* and the briefer changes are termed *events*. The Pleistocene is spanned by two epochs, the Brunhes Normal Epoch which includes the present, and the Matuyama Reversed Epoch. Within each of these are briefer reversal events as shown on Figure 11.11. The Matuyama-Brunhes boundary, slightly under 700,000 years ago, is now being taken as a convenient dividing line between the Lower and Middle Pleistocene.

In the past there has been little agreement on the date or the event to choose as the beginning of the Pleistocene. The original criterion used was the spread of cold-adapted microscopic marine animals throughout the northern hemisphere. Later investigators decided also to use the presence of certain land-dwelling mammals as indicators of the Pleistocene. It appeared that the spread of these land-dwelling forms (modern genera of elephant, cattle, horse and camel) occurred at the same time as the marine cold fauna. It

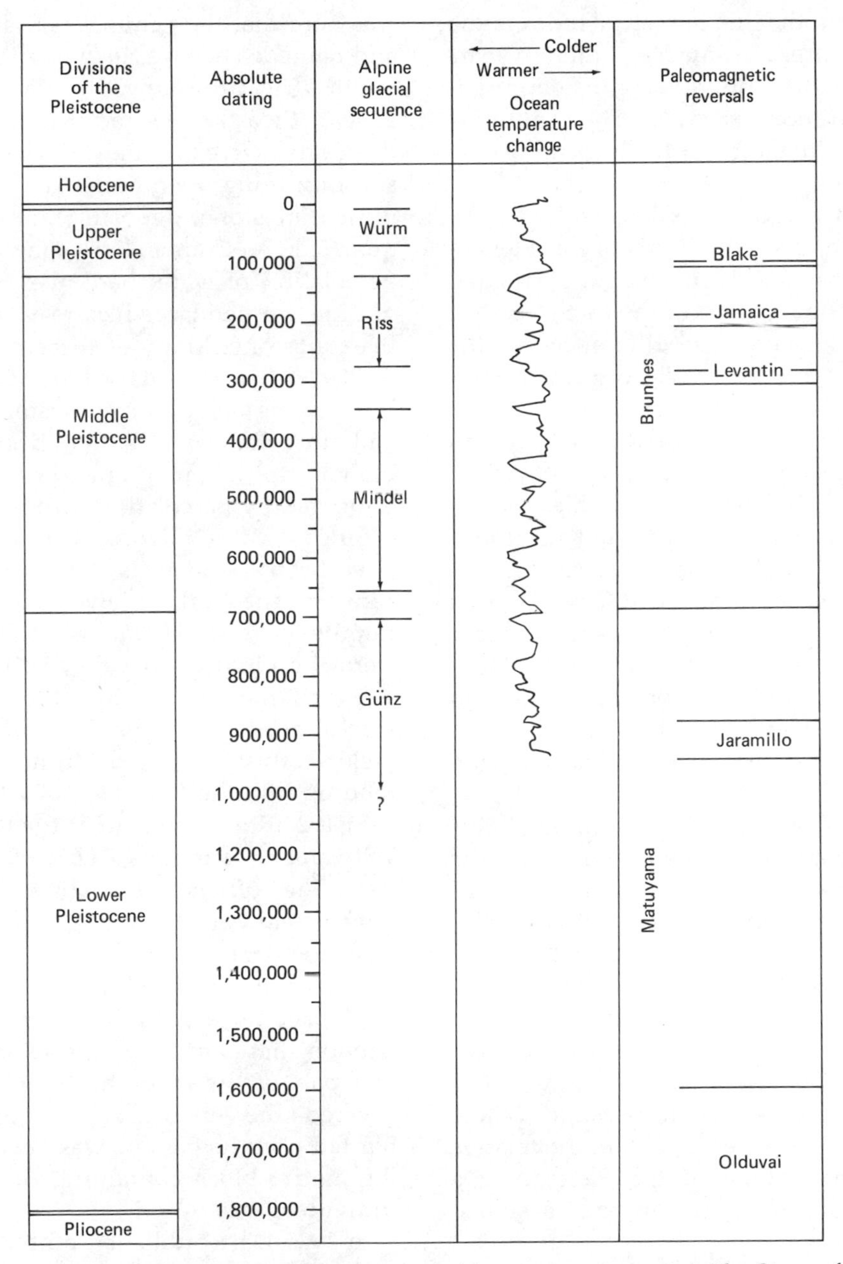

FIGURE 11.11 Divisions of the Pleistocene. (*Source: Man-apes or Ape-men? The Story of Discoveries in Africa,* Sir Wilfrid E. LeGros Clark. Copyright © 1967 by Holt, Rinehart and Winston, Inc. Reproduced by permission of Holt, Rinehart and Winston, Inc.)

has turned out that the terrestrial index fauna began their spread as much as 3 million years ago while the marine forms spread approximately 1.8 million years ago. The trend now is to use only the latter as the beginning date of the Pleistocene.

Many years ago a Russian geologist, M. Milankovitch, presented a theory of Pleistocene glaciations which said that regular changes in the position of the earth, relative to the sun, resulted in differences in the amount of solar energy reaching the earth's surface. Periods of less solar energy resulted in glaciations. This theory was used by an American geologist, Zeuner, to provide a chronological framework for the Pleistocene. Many anthropologists adopted Zeuner's chronology. They, and Zeuner, were accused of using a theory, rather than facts, to order the Pleistocene. Ironically enough, new facts, the alternation of cold and warm periods revealed in deep-sea drill cores, now confirm the Milankovitch theory. The warm-cold variation shown by deep-sea cores is shown in column 4 of Figure 11.11.

Despite the apparent vindication of this theory of Pleistocene climates, it remains the case that evidence of cold-warm fluctuations on land are not well correlated with the cold-warm fluctuations seen in deep-sea sediments. And cold-warm fluctuations in one continental region are often not well associated with the sequence of fluctuations in another region. Therefore we utilize, by preference, absolute dates where possible. Where dates of fossils are inferred from the relative (climatic) chronology of the Pleistocene a fairly large margin of error must be assumed for each date.

AUSTRALOPITHECUS

The earliest large collection of fossils of *Australopithecus* comes from South Africa. It was here that the genus was first discovered and named. The five *Australopithecus* sites in South Africa are Taung, Sterkfontein, Makapansgat, Swartkrans, and Kromdraai. All of these sites were originally discovered as a result of mining operations. They are in dolomitic limestone deposits that were being quarried. As in most limestone deposits the percolation of water had, over long periods of time, formed large fissures and caverns in the limestone. Many of these openings had refilled with a secondary deposit of limestone *breccia*, angular bits of limestone plus sand and other materials that had been washed or blown into the opening and had become recemented by percolating carbonates.

Since 1959 a number of *Australopithecus* have been found in Olduvai Gorge in Tanzania by the Leakeys. Even more recently a number of *Australopithecus* fossils have come from the Omo River valley in Ethiopia and an even larger number from the Lake Turkana area of northern Kenya. The latter two areas yielded *Australopithecus* from deposits extending into the Pliocene.

In 1924 Raymond Dart at the University of Witwatersrand received a box of fossils saved by mine officials operating a limestone quarry near Taung. Among this material was an endocranial cast, a natural cast of the endocranial cavity of a skull, which in Dart's opinion looked far too hominid to be from a fossil baboon that Dart had expected to find. The rear part of the cranium, which must have covered the endocast prior to its being blasted from its fissure, was never found but the matrix block containing the face of the individual was found.

In 1925 Dart published a note on his find and named it *Australopithecus africanus*, the "southern ape" of Africa. *Australopithecus africanus* was, judging from tooth eruption, a child of approximately five years of age. From an analysis of its features Dart concluded

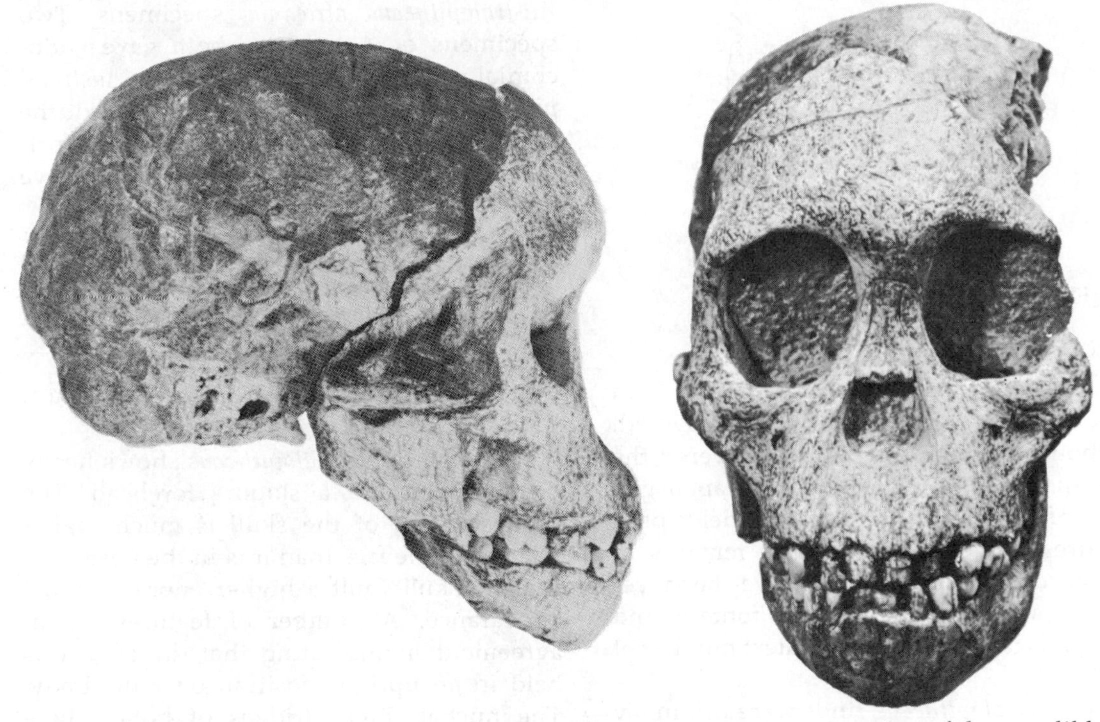

FIGURE 11.12 *Australopithecus africanus*, side and front views. Lower rear portion of the mandible and the back of the skull are missing. The endocast of the missing portion of the skull is in place. (*Source: Man-apes or Ape-men? The Story of Discoveries in Africa*, Sir Wilfrid E. LeGros Clark. Copyright 1967 by Holt, Rinehart and Winston, Inc. Reproduced by permission of Holt, Rinehart and Winston, Inc.)

that *Australopithecus africanus* was intermediate between the pongids and hominids.

The Taung child included much of the mandible, the bones of the face including most of the frontal bone, and the endocranial cast that permitted a reasonable reconstruction of the contours of the remainder of the skull (Figure 11.12.)

In London, the established authorities on fossil man did not think highly of Dart's view that *Australopithecus africanus* was a transitional early-man form. They chose to view *africanus* as a deviant ape and nothing more. They did not visit the discovery site nor view the fossil. A manuscript of a monograph providing a more definitive description of *africanus* was refused publication. Dart and a colleague, Dr. Robert Broom, continued their work on the South African "man-apes," and only after a lifetime of work in accumulating additional evidence were the early views of Dart on the taxonomic affinities and cultural abilities of *africanus* largely vindicated.

Robert Broom discovered and described the first adult specimen of *Australopithecus africanus* in 1937. This came from the site of Sterkfontein. Broom soon discovered another site at Kromdraai, only a mile distant from Sterkfontein (Figure 11.13). In 1948 the

FIGURE 11.13

Swartkrans site, only a short distance from Sterkfontein and Kromdraai, was discovered. At about the same time Dart discovered the fifth *Australopithecus* site at Makapansgat. Some of these sites continue to yield fossil treasures. To date fragmentary remains of several dozen individuals have been recovered, with Swartkrans, Sterkfontein, and Makapansgat yielding the greatest number of finds.

The *Australopithecus* finds were originally given a number of names designating different species and genera. Today most authorities will agree that only two types are involved. These shall be designated *Australopithecus africanus,* the slender form, and *Australopithecus robustus,* the heavier form. The species name *Australopithecus robustus* comes from a Kromdraai fossil first designated by Broom as *"Paranthropus robustus."* In the following paragraphs we will first review the general characteristics of *Australopithecus.* After that we will note the ways in which *africanus* differs from *robustus.*

Australopithecus had a very small brain compared to modern man. The Taung child had an endocranial capacity of approximately 405 cc. Judging the Taung specimen from tooth eruption to be around five years of age, the adult individual would have had a cranial capacity of 440 cc. Holloway (1970) has given an estimated mean of 442 cc based on six *Australopithecus africanus* specimens. Two specimens of *A. robustus* both gave endocranial values of 530 cc. This places *Australopithecus'* brain size at the upper end of the range of gorilla brain sizes. Among large primates today we find the adult males to have the following endocranial capacities:

Man	1100–1700 cc
Gorilla	350–650 cc
Chimpanzee	300–480 cc
Orangutan	300–500 cc

The largest capacity recorded for a gorilla is 752 cc.

The skull of *Australopithecus* shows heavy brow ridges and a sloping forehead. The highest point of the skull is much higher relative to the face than it is in the apes, giving the skull vault a higher, more rounded appearance. A number of features are in agreement in indicating that the skull was held in an upright position over the body. The nuchal ridges (ridges of bone where neck muscles attach at the back of the skull), sometimes raised in a heavy torus of bone, are low on the back of the skull as compared to the condition in apes. The occipital condyles, the points of articulation of the skull with the vertebral column, are located much further under the skull than in apes. The foramen magnum (the opening through which the spinal cord enters the skull) opens downward as it does in *Homo* and not down and back as in pongids. The face of *Australopithecus* is quite heavy, especially in the jaws, and the center of gravity of the complete skull is further forward from the occipital condyles than it is in modern man.

The face is long, massively constructed, and projecting in the tooth region. It does not project as much as in pongids, and relative to the pongids the face is somewhat more under the frontal lobes of the brain. Like pongids, because of the protrusion of the

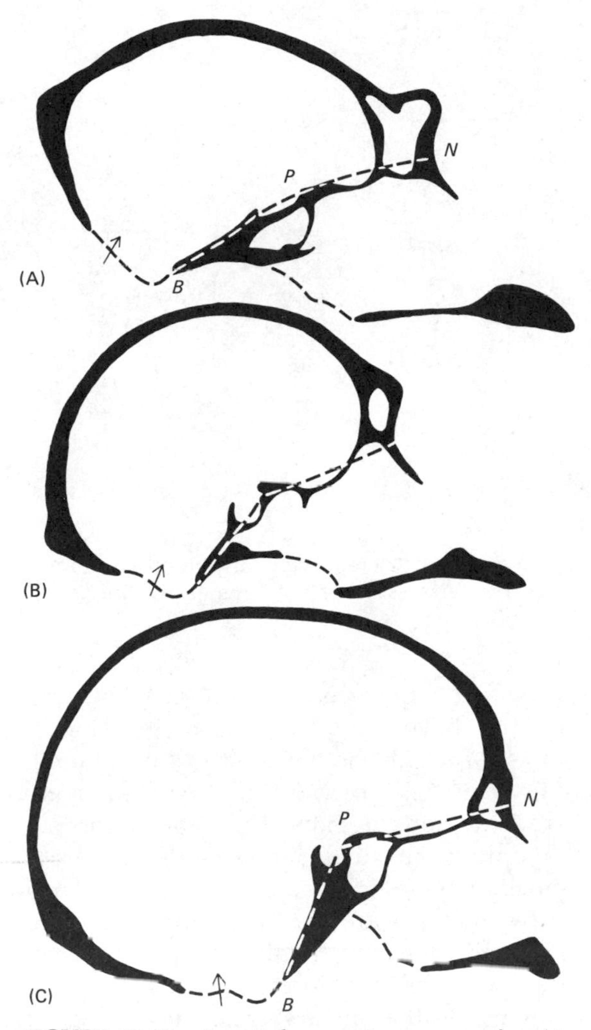

FIGURE 11.14 Saggital craniograms of (A) gorilla (female), (B) *Australopithecus,* and (C) *Homo.* The position and axis of orientation of the foramen magnum is shown by the arrow. Flexion of the cranial base (*BPN*) indicates the extent of folding of neurocranium over the face. (*Source: Man-apes or Ape-men? The Story of Discoveries in Africa,* Sir Wilfrid E. LeGros Clark. Copyright © 1967 by Holt, Rinehart and Winston, Inc. Reproduced by permission of Holt, Rinehart and Winston, Inc.)

maxilla, the nasal bones do not project out from the frontal plane of the face as they do in *Homo.* Due to the retreat of the face, but even more due to the transposition of the foramen magnum, the sphenoidal angle (indicated by the points *BPN* in Figure 11.14) is intermediate between what is found in apes and in man.

The shape of the mandibular fossa, where the mandible hinges with the base of the skull, is hominid. The mastoid processes, the pyramidal bony extensions projecting down from either side and to the rear of the ear opening, are well developed. Poorly developed mastoid processes sometimes form on male gorilla skulls. In hominids they are a critical feature for the attachment of muscles involved in the rotation and balance of a skull held in an upright position.

The mandible of *Australopithecus* is heavy, and the ascending ramus—the rear, nearly vertical, branch of the jaw—is very long. The bone of the symphyseal region slopes backward as it does in the apes. In other words it has no protruding chin. But it also has no simian shelf, the internal shelf of bone that buttresses the symphysis of the two halves of the mandible in modern pongids. The tooth row of the mandible is arcuate in form, a hominid trait. Pongids display a rectangular tooth row with the large canines at the anterior corners of the rectangle. The arcade of the upper dentition corresponds to that of the jaw (Figure 11.15). It can be noted that although clearly hominid, *Australopithecus* does not have quite the open parabolic form of the tooth row found in modern man. Its tooth row might instead be described as hyperbolic, in contrast to the parabolic form exhibited by modern man.

The teeth of *Australopithecus* are hominid in almost every respect. When Dart separated the upper and lower teeth of *Australopithecus africanus* a few years after its dis-

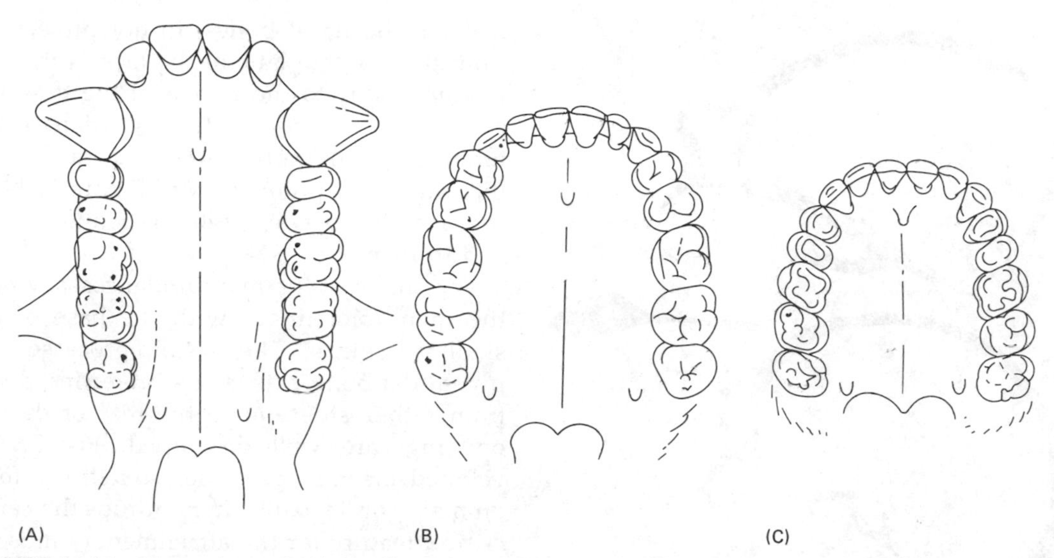

FIGURE 11.15 Upper jaw of (A) gorilla, (B) *Australopithecus*, and (C) relatively large-toothed modern man, occlusal view, two-thirds size. (*Source:* Clark, 1967. By permission of the Trustees of the British Museum, Natural History.)

covery, he clinched its hominid status. The incisors are small as in *Homo*, not large as in the pongids. The canines are not round and tusklike as in pongids. They are small, somewhat spatulate in shape, and tend to grow only slightly beyond the plane of the other teeth. But, also as in modern man, the canines wore down such that they did not normally project beyond the other teeth at all. The canines had become, in function, assimilated to the incisor teeth. Compared to modern man's, the premolars of *Australopithecus* are surprisingly large. But they differ more importantly in their shape from the premolars of pongids. The first lower premolar, P_3, of apes is a sectorial tooth shearing against the upper canine. In *Australopithecus*, as in all hominids, the premolars are all bicuspid. In *Australopithecus* the bicuspids have a larger crown area and are more cuboidal in shape than they are in modern man.

The molar teeth of *Australopithecus* are very large compared to modern man's and tend to have a relative size progression of $M_1 < M_2 < M_3$, as opposed to the common $M_1 > M_2 > M_3$ or $M_2 > M_1 > M_3$ in small-toothed modern man. The molars show a dryopithecine cusp pattern and the larger molars have a small sixth cusp. The fissures between cusps are narrow and the cusps are rounded. Both are hominid traits that contrast with the pongid condition. Finally the crowns of the molars wear down in flat planes as in man.

The postcranial anatomy of *Australopithecus* is even more clearly hominid than the skull. The picture is overwhelmingly that of an animal orthograde in posture and bipedal in locomotion. Vertebrae and a sacrum from Sterkfontein reportedly show a lumbar curve, a trait associated with orthograde posture (Figure 10.9). At Kromdraai a talus (ankle bone) that is predominantly hominid in form was found. The superior articular surface—

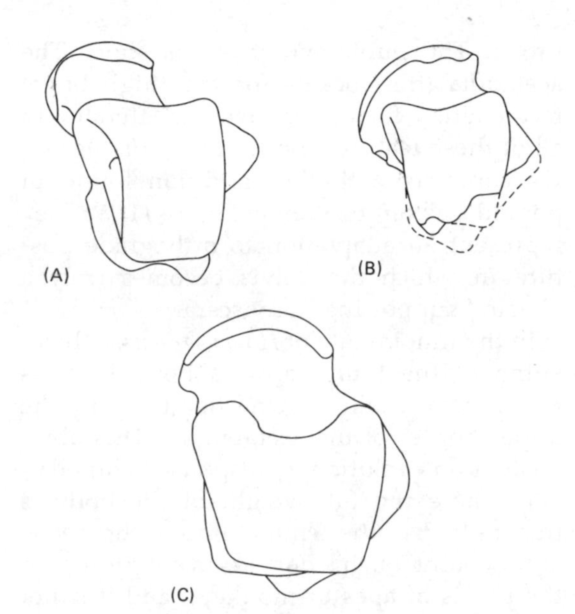

FIGURE 11.16 Right talus (ankle bone) of (A) chimpanzee, (B) *Australopithecus robustus*, and (C) *Homo sapiens* (top view). (*Source:* Clark, 1947. Used with permission.)

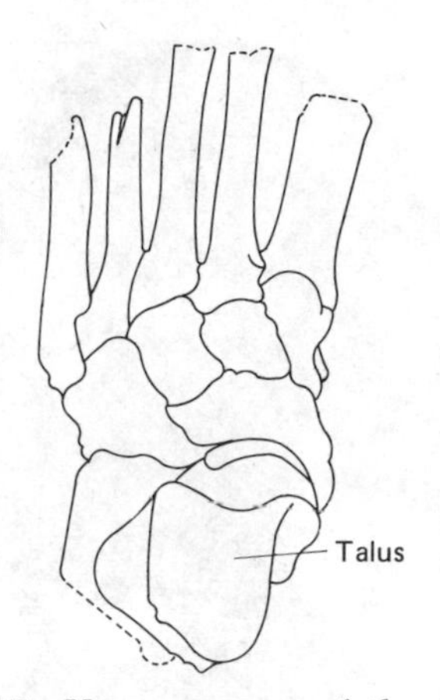

FIGURE 11.17 Upper aspect of the left foot of an *Australopithecus* skeleton from Bed I, Olduvai Gorge, Tanzania. Toes and metatarsal ends are missing, as is the rear of the calcaneus (bottom left). (*Source:* Clark, 1964. Copyright © 1955 by the University of Chicago Press. Used by permission of the University of Chicago Press.)

the surface in contact with the tibia (Figure 11.16) – is hominid. The angle between the axis of the neck and the main axis of the talus is intermediate between that of apes and man. This fact, plus the broad area of weight transmission on the forward articular surface, suggests that the main axis of weight transmission in completing a step was not through the first toe as it is in modern man but was somewhere lateral to the first toe. This is confirmed by the find of an almost complete foot, minus toes, from Bed I, Olduvai Gorge in East Africa (Figure 11.17). Day and Napier have noted that the talus of the Olduvai foot is much like the talus from Kromdraai (Day and Napier, 1964). They show that the third metatarsal (to the middle toe) is unusually strong and long. This further confirms that weight transmission was not as in modern man. Apes in walking tend to curl the foot slightly and to walk on the outer edge of the foot. *Homo sapiens* transmits his weight nearer the inner edge of the foot, using the outer edge for control and balance in the middle of the stride. Again *Australopithecus* seems to be intermediate.

Olduvai bones of the lower leg, the tibia and fibula, show at their articulation with the talus that they were adapted to completely plantigrade (both heel and sole squarely on the ground) bipedal walking (Davis, 1964). But the places of insertion for some muscles of the leg differ from both modern pongids and modern man and suggest that their gait was not exactly as found in modern man. Upper leg bones, femurs, from Olduvai, Kromdraai, and Sterkfontein are consistent with the lower leg bones in showing bipedalism but with muscle attachments differing slightly from the present condition in *Homo*.

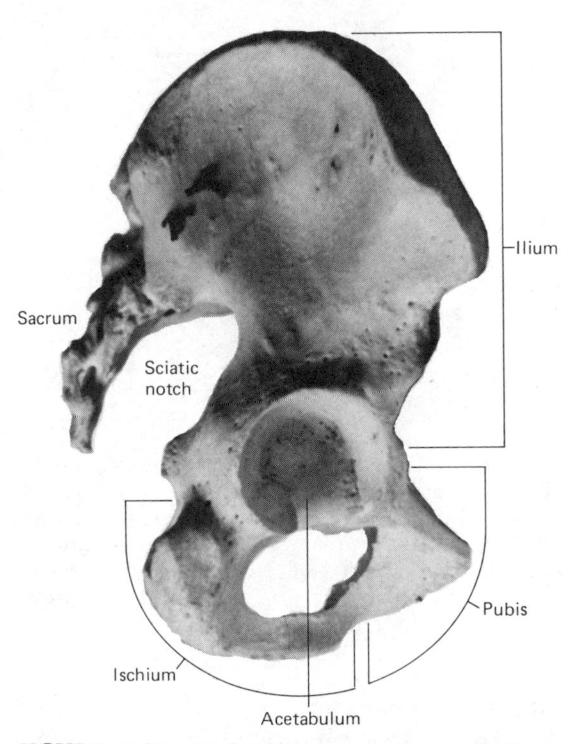

FIGURE 11.18 Right innominate bone of *Homo sapiens*.

The pelvic girdle consists of two innominate bones (Figure 11.18) joined at the front by the pubic symphysis, and a sacrum that connects the posterior borders of the innominate bones (Figure 10.12). Each innominate is in turn formed of three bones: the ilium, ischium, and pubis. *Australopithecus* innominates have been recovered at Sterkfontein, Swartkrans, and Makapansgat. The Sterkfontein pelvis includes the upper portion of the sacrum. The following features (Figures 11.19 and 11.20) are common to all the *Australopithecus* pelvises and to modern man: The ilium is not high. The wing of the ilium is broad and is expanded far to the rear producing, by flexion of the posterior margin, a deep sciatic notch. The iliac crest (the upper border) exhibits a sigmoid curve when viewed from above. The sacrum is short and broad. The pubic symphysis is short. The acetabula (the sockets for the thigh bone) face slightly forward as well as laterally. In all of these features the *Australopithecus* pelvis contrasts with the condition found in pongids, as can be seen in Figure 11.19. They represent an adaptation to orthograde posture in which the pelvis becomes truly a "basin" supporting the viscera.

In the innominate of *Homo sapiens*, a thickening of the ilium forms a bony buttress running vertically down the ilium to the upper border of the acetabulum. This obviously is an evolutionary response to bipedalism where the full weight of the body is transmitted to the femur through the pelvis at this point. Such a buttress is not present in the pelvis of apes or monkeys, and it is not present in *Australopithecus*.

There are differences between these pelves and those of modern man. The *Australopithecus* innominate does not curve around to the front of the body as much as the innominate of modern man does. Consequently it shows a more lateral flare of the ilium. This could have reduced the stability of the upright stance by shifting the point of origin of muscles important in lateral stability in man today. But this condition is compensated by an unusual extension of the anterior-superior iliac spine (Figure 11.20). It has further been suggested that the ischium was apelike in being relatively long. This is particularly true in the Swartkrans innominate where the ischium is long and the roughened area where heavy muscles attach (Figure 11.19) flares outward. This has led authorities to suggest that *Australopithecus,* with a different orientation of the hamstring muscles, had not achieved the efficient striding walk of *Homo sapiens* even though he was fully bipedal (Washburn, 1960; Napier, 1964; Clark, 1967).

The skeletal remains of *Australopithecus* have been discussed in greater detail than

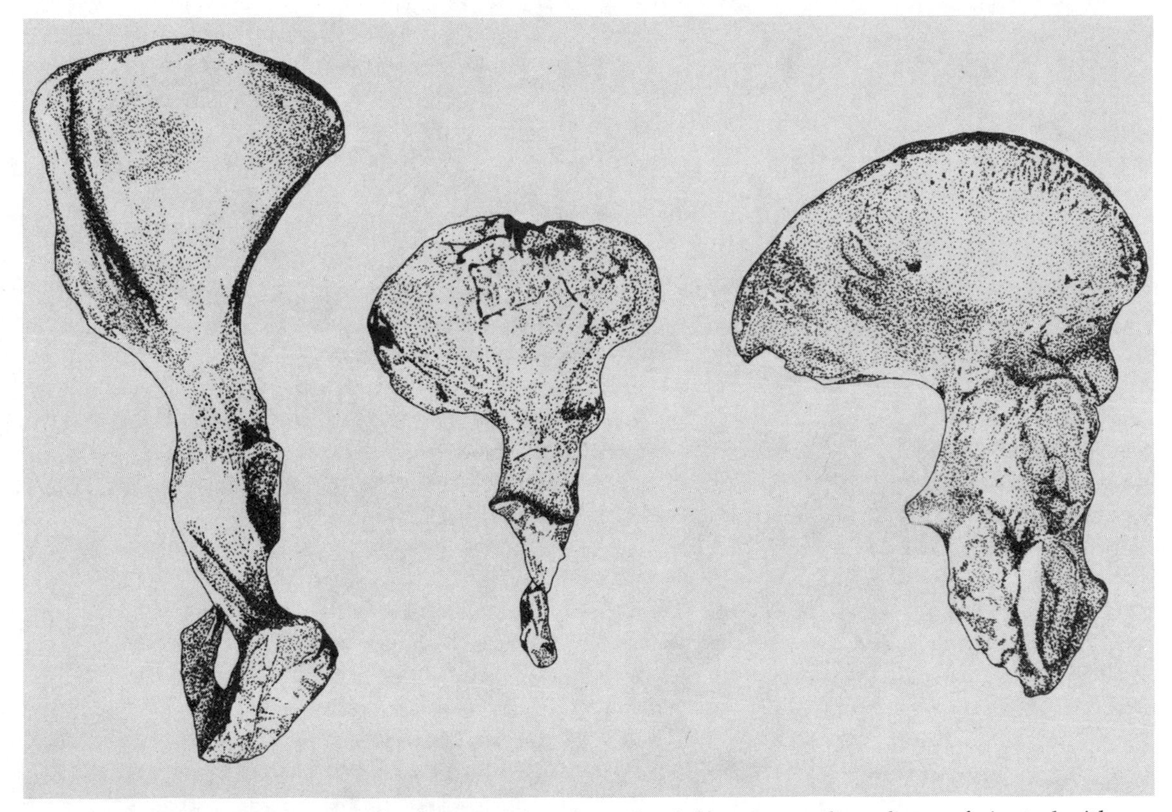

FIGURE 11.19 The pelvic bone (innominate) of gorilla (left) compared to those of *Australopithecus* (center) and modern *Homo sapiens* (right).

will be the case for later fossils. The reason for this is that the *Australopithecus* morphology can now provide a base line for comparative statements about later fossils. Also we have more remains of *Australopithecus* than we have for any other fossil man taxon until we come to Neandertal man in Upper Pleistocene times. Furthermore the position of *Australopithecus* is critical in that this is the first group certainly known to have made the full transition to bipedal, terrestrial life. And he was the first, as will be shown later, to have made stone tools and to have hunted other animals.

There are important differences within the genus *Australopithecus*. These are usually designated as a species distinction, *Australopithecus africanus* being the smaller, slender form and *Australopithecus robustus* being the heavier form. This distinction between a large and small form of *Australopithecus* was first made on the basis of the South African finds. *Australopithecus africanus* includes most of the Sterkfontein and Makapansgat finds and, perhaps, Taung. *Australopithecus robustus* includes the finds from Kromdraai and Swartkrans.

The many East African finds that have come to light since 1959 include: fossil specimens which represent both *A. africanus* and *A. robustus;* a few finds which seem intermediate between these two; and some finds

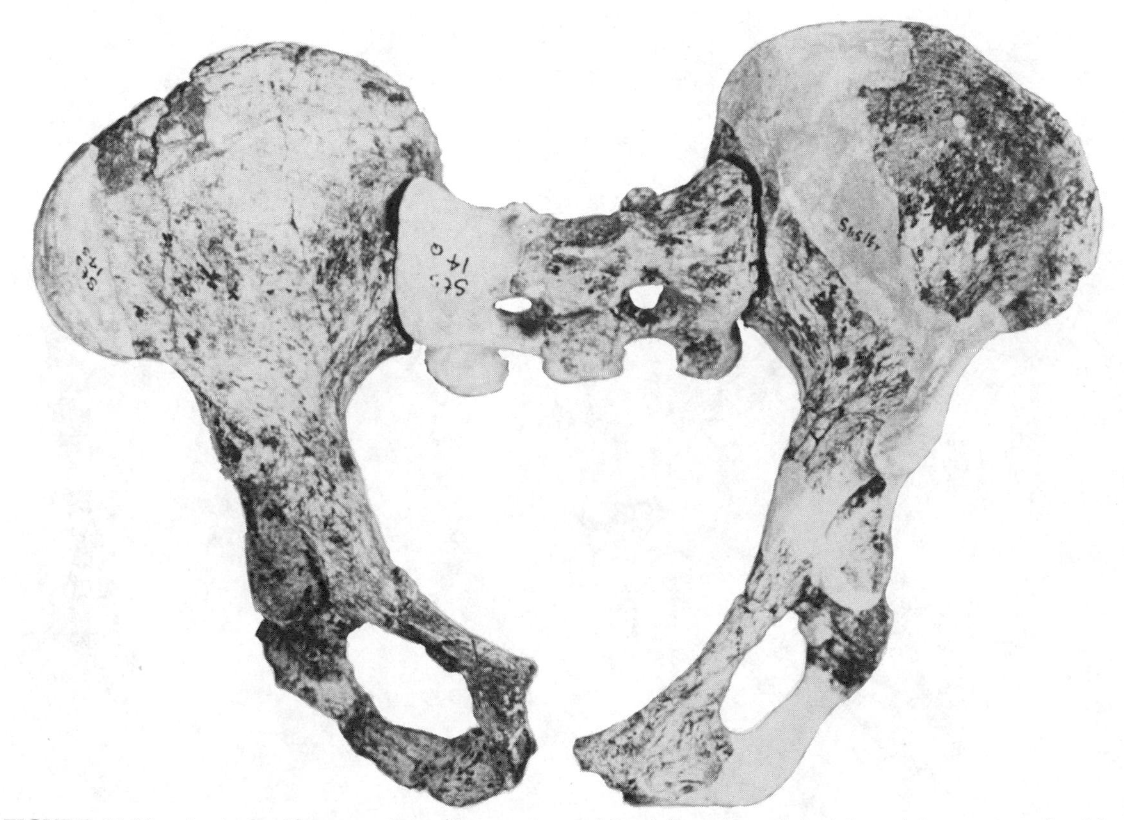

FIGURE 11.20 *Australopithecus* pelvis. Front view of Sterkfontein pelvis (above) is compared with a larger pelvis of American White (opposite page). (*Source:* Sterkfontein pelvis, Bernard G. Campbell.)

which are more advanced, that is, more like later representatives of the genus *Homo*. Most of the East African finds come from Olduvai Gorge in Tanzania, the Omo River basin of Ethiopia, the Afar region of Ethiopia, and the East Rudolf region of Kenya.

Most of the differences between these two forms can be understood as a reflection of the greater size and muscularity of *robustus*. But, it should also be noted, that the early literature tended to overemphasize the magnitude of the differences between the two species. Some views on human phylogeny are still influenced by an overdrawn picture of differences.

In a careful study of postcranial remains, McHenry (1974) has estimated that the South African *africanus* forms averaged about 4 feet 9 inches (1451 mm) in height; the *robustus* forms, 5 feet (1527 mm) in height. The East African hominids averaged around 5 feet 4 inches (1630 mm) in height. Estimates on weight are less secure. For the South African finds *africanus* probably averaged 50 to 60 pounds.

The heavier musculature of *robustus* is reflected in the skull in a generally more rugged bony architecture. The zygomatic arch (the arch formed by the backward extension of the cheekbone) is heavier and displays a

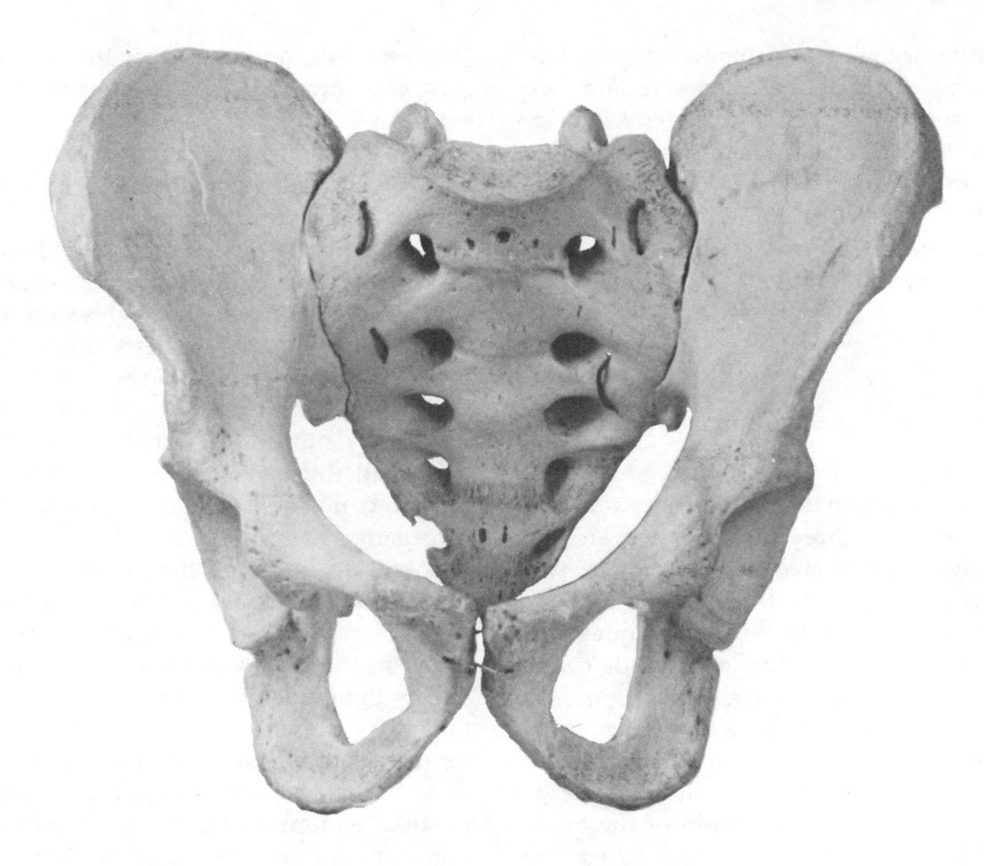

more prominent malar root, where the cheekbone joins the upper jaw. The entire masticatory apparatus of *robustus*—the palate, the jaws, and the cheek teeth—are larger than they are in *africanus*. The occipital torus and the supraorbital torus are more pronounced in *robustus*. In a few of the *robustus* specimens, a sagittal crest is evident.

The slightly larger cheek teeth of *robustus* are in themselves of little taxonomic significance. But in some respects the dental differences are more than that of relative size. The lower canines of *robustus* are smaller than the canines of *africanus*. The premolars are remarkably large in both forms, but in *robustus* they are larger in the mesiodistal diameter (the diameter measured parallel to the tooth row) than they are in *africanus*. The upper premolars of *robustus* are usually three rooted, those of *africanus* usually two rooted, as is P^3 of modern man.

DATING OF AUSTRALOPITHECUS

Because of the nature of the deposits, there has been up to the present no way of getting absolute dates on the South African cave deposits. In East Africa the situation is quite different. There *Australopithecus* fossils have been found in deposits dateable (usually by potassium-argon) from somewhat over five million years up to between one and two million years. In other words, *Australopithecus* is a Pliocene form which continued into the beginning of the Pleistocene.

On the basis of faunal and soil analysis, the five South African sites have been suggested to show a relative age sequence of

Kromdraai	Later
Swartkrans	
	↑
Makapansgat	Earlier. Perhaps overlapping.
Sterkfontein	
Taung	

This sequence has phylogenetic implications since the three earlier sites are the *Australopithecus africanus* sites and the two later sites are the *Australopithecus robustus* sites. (The placement of Taung is questionable both temporally and morphologically.) On the other hand, *robustus* and *africanus* forms are contemporaneous or, at least, overlapping in time and space in all of the major *Australopithecus*-yielding regions of East Africa. A careful consideration of the available dates indicates that, in East Africa, the slender form appears earlier than the robust form which is not found at much over 2.1 million years ago (Coppens and others, 1976). One hypothesis advanced to account for this situation is that an *africanus* basal population diverged into two separate lines, one of them maintaining the *africanus* characteristics until it evolved into *Homo erectus* and then into *Homo sapiens,* and the other line leading to *Australopithecus robustus,* which then became extinct.

J. T. Robinson noted that the *robustus* molar crowns showed fine chipping of the enamel surface of a kind that would be produced by sand or grit particles. From this, and the generally heavier chewing apparatus, Robinson suggested that *robustus* had evolved as a vegetarian species that ate tubers, rhizomes, and other vegetal material in contrast to *africanus,* which had a more omnivorous diet supplied in part by hunting. Later, however, Tobias examined the molar teeth of *africanus* carefully and he found comparable chipping of the enamel.

Therefore, although the prevailing view is that *africanus* and *robustus* represent hominid speciation, there are alternative possible hypotheses that have not been completely ruled out by the evidence. One such alternative is that the difference between the two forms represents the evolution of subspecific variation, allopatric (nonoverlapping homelands) in some regions and sympatric (overlapping homelands) in others. Another alternative is that this difference represents the evolution of sexual dimorphism within a single species. The differences in skulls in East Africa and in the latter two South African sites (Figure 11.21) are suggestive of the differences that can be seen, generally due to sexual dimorphism, in the skulls of gorillas, chimpanzees and orangutans [Figure 11.22(B)].

FIGURE 11.21 Variability in *Australopithecus* shown in the reconstruction of skulls from (A) Sterkfontein, (B) Swartkrans, and (C) Olduvai "Zinjanthropus."

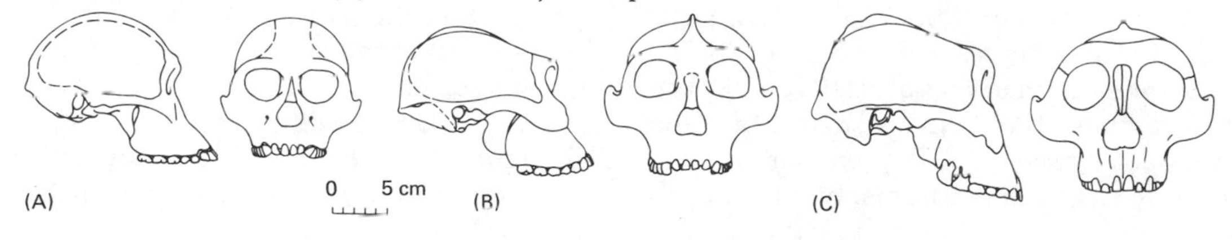

Throughout the preceding descriptions the major distinguishing features of *Australopithecus* have been discussed. But this does not mean that the South African forms and the East African forms were identical. A notable difference is that some East African individuals are more robust than *robustus* individuals from South Africa.

The first of these "hyperrobust" individuals was discovered at Olduvai in 1959. The name originally suggested by L. S. B. Leakey for this specimen was "*Zinjanthropus boisei.*" But it and later, similar finds were clearly not different enough to be given a new generic name. The question is whether it should even be considered a different species from *Australopithecus robustus.* Some do regard these specimens as sufficiently different and now designate them as *Australopithecus boisei,* a third species of this genus.

Others, including the author, take a somewhat different view and ask, is this difference greater than you would expect as geographic variation within a single hominid species? This cannot be answered directly on a species that lived two million years ago. But if we look at living human populations today, it seems that this much variation is easily encompassed. Indeed, if we look at populations in these same general regions of Africa today we can find as much difference in morphology (although not necessarily in the same features) as between the robust South African forms and the hyperrobust East African forms. Assuming comparable variability in earlier hominids, we would have to say there is not sufficient difference to justify a separate species name for the very robust East African specimens.

Most investigators in this area believe that *africanus* and *robustus* were true species, that *africanus* evolved into later forms of man, and that *robustus* became extinct somewhere around 1.1 million years ago. In addition, a number of workers differentiate a third, more advanced, species in the East African and South African deposits. Some refer to this more advanced form as *Homo* sp., a way of saying it is a member of the genus *Homo* but avoiding the question, for the present, of species assignment. Others have called it a new species, *Homo habilis.*

Professor L. S. B. Leakey introduced the name *Homo habilis* to designate some of the fossils from Bed I and Bed II (the lower two major depositional beds) of Olduvai Gorge. Some of the dentition and some of the cranial fossils had somewhat more modern features than others. The more advanced forms were put in the taxon *Homo habilis.* The major features of this distinction included a less massive mandible, a more open dental arcade, smaller cheek teeth with less mesiodistal crowding, and a more rounded skull vault. Some later finds at other East African sites, principally dental fragments, have been referred to as *habilis.*

In South Africa, in later deposits at both Swartkrans and at Sterkfontein, more advanced forms have been found which are now referred to the genus *Homo.* At Swartkrans *Homo* is represented by two mandibles, a maxilla, and part of a skull which, some time after its discovery, was found to articulate with the maxillary piece. Originally these jaw fragments were referred to as "*Telanthropus*" but this name has been dropped. A recent find of a maxilla, facial and frontal bones, and other fragments of the skull of a single individual at Sterkfontein is similarly ascribed to *Homo.* At both of these sites fossil bearing deposits could be as old as two million years, or it could, in both cases, be as young as half that age.

Whether "*Homo habilis*" comes to be treated as a valid and widely used taxon depends on conventional agreements which are still developing. But name difficulties should

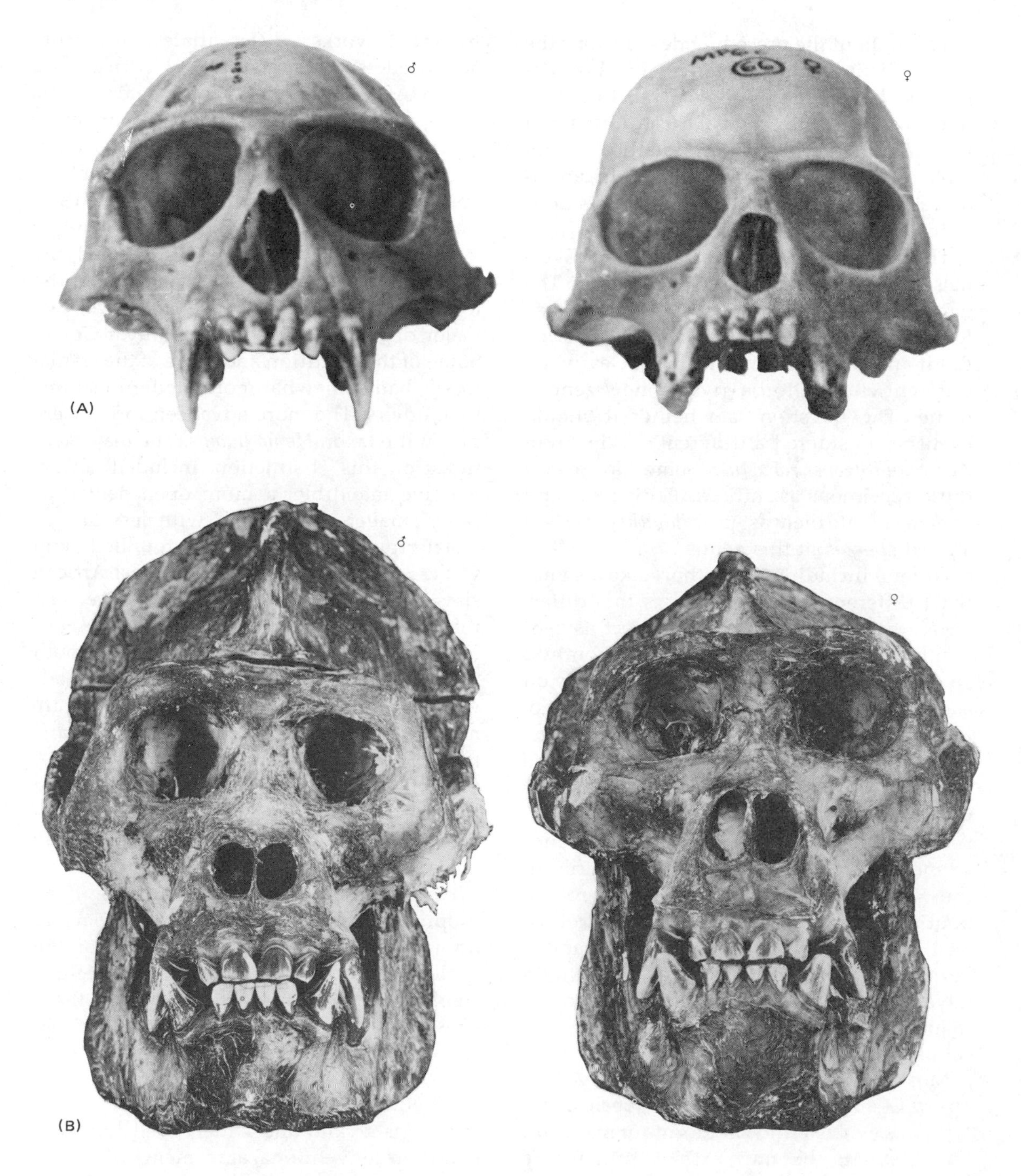
(A)
(B)

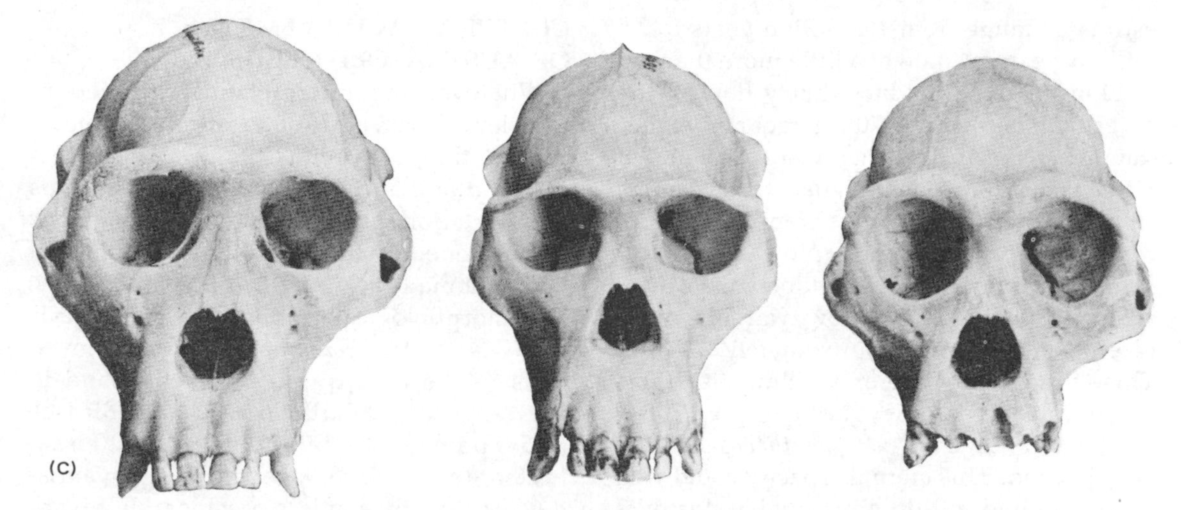

FIGURE 11.22 Variability in the skulls of living primates: sexual dimorphism in (A) monkey *Colobus* and (B) gorilla, and (C) variability among males in West African chimpanzees (*Pan*).

not obscure the overall features of the evolutionary record which are, in this case, fairly clear. As will be shown in the next chapter there is a well-known fossil species, *Homo erectus,* which extends in time from the Lower Pleistocene into much of the Middle Pleistocene. There is a well-documented fossil series leading from *Homo erectus* into *Homo sapiens.* Similarly the fossil record is now rich enough to show that *A. africanus* evolved into *Homo.*

This process is called anagenetic speciation, the evolution of subsequent species out of a preceding species. In such cases, the point at which the new species designation is made is quite arbitrary and a matter of convenience and common agreement, hopefully. So the only question on *Homo habilis* is whether or not there is room to usefully subdivide this evolving lineage into another species. Probably not. As will be documented in the next chapter, the South Asian fossils which have long been designated *Homo erectus,* appear now to go back almost as far in time as do the *habilis* fossils.

There is another consideration in the interpretation of fossils which will be dealt with more systematically in Chapter 13. It is simply this: within a species at any one time there will be variation. If, at a later time period, the descendants of this species are recognized as a different species, because of the variation still present, some specimens can be more "primitive" than some members of the earlier species. This can lead to great problems in assigning every fossil to an appropriate species on the basis of morphology alone. Such problems only arise as the fossil record becomes more complete. A possible case of this is represented by two finds from East Africa.

In 1973 Richard Leakey announced the find of an advanced hominid from early deposits of East Rudolf, Kenya. This was a skull cataloged as KNM-ER 1470. The "1470" skull was reported as having a much larger endocranial capacity than *Australopithecus* skulls. But 1470 may be older than many of the smaller *Australopithecus* skulls. It comes from a level below a volcanic deposit on which K-Ar age estimates range from 1.4 million years to 2.6 million years. A date of a little more than 1.8 million years seems most likely now. If the older date is correct 1470 is larger and earlier than the specimens that were designated *Homo habilis.* It was assigned to the genus *Homo.* Another recently reported skull, KNM-ER 1590, is similar.

Initial estimates on the endocranial capacity have been revised downward in more careful studies, to approximately 775 cc. This, though no longer startling, is still a larger value than other *Australopithecus.* But the known range of *Australopithecus* skulls is very limited. This cranial capacity could well be considered within a reasonable *Australopithecus* range as well as that of an early *Homo;* a recent paper by Walker (1976) shows a number of morphological features and proportions, not previously emphasized, to be like *Australopithecus.*

FIGURE 11.23 Pebble tools from Swartkrans, South Africa. (*Source:* M. D. Leakey, 1970. Used with permission.)

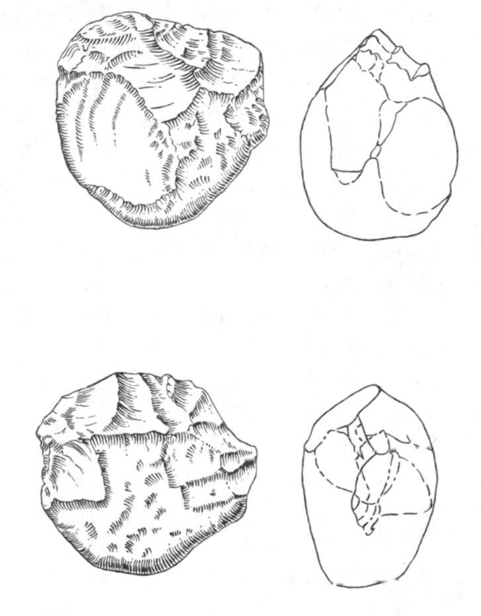

CULTURAL ACHIEVEMENTS OF AUSTRALOPITHECUS

Whatever the interpretation attached to the slender versus robust forms, it appears certain that the genus *Australopithecus* is hominid and ancestral to modern man. This interpretation, based on morphology, is strengthened by the "fossilized" evidence of behavior made available through archeology. The morphological evidence itself suggests that *Australopithecus* habitually employed tools in the food quest, in defense, and in aggression. As Bartholomew and Birdsell (1953) pointed out, bipedalism cannot be expected to be advantageous to a savanna-dwelling primate unless considerable advantage accrues from having the hands freed for functions other than locomotion. Also it is apparent that the daggerlike canine teeth of ground-dwelling primates had been replaced in function by something else, presumably tools.

The earliest tools used by hominids were probably made of perishable materials, wooden digging sticks and clubs. Direct evidence of such tools may never be recovered. But stone tools, once they are modified enough to be recognizable, leave a permanent record of man's activities. Simple chipped stone tools of this kind have been found dating back 2.0 to 2.6 million years in East Africa and at Sterkfontein and Swartkrans (Figure 11.23) in South Africa. These are often referred to as "pebble tools," as they usually represent river pebbles that have been fractured in such a way that a cutting edge is produced (Figure 11.24).

In addition to these stone tools, Raymond Dart has postulated the presence of an "osteodontokeratic" culture (Dart, 1957). By this Dart means that *Australopithecus* utilized "bones, teeth, and horns" as tools. He has suggested that much of the bone accumulation of the breccias is a result of the activities

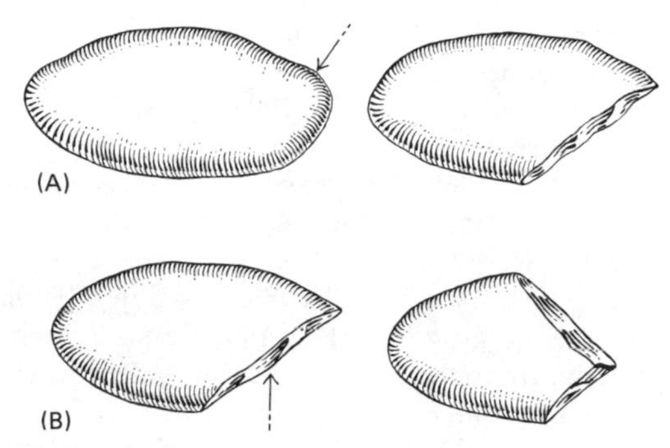

FIGURE 11.24 Production of crude cutting edge on a river pebble by (A) unifacial flaking or (B) bifacial flaking.

of *Australopithecus.* Although some of the specific uses Dart postulates for particular fragments of bone are far beyond what can be confirmed, he has convinced many experts that hominid manual dexterity was involved in some of the bone fracturing apparent within the cave deposits. Most convincing has been his argument that most of the baboon skulls show evidence of bludgeoning while the bone was fresh. A collection of antelope leg bones showing battered ends could have been the clubs which produced the double depressed fractures found on a number of the skulls (Dart, 1949).

The sites excavated by Mary Leakey and L. S. B. Leakey in Olduvai Gorge, Tanzania have been most important in providing evidence of the cultural capacities of *Australopithecus.* Some of the finds of Olduvai are occupation sites, the places where *Australopithecus* lived, which were later covered with mud by a seasonally fluctuating Pleistocene lake. One of the earliest sites in the gorge, DK I, contains a ring of stones, perhaps nine feet in diameter, that appears to have formed the base of a hut or a windbreak. At the "Zinj" site, considerably higher in Bed I, all of the bone except that of the hominids had been broken open to extract the marrow. This is a common practice among ethnologically known hunting groups. Olduwan pebble tools show a progressive development into the handaxes common to Middle Pleistocene times. There is also a biological change evident in succeeding strata at Olduvai. The change is sufficient to justify a different designation for the fossils—*Homo erectus.* Since this taxon was first and most extensively described on the basis of Asian fossils it is there that we now turn our attention.

REFERENCES CITED

Batholomew, G. A. and J. B. Birdsell 1953 Ecology and the protohominids. *American Anthropologist* 55:481–498.

Brain, C. K. 1958 The Transvaal Ape-Man Bearing Cave Deposits. *Transvaal Museum Memoir* no. 11. Pretoria: Transvaal Museum.

Broom, R. 1943 An Ankle-Bone of the Ape-Man, *Paranthropus robustus. Nature* 152:689–690.

Butzer, K. W. 1964 *Environment and Archeology.* Chicago: Aldine.

Clark, W. E. Le Gros 1947 Observations on the Anatomy of the Fossil Australopithecinae. *Journal of Anatomy* 81:300–333. 1964 *The Fossil Evidence for Human Evolution.* Chicago: University of Chicago Press. 1965 *History of the Primates,* 9th ed. London: Trustees of the British Museum (Natural History). 1967 *Man-Apes or Ape-Men? The Story of Discoveries in Africa.* New York: Holt, Rinehart & Winston.

Dart, R. A. 1925 *Australopithecus africanus:* The Man-Ape of South Africa. *Nature* 115:195–199. 1949 The Predatory Implemental Technique of *Australopithecus. American Journal of Physical Anthropology* 7:1–38. 1957 The Osteodontokeratic Culture of *Australopithecus promethesus. Transvaal Museum Memoir,* no. 10. Pretoria: Transvaal Museum.

Davis, P. R. 1964 Hominid Fossils from Bed I, Olduvai Gorge, Tanganyika: A Tibia and Fibula. *Nature* 201:967–968.

Day, M. H. and J. R. Napier 1964 Hominid Fossils from Bed I, Olduvai Gorge, Tanganyika: Fossil Foot Bones. *Nature* 201:969–970.

Ewer, R. F. 1957 Faunal evidence on the dating of the Australopithecinae. *Proceedings of the Third Pan-African Conference on Prehistory, Livingstone, Northern Rhodesia, 1955,* pp. 135–142. London: Chatto and Windus.

Goodall, J. 1965 Chimpanzees of the Gombi Stream Reserve. *In* Irvin DeVore (ed.), *Primate Behavior,* pp. 425–473. New York: Holt, Rinehart & Winston.

Holloway, R. L. 1970 New Endocranial Values for the Australopithecines. *Nature* 227:199–200.

Köhler, W. 1927 *The Mentality of Apes,* 2d ed. London: Routledge & Kegan Paul.

Leakey, L. S. B. 1968 Bone Smashing by Late Miocene Hominidae. *Nature* 218:528–530.

Leakey, M. D. 1970 Stone Artefacts from Swartkrans. *Nature* 225:1222–1225.

McHenry, H. M. 1974 How Large Were the Australopithecines? *American Journal of Physical Anthropology* 40:329–340.

Napier, J. R. 1964 The Evolution of Bipedal Walking in Hominids. *Archives de Biologie* 75:673–708.

Napier, J. R. and P. R. Davis 1959 The Fore-limb Skeleton and Associated Remains of *Proconsul africanus. Fossil Mammals of Africa* 16:1–78.

Oakley, K. P. 1951 A Definition of Man. *Science News* 20:69–81. Reprinted and revised. *In* M. F. A. Montagu (ed.), *Culture and the Evolution of Man,* pp. 3–12. New York: Oxford University Press.

Penck, A. and E. Brückner 1909 *Die Alpen im Eiszeitalter.* Leipzig: C. H. Tauchnitz.

Simons, E. L. 1960 New Fossil Primates: A Review of the Past Decade. *American Scientist* 48: 179–192.

Simons, E. L. 1961 The Phyletic Position of *Ramapithecus. Postilla* 57:1–9. 1962 Fossil Evidence Relating to the Early Evolution of Primate Behavior. *Annals of the New York Academy of Sciences* 102:282–295. 1963 A Critical Reappraisal of Tertiary Primates. *In* J. Buettner-Janusch (ed.), *Evolutionary and Genetic Biology of Primates,* vol. 1, pp. 65–129. New York: Academic Press. 1964 On the Mandible of *Ramapithecus. Proceedings of the National Academy of Sciences* 51:528–535.

Simons, E. L. and D. R. Pilbeam 1965 Preliminary Revision of the Dryopithecinae (Pongidae, Anthropoidea). *Folia Primatologia.* 3:81–152.

Tobias, P. V. 1963 Cranial Capacity of *Zinjanthropus* and Other Australopithecines. *Nature* 197: 743–746.

Walker, A. 1976 Remains Attributable to *Australopithecus* in the East Rudolf Succession. *In* Coppens, Y., F. C. Howell, G. Isaac, and Leakey, R. E. F. (eds.) *Earliest Man and Environments in the Lake Rudolf Basin,* pp. 484–489. Chicago: University of Chicago Press.

Washburn, S. L. 1960 Tools and Human Evolution. *Scientific American* 203:63–75.

12

Middle Pleistocene Ancestors

The next species in our succession has been named *Homo erectus.* The species name, *erectus,* is in recognition of the upright stance of this ancestor, a point that was debated for many years before the discovery of *Australopithecus africanus. Homo erectus* weighed about the same as the average individual in many human populations today. In all other ways *Home erectus* was intermediate between *Australopithecus africanus* and *Homo sapiens.*

HISTORY OF THE FINDS

The separate stories of the finds of *Homo erectus* in Java and in China both read like science fiction stories with improbable outcomes. A part of the story of the Java finds has been recounted by Ralph von Koenigswald (1956), who is himself a part of the story. In 1887 Eugene Dubois, a young Dutch physician, took a government position in Indonesia, then the Dutch East Indies. Dubois was influenced by the writings of Darwin and others to the extent that he was convinced that man had evolved in the tropics where the great apes live today. The East Indies was that part of the specified region controlled by the Dutch. This is where Dubois went to search for what was at that time an imaginary "missing link" that had been dubbed *"Pithecanthropus alalus,"* the speechless ape-man.

In 1891 near the village of Trinil on the Solo River of central Java, Dubois found what he

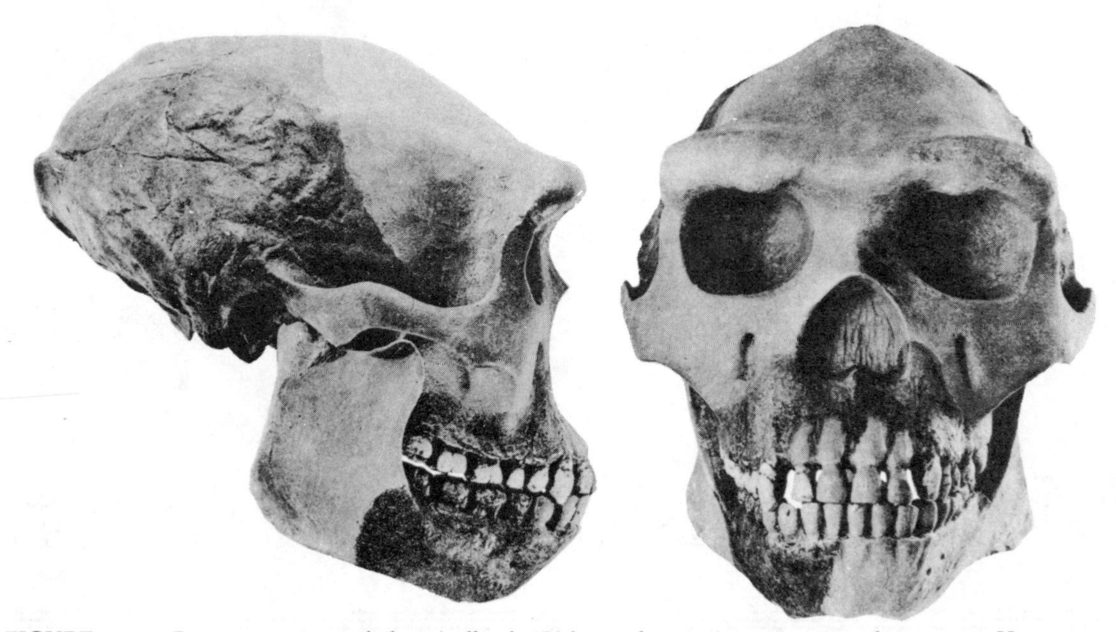

FIGURE 12.1 Reconstruction of the skull of *"Pithecanthropus" erectus* now known as *Homo erectus erectus.* (*Source:* Weidenreich, 1945.)

was looking for. The first find at Trinil was only a molar tooth, but then a skull cap, the calotte, was found. In the next year, 1892, he found a femur. He published a description of his finds in 1894 and named the fossil species *Pithecanthropus erectus,* as the femur was that of an individual who had upright posture. Most anthropologists today consider this and closely related forms to be early members of the genus *Homo.* Therefore the originally assigned genus name is dropped and Dubois' find becomes *Homo erectus* (Figure 12.1).

A controversy began immediately over the finds Dubois had announced. Some authorities assured themselves that this was a fossil ape. Others were convinced that the femur, difficult to distinguish from that of some present-day femurs, could not have been attached to the same kind of individual who had the low vaulted skull with its shelving brow ridges. Others agreed with Dubois. Dubois wavered. He, living then in Haarlem, became reclusive and was unwilling to show or discuss his finds. His original finds were housed in the museum in Leiden. Other finds were housed, reputedly, under the floor of his house. Dubois did not return to Java and nothing more was known of *"Pithecanthropus"* until 1936.

In 1936 G. H. R. von Koenigswald was conducting geological surveys near the village of Modjokerto in eastern Java. An assistant found a small, extremely thin skull that von Koenigswald identified as a hominid and judged to have been two years of age or less at death. He initially called this *"Homo modjokertensis"* (Figure 12.2). He believed it to be an infant of the same species discovered by Dubois, but Dubois was by now very sensitive about anyone's use of the term *"Pithecanthropus."*

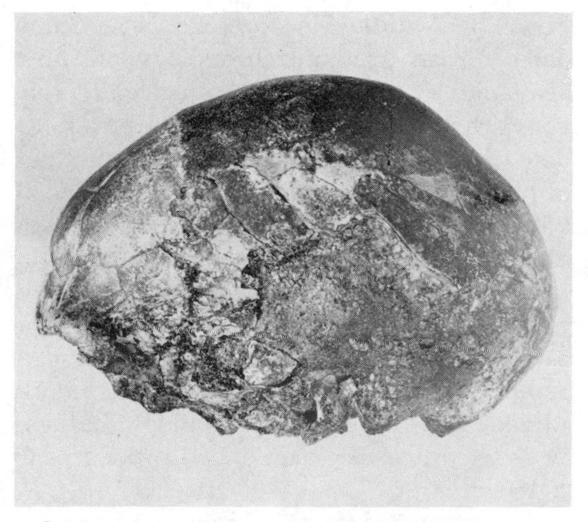

FIGURE 12.2 Infant skull of "*Homo modjokertensis.*" (*Source:* G. H. R. von Koenigswald.)

Soon after this, near the village of Sangiran, von Koenigswald and his assistants found part of a very heavy mandible. Next they found another calotte (*P* II) a little more complete than the one (*P* I) found originally near Trinil (Figure 12.3). Prior to the outbreak of World War II, two more skulls, a maxilla and another mandible, were found in the area.

When Japanese troops arrived in Java, von Koenigswald had substituted well-made casts for many of the original fossils that he had distributed among his friends for safe keeping. Some were even reburied. At the end of the hostilities von Koenigswald was released from a prisoner-of-war camp and surprisingly enough was able to recover all the original fossils. Since that time two more mandibles, three calvaria, and other fragments have been found (Jacob, 1972).

No such happy outcome can be reported for the Chinese *Homo erectus* finds. This very important group of fossils came from caves in a limestone hill near the village of Chou-k'ou-tien. In 1921 Swedish researchers recognized this area as an important fossil site and

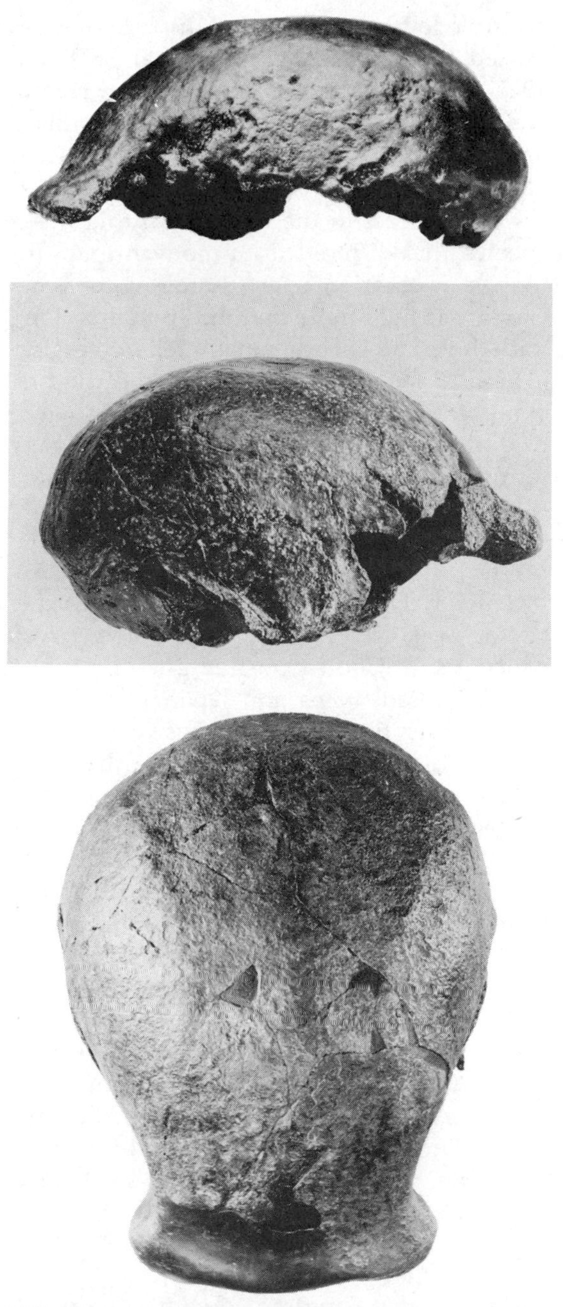

FIGURE 12.3 Skulls of Java man: (above) "*Pithecanthropus*" I, side view; (center and below) "*Pithecanthropus*" II, side views. (*Source:* G. H. R. von Koenigswald.)

possible locale of early man. In 1926 they reported the finding of two human teeth. In 1927 a third tooth was found. After careful study of this one specimen Davidson Black designated this a new species and called it "*Sinanthropus pekinensis.*" Black was proven correct, at least to the extent that this was a species that differed from modern man. Intensive excavation over the following years brought to light more hominid remains along with thousands of other mammalian remains.

In 1929 Dr. W. C. Pei discovered the first skull of "*Sinanthropus.*" Dr. Pei continued to find the remains of what was colloquially termed Peking Man, principally in a large filled-in fissure designated as Locality 1 and known locally as Ko-tzu-t'ang, the Chamber of the Pigeons. Black continued analyzing the material until his death in 1934. His successor was Dr. Franz Weidenreich. Work continued at Chou-k'ou-tien until 1941.

The United States and Japan were not at war at this time. The Japanese army controlled this region of China, but they permitted the excavations, which were supported by the United States but directed by the Geological Survey of China, to continue. As the international situation deteriorated it was decided that the fossils should be shipped to the United States. They were packed and entrained with a small United States Marine contingent on December 5, 1941. They were to reach the port on December 7. The American ship was captured on that day as were the marine guards on the train. The irreplaceable fossils of "*Sinanthropus pekinensis*" have not been seen since.

JAVA MAN

The finds at Sangiran come from two distinct levels. Strata of the upper level contain a faunal assemblage like the one found at Trinil. The lower level is referred to as the Djetis faunal beds. These fauna appear to be Lower to Middle Pleistocene in age. Some dates for an absolute chronology are now available. Volcanic material considerably above the Trinil fauna have indicated a K-Ar date of 500,000 years. Tektites, glass nodules of meteoritic origin, have indicated dates of 730,000 years. These are fairly closely associated with the upper strata of the Trinil beds, and this date will be taken as a fair approximation of the age of the later Java finds. The underlying Djetis beds are much earlier. A recent K-Ar date for the lower Djetis is just under 2 million years. This would make these beds as early as the base of Olduvai Gorge in East Africa.

All but one of the mandibles of "*Pithecanthropus*" come from the Djetis level. Dubois found the other, known as the mandible of Kedung Brubus, in 1890, but it was not mentioned until 1924. The mandibles are large and the bone of the body is very thick. In the teeth there are striking resemblances to the teeth of *Australopithecus* and "*Homo habilis*" (Robinson, 1953; Tobias and von Koenigswald, 1964). This can be seen in Figures 12.4 and 12.5. The upper incisors show some "shovelling," a marked concavity in the lingual surface that is present on incisors attributed to *Ramapithecus* (*Kenyapithecus*), on the incisors of most Neandertals, on Asian and Amerindian populations today, and occasionally in all populations. Based on a small sample (two mandibles, one maxilla) the third molar tends to be smaller in crown area than the second molar.

Unfortunately none of these mandibles includes the ascending ramus. The one maxilla from the Djetis beds indicates that the face was not as long, relative to the remainder of the skull, as it was in *Australopithecus.* The palate of this maxilla is very long (ca 72 mm) but shorter than that of "Zinj." Interestingly enough a diastema is present between the lateral incisor and the canine. This trait is usually considered as indicative of pongid

FIGURE 12.4 Comparison of (left) the right maxilla of Olduvai Hominid 13, attributed to *"Homo habilis"* from lower Bed II, and (right) the *"Pithecanthropus"* IV maxilla from Java. (*Source:* Tobias and von Keonigswald, 1969. Redrawn by permission from *Nature*.)

rather than hominid status. But it must be noted that such a diastema occurs occasionally in modern populations of man.

There is quite a bit of variation in size of the mandibles, much more so than in the skulls. Mandible D, found quite near Mandible B in the Djetis beds, is approximately 50 percent larger in size of the bone.

The bone of the skull vault is quite thick. A prominent brow ridge forms a shelf of bone above the orbits. There is a postorbital constriction that is less marked than in *Australopithecus* but more than in any of the later hominid fossils. Strongly marked nuchal lines and a heavy nuchal torus are present. Sagittal and nuchal crests are not present but there is some "keeling" of the skulls. The only skull from the earlier Djetis level is thicker than the later skulls and shows a median torus of bone but does not have the sharper configuration that would be considered a sagittal crest. Estimates of endocranial capacity of these skulls range from 750–900 cc. They are broadest near the base of the skull (on the temporal bone) instead of on the parietals as is the case in modern man.

The femurs are little different from strongly built femurs of modern man. There are no other postcranial remains and there are no tools found in association with Java Man. This is due to the fact that the finds are from fluviolacustrine (partially in flowing water

FIGURE 12.5 Comparison of (left) the mandible of Olduvai Hominid 13, attributed to "*Homo habilis*" from lower Bed II, and (right) the Sangiran B mandible from Java. (*Source:* Tobias and von Koenigswald, 1964. Redrawn by permission from *Nature*.)

and partially in lake water) deposits formed by streams into which Java Man fell or had his remains washed into after death. Indeed, one of the recently discovered mandibles shows the marks of having been bitten into by a crocodile while still fresh. The Java representatives of *Homo erectus* have yet to be found on their living sites. On the other hand, "*Sinanthropus*," the Chinese *Homo erectus*, was found right at home, or perhaps in the home of a hungry neighbor.

PEKING MAN

Although now classified by most experts as being of the same species as Java Man, the Peking finds from Locality 1, Chou-k'ou-tien, are considerably later in time and correspondingly more advanced than the Java specimens. The best current estimates are that the occupants of this cave lived approximately 400,000 years ago. This date is based on mammalian remains and on fossil pollens that indicate that human occupation began

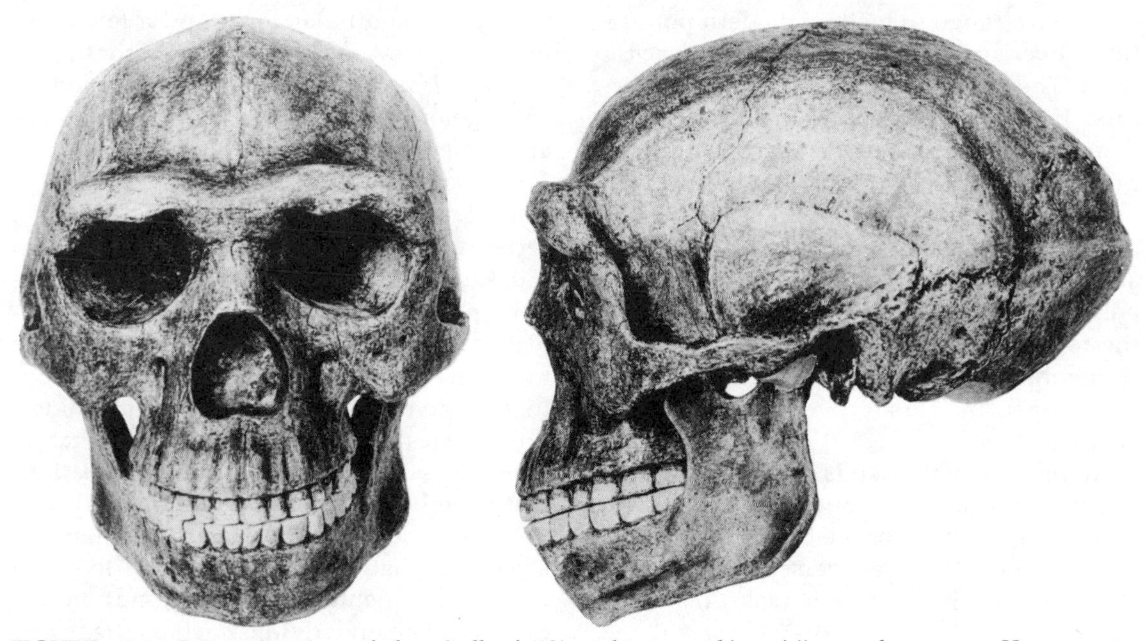

FIGURE 12.6 Reconstruction of the skull of "*Sinanthropus pekinensis*" now known as *Homo erectus pekinenesis.* (*Source:* Weidenreich, 1945.)

in Mindel I times and extended into Mindel II. This means there is as great a time span between Java Man and Peking Man as between Peking Man and modern man.

Although the original fossils have been lost, there are excellent casts and published descriptions of the Chou-k'ou-tien skulls. We will now refer to these skulls as *Homo erectus pekinensis* in contrast to *Homo erectus erectus* of Java. It has been estimated that approximately 40 individuals were represented in the finds of 14 skulls, 147 teeth, and a few postcranial fragments that were found and described in 1941.

The skull vault of *pekinensis* is slightly higher than that of the Java *erectus.* The brow ridges tend to curve, conforming to each orbit rather than forming a straight shelf of bone. Seen from the side *pekinensis* appears to have even more pronounced brow ridges, but this is due to the fact that a forehead rises at a steeper angle in *pekinensis* than in the *erectus* subspecies (Figure 12.6). A keeled frontal region is present as it was in the earlier forms. The endocranial capacity of the skulls ranges from 915 cc (for a juvenile) to 1225 cc, overlapping the range in modern man. The mean endocranial capacity has been estimated as 1075 cc.

The jaw is less massive and the teeth smaller than in the Java specimens. Like the Java *erectus,* the ascending ramus is only moderately long and correspondingly the face is not particularly long as it was in the earlier African *Australopithecus.* No projecting chin is present. All the jaws have a mandibular torus. This is a pronounced thickening of dense bone along the upper, inside edge of the mandible in the region of the canine and premolars. All the jaws have multiple mental foramina as do some *Homo erectus* jaws from North Africa.

The size range in teeth of modern populations overlaps the size range of *pekinensis* teeth. The size of the molar crowns as measured by Weidenreich's "robustness" index (length × breadth) is like modern man in the progression $M^1 > M^2 > M^3$, but in the lower molars M_2 is the largest, unlike the average condition today. The molars and premolars are "taurodont," meaning that they have an enlarged pulp cavity extending down into the roots of the teeth. This condition is also common in Neandertal Man of later times. The upper incisors are commonly shovel shaped.

All but one of the skulls having the occiput present show an Inca bone. This is a triangular bone at the apex of the occiput formed by the presence of a transverse cranial suture (Figure 12.7). Weidenreich took this, along with the mandibular torus and a few other features, as evidence of a genetic continuity between *Homo erectus pekinensis* and modern "Mongoloid" populations including Asian and Amerindian groups.

FIGURE 12.7 Skull, seen from the rear, of an early California Amerindian showing the presence of an Inca bone.

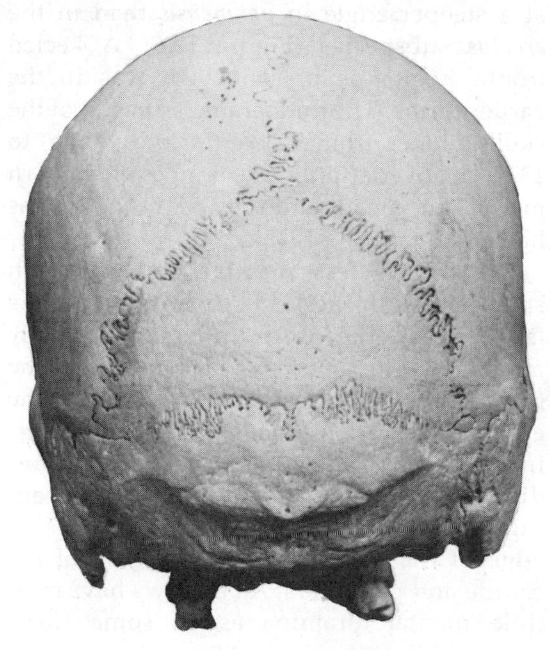

Undoubtedly *Homo erectus* was ancestral to *Homo sapiens,* and it is quite believable that certain traits that were more common in Asia 400,000 years ago continue to be more common in Asian populations than in other populations. This has led others, however, to the unjustified conclusion that the Chou-k'ou-tien residents were "primitive Mongoloids." This carries with it the implication that all other traits typical of those living populations that have been classified as Mongoloid were present in *Homo erectus pekinensis.* This is an aspect of what was earlier referred to as the "archetypal mentality": the assumption that there is some underlying, unchanging reality or "essence" to classificatory categories.

We know that traits influenced by genes subject to crossing over and to genetic reassortment in meiosis can be acted upon independently by the forces of evolution. Therefore the continued high frequency of an observable trait says nothing about the constancy of trait frequencies not observable on the skeleton. As a corollary to this, it follows that such evidences of genetic continuity within a given region of the world do not imply that there has been any insuperable barrier to gene flow between different human populations at any given time in the Pleistocene. A legitimate inference is that prior to the Holocene, gene flow between major regions of the Old World was not great enough to swamp differences that may be maintained by only small regionally adaptive selective factors.

According to Weidenreich, the variability in the teeth of *pekinensis* was greater than it is in present-day populations of man. He as-

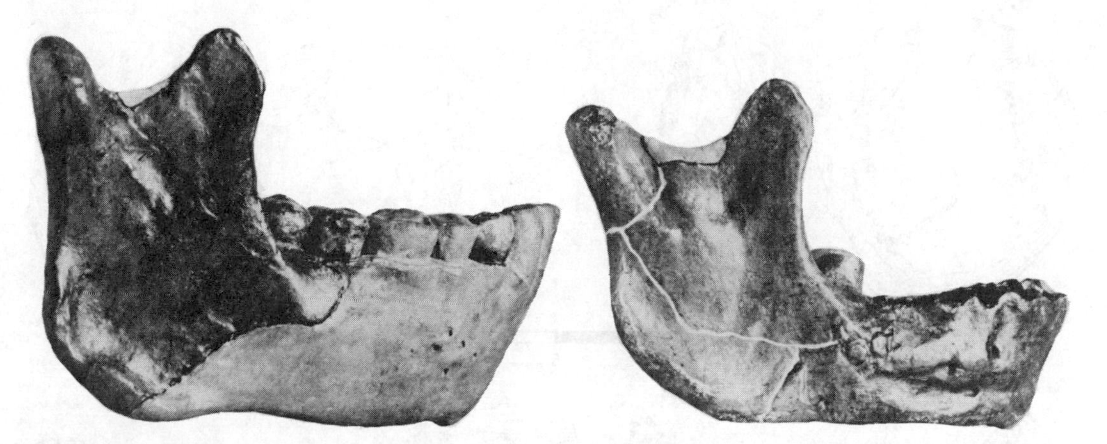

FIGURE 12.8 Restored mandibles of "*Sinathropus*" which, according to Weidenreich are (left) male, and (right) female. (*Source:* Weidenreich.)

cribed this to greater sexual dimorphism in *Homo erectus* than in modern man. In fact Weidenreich felt that he could divide teeth, jaws, and crania into two distinct groups and that these groupings represented a sex difference. Two mandibles, ascribed to male and female, are shown in Figure 12.8

Locality 1, the Chapter of the Pigeons, was an occupation site. Occupational debris and other material had accumulated to a depth of approximately 140 feet in the cave. Throughout this deposit—and beginning in strata slightly below where the earliest human remains are found—are stone tools. These are predominantly flake tools, although core tools also occur. The distinction between flake and core tools is fairly simple. If the toolmaker fractures a spall from a stone nodule, then utilizes that flake or spall, it is a flake tool. If he is left with a sharp edge on the nodule he may utilize it as a core tool. All stone toolmakers have used both flake and core tools with varying frequencies, different techniques, and varying degrees of sophistication.

The stone tools at Locality 1 represent a small but clear advance over the tools of *Australopithecus.* The majority were flakes that were knocked off a core with a hammerstone, and there was apparently little regard for the shape of the resulting core. Flakes were used as they were or were retouched by finer flaking along the edge. The flake tools do not fall into neat categories that would indicate functionally differentiated tools. They were generalized cutting and scraping tools (Figure 12.9).

The core tools have been called choppers when flaked from one side, and chopping tools when bifacially flaked. The distinction between chopper and chopping tools has no functional significance. Whether the tool was unifacially or bifacially worked probably depended on the thickness of the blank.

A most important cultural achievement of *Homo erectus* first appears at Chou-k'ou-tien. This is the use of fire. Burned areas and lens-shaped ash deposits occur at different levels in the fill indicating that for the first time man had some control over fire. A recently discovered site in Hungary, Vértesszöllös, has been found to contain fire hearths and is

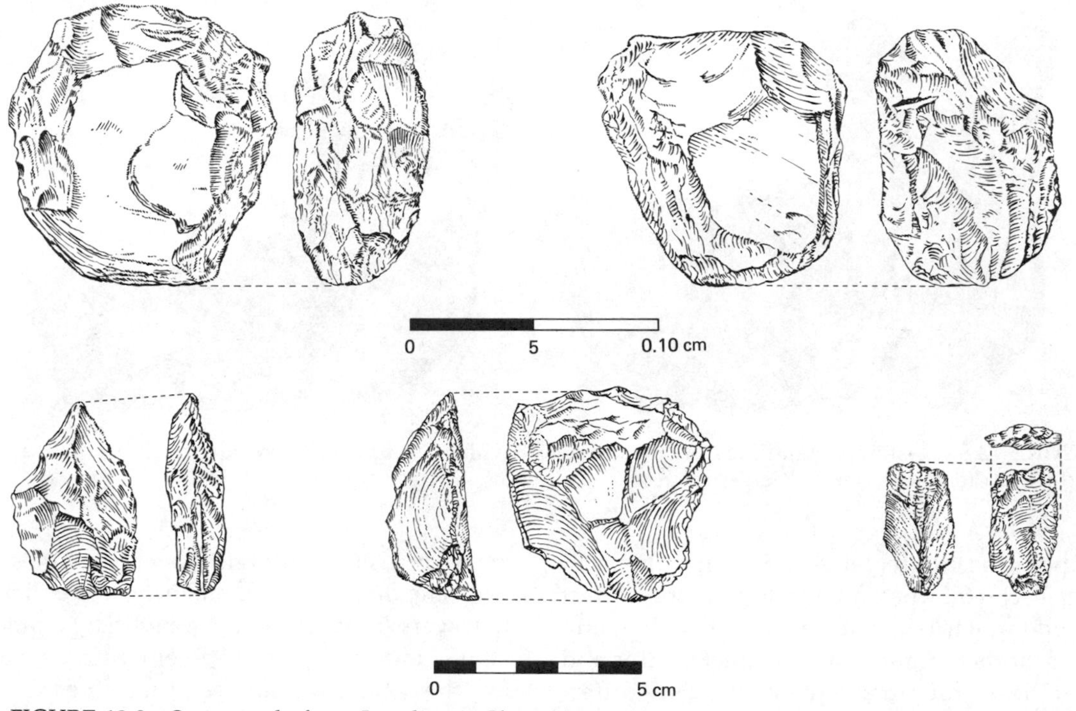

FIGURE 12.9 Stone tools from Locality 1, Chou-k'ou-tien. (*Source:* Chang, 1968.)

about the same age as the deposits at Locality 1, Chou-k'ou-tien.

Many bone fragments in the cave are charred and suggest that *pekinensis* was cooking meat at least part of the time. He was an accomplished hunter although we cannot be sure that all the mammalian bone in the cave, ranging from elephant to horse, was a result of his hunting activities. Certainly he preyed on deer and 70 percent of the animal remains are deer. The hominid skulls are probably of those who fell victim to the cave dwellers. To quote Hooten:

> For the most part, only the skulls of Sinanthropus seem to have been brought into the caves at Choukoutien and, with the exception of the few fragmentary postcranial parts mentioned above, there are simply no long bones, vertebrae, etc. in the deposits. It appears that these skulls were trophies of head hunters, and, furthermore, that said hunters usually bashed in the bases of the skulls when fresh, presumably to eat the brains therein contained. Many crania also show that their owners met their deaths as a result of skull fractures induced by heavy blows. (Hooten, 1946, p. 304.)

In 1950 Chinese investigators found another mandible at Locality 1. In 1963 a mandible, and in 1964 a skull, were found at different sites near Lan-t'ien in east-central Shensi province. The Lan-t'ien skull (Figure 12.10) was somewhat more primitive than Chou-k'ou-tien finds, having an endocranial capacity estimated at only 780 cc and a thicker vault. In all respects the Lan-t'ien fossils appear to be like the Javanese *Homo*

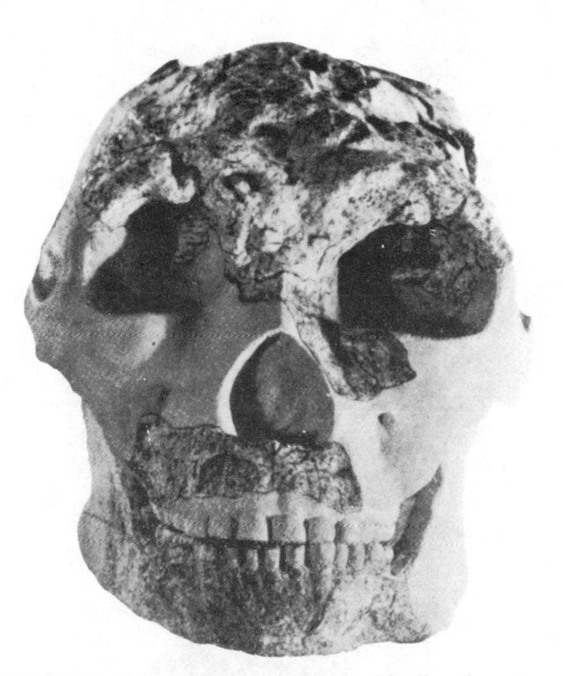

FIGURE 12.10 The Lan-t'ien skull showing close resemblances to *Homo erectus erectus*. (*Source:* Woo, 1966.)

erectus. The associated fauna indicates a warm climate and probably a much earlier time than Chou-k'ou-tien.

GIGANTOPITHECUS

Before discussing the range of *Homo erectus* outside of Asia we should for historical purposes consider an interesting primate that lived in South Asia contemporaneously with *Homo erectus*. This is *Gigantopithecus blacki*. The species was established on the basis of giant hominid-appearing molars that G. H. R. von Koenigswald acquired in a Chinese drugstore. Until recently, and even to an extent at present, fossils have been sold as "dragon bones" and "dragon teeth" in Chinese drugstores. Professor von Koenigswald, not one to pass up such a possibility, usually checked the local Chinese drugstores. These particular molars had a dryopithecine cusp pattern and wear facets like hominid molars. Professor Weidenreich, knowing of the decrease in tooth size from *Homo erectus* to *Homo sapiens*, concluded that this trend must have existed for some time in the past and that *Gigantopithecus blacki* must represent the giant ancestor preceding *Homo erectus*. More finds and better dating have proven this untrue. Three mandibles of *Gigantopithecus* have now been found in South China and one in North India. *Gigantopithecus* was a large and unusual ape. He was perhaps a cave dweller. Although there was some interlocking of the canines in the males, they maintained a rotary chewing motion as in hominids (Figure 12.11). The new fossils found in China were associated with a Middle Pleistocene fauna and were thus contemporaneous with *Homo*. The Indian find is from Pliocene deposits.

HOMO ERECTUS IN AFRICA AND EUROPE

From North Africa four mandibles and possibly a fifth are referrable to *Homo erectus*. Camile Arambourg has described three mandibles from a site near Ternifine, Algeria (Figure 12.12). These came from what had been the bottom of a Pleistocene artesian lake. There are multiple mental foramina in I and II as in *Homo erectus pekinensis*. In all specimens the mandible slopes backward in the symphyseal region. Interestingly enough the smallest molars are in the largest mandible. A parietal bone, associated with the mandibles, shows on its internal surface the impress of blood vessel patterns more primitive than in *Homo sapiens* and much like *Homo erectus pekinensis*. Faunally the site is early Middle Pleistocene. Stone tools are present, including scrapers and crude handaxes. The tools are of a type identified in

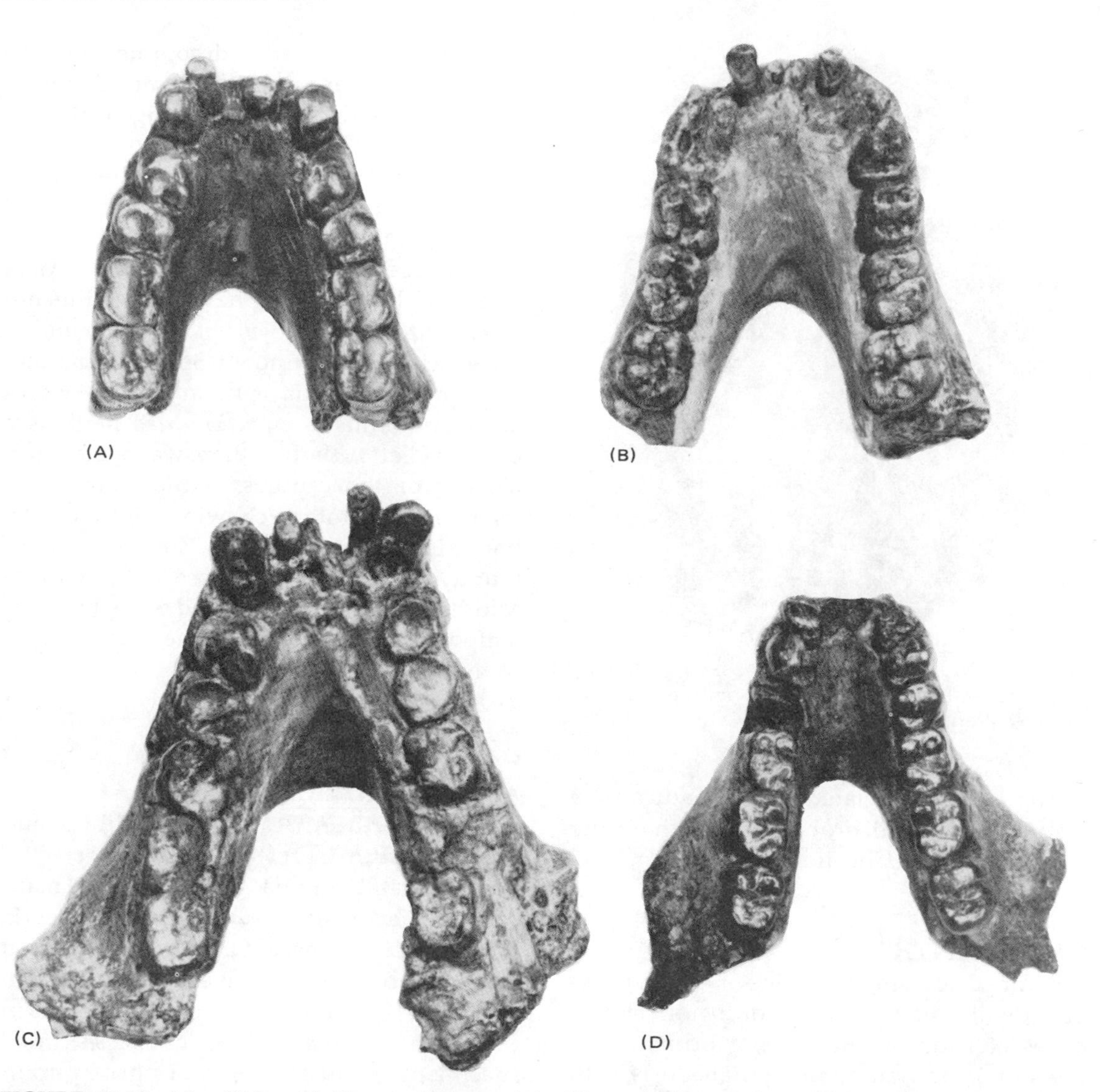

FIGURE 12.11 Mandibles of *Gigantopithecus:* (A), (B), and (C) are from China, (D) is from India. (*Source:* Drawn from photographs by Boltin; Simons and Ettel, 1970.)

Europe and Africa as *Acheulian* after a site in France.

From "Littorina Cave" at Sidi Abderrahman near Casablanca, Morocco, come two mandibular fragments that are also *Homo erectus*. They are associated with a Middle Acheulian industry. Based on marine terrace changes they are late Middle Pleistocene. Another mandibular fragment comes from Rabat, Morocco and is roughly contemporaneous with the Casablanca mandibles. Some have suggested that this mandible is

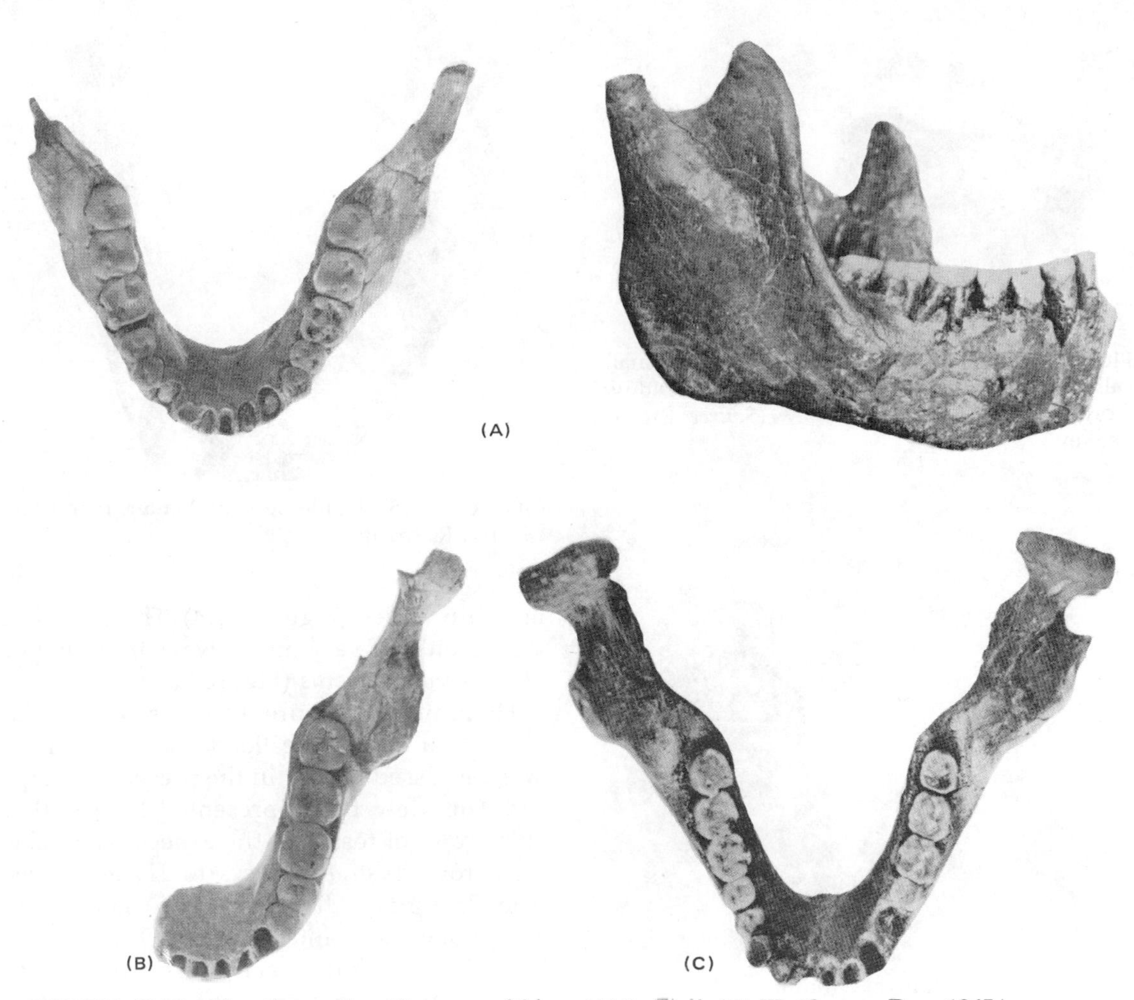

FIGURE 12.12 The three Ternifine mandibles: (A) I, (B) II, (C) III. (*Source*: Day, 1965.)

Homo erectus. Others have suggested that it is an early form of *Homo sapiens* known as Neandertal. Assignment to one or the other is arbitrary, since the fossil seems truly transitional.

In Tanzania, East Africa, there are two skulls from Bed II Olduvai Gorge that are clearly *Homo erectus*. It may be that all the Bed II fossils from Olduvai are referrable to *Homo erectus*. As mentioned earlier, the Bed I fossils are not distinguishable from *Australopithecus*. Bed II by no means represents a single depositional cycle. Bed II is 50–90 feet in thickness. A K-Ar measurement from near the base of Bed II gave a 1.1 million year date. A faunal break occurs within Bed II. The upper part of the bed is not well dated. An earlier date of 490,000 years has been called into question. A skull from 15–20 feet below the top of Bed II, originally called the "Chellean skull" by Leakey (or Hominid 9), has not been well described as yet, but pho-

FIGURE 12.13 Olduvai hominid 9 originally called "Chellean man" from upper Bed II, Olduvai Gorge, Tanzania, left lateral view. (*Source:* Richard Leakey.)

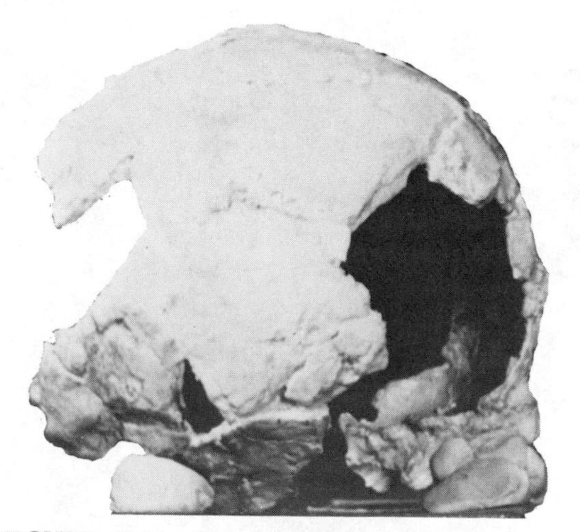

FIGURE 12.14 Olduvai Hominid 13 skull, occipital view. (*Source:* Richard Leakey.)

tographs (Figure 12.13) show it to be very similar to the Chinese *Homo erectus.*

From the lower part of Bed II comes "George" (Hominid 16), a skull that is small and that, also, has not been definitively described but which appears to be a good predecessor of the later Hominid 9. Another specimen, Hominid 13, consists of parts of the cranium, a maxillary fragment, and a mandible. The skull is small (appearing to be within the *Australopithecus* range) but well rounded

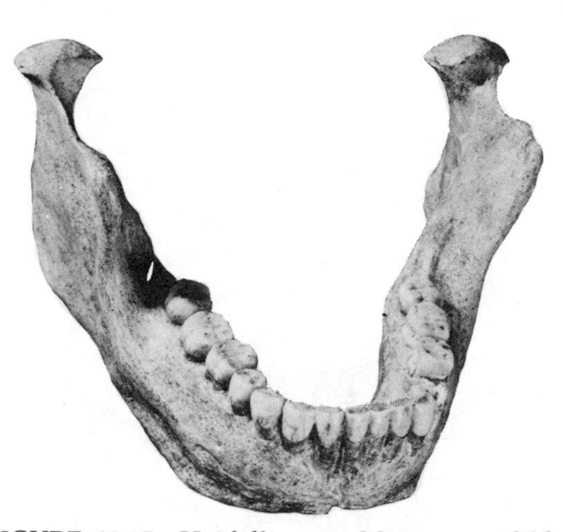

FIGURE 12.15 Heidelberg or Mauer, mandible. (*Source:* R. Kraatz.)

and thin walled (Figure 12.14). The dentition differs only in very minor ways from that of *Homo erectus erectus* (Figure 12.5).

Hominid 13 (Figure 12.14) is one of the skulls that has been called *habilis.* This taxon was discussed briefly in the preceding chapter. The view being presented here is that this group of fossils is the expectable transition from *Australopithecus* to *Homo erectus.* The data fit such a transition temporally, morphologically, and culturally. Others, however, find it useful to segregate a *habilis* group within the evolving continuum at this time.

From Europe comes one of the early and most famous finds, the Heidelberg jaw, also called the Mauer mandible (Figure 12.15), which is a European *Homo erectus.* It has multiple mental foramina. It lacks a chin (*mentum osseum*) but is extremely thick in the symphyseal region. The ascending ramus is low and exceptionally broad. As in other *Homo erectus* specimens the roots of the anterior teeth are larger, relative to crown size, than in *Homo sapiens.* Tooth size in the Mauer jaw falls well within the range of *Homo erectus pekinensis* and is very similar to Ternifine III.

The Mauer mandible was found in 1907 in a sand and gravel quarry near the village of Mauer, Germany. There is some question as to whether the jaw was seen in place in the exposed face of the pit, but it reportedly came from a point approximately 65 feet below the upper edge of the pit. The "Mauer Sands" of this stratum contain a warm-climate fauna of identifiable age. It appears that Mauer is comparable in time to Peking man and slightly earlier than a Hungarian find that shares with Mauer the honor of being the earliest known man of Europe.

At Vértesszöllös in Hungary recent excavations turned up tooth fragments of an infant and the occipital bone of an adult who lived in Elster (Mindel) times. The occipital bone is thick and has an occipital torus at the most posterior point of the skull as in *Homo erectus*. The skull was very broad at the base. It is slightly larger and more rounded than the largest of the Peking skulls. Like Peking, and unlike Mauer, Vértesszöllös was an occupation site. There is evidence of fire and numerous small tools of a pebble "chopper/chopping tool" industry are present (Kretzoi and Vértes, 1965).

The tendency toward a more rounded contour of the rear of the skull, presaged in Vérteszöllös, is found in later European fossils of the Middle Pleistocene. These are specifically Swanscombe, Steinheim, and Fontechevade (of later date). These are later in time than the specimens we have been discussing, and it might be expected that they would have somewhat more filled-out cranial skeletons. Much significance has been attributed to this trend because in later Europeans, the Neandertals, the skull base is long with consequent extension of the occipit into a "bun-shaped" occipital form. But in Neandertal, unlike early *Homo erectus*, the inion, where the superior nuchal lines meet in the rear of the skull, is well below opisthocranion, the most posterior point on the skull.

POSSIBLE EARLY HOMO SAPIENS

The Swanscombe skull consists of an occiput and both parietal bones (Figure 12.16). These three separate bones were found at different times in a gravel pit to the south of the Thames River, Swanscombe, England. This is one of the best dated of the early European fossils. The gravels in which the skull was found date from the Hoxnian warm period. This would be roughly 250,000 years ago. These gravels also contain abundant flake tools and hand axes classified as Middle Acheulian (Acheulian III). Bits of charcoal and fire-scarred flint have also been found in the same level.

The bone of the Swanscombe skull is thinner than that of Peking Man but thicker than is modern man's. It is the first skull in which the inion is below opisthocranion. This is in accord with the generally well-rounded contours of the rear of the vault. The pattern of the midmeningeal vessels, seen as impressions on the inner surface of the parietals, is more complicated than in the Peking or Ternifine remains. Maximum breadth is not as low on the skull as in Peking Man, and the sides of the skull vault are nearer the vertical condition found in modern man than can be seen in any earlier skulls.

Much controversy has raged in the past over different views of Swanscombe. Some have depicted this as almost modern man. Others have seen Swanscombe as a good ancestor of Neandertal. Perhaps the best way to resolve this is to look at Steinheim—which is much like Swanscombe in the rear of the vault but which also has much of the facial skeleton present. Unfortunately, Steinheim,

FIGURE 12.16 Swanscombe skull: (A) lateral, (B) occipital, and (C) top views. (*Source:* Trustees of the British Museum, Natural History.)

like too many important hominid fossils, has not received definitive treatment in print. The Steinheim skull also comes from a gravel pit. This one is at the village of Steinheim, 12 miles north of Stuttgart in Germany. The deposit in which the skull occurred is of the same interglacial age as Swanscombe, although perhaps slightly later in the interglacial.

The skull is that of an adult, probably female and small (Figure 12.17). Endocranial capacity has been estimated as 1150–1175 cc (Howell, 1960) as compared to approximately 1300 cc for Swanscombe. The bone is thinner than in Swanscombe and well within the range of *Homo sapiens.* The skull is long and narrow but has been quite distorted by the crushing of the left side. The crushing was due to earth pressure on the post mortem skull but may have been facilitated by a wound to the left side of the individual at death. Maximum breadth is at the same point as it is in Swanscombe and the sides of the vault appear to have been quite vertical. Like Swanscombe the occiput is rounded and the nuchal ridges do not form a prominent toric structure (an occipital torus) as in Peking Man or even as in Vértesszöllös Man.

The facial skeleton is not that of modern man. It is not as much under the frontal region of the brain as in modern man. A supraorbital torus is formed by frontal air sinuses that are larger than in Peking Man. This bony torus arches sharply over the orbits that are set wide apart. A broad nasal opening and little or no canine fossa are present. There is a marked postorbital constriction of the frontal bone. In facial features Steinheim could indeed be a good ancestor for the later Neandertals.

Another early European to be considered here is Fontéchevade, which is probably not Middle Pleistocene but Upper Pleistocene. Faunal associations suggest a Riss-Würm (Eemian) interglacial age for Fontéchevade, but it is considered at this point because of its similarity to Swanscombe and Steinheim. Fontéchevade I is a fragment of the frontal bone of a young individual. Fontéchevade II includes most of both parietals and the squamosal (or upper) portion of the frontal bone.

FIGURE 12.17 Steinheim skull, right lateral view (from a cast.)

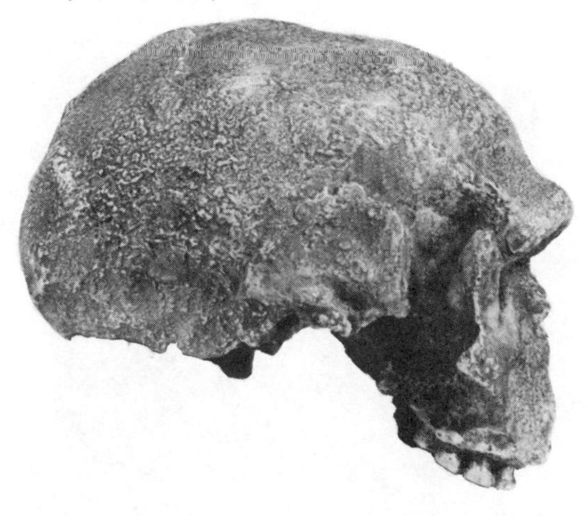

Again there has been extensive argument over what was "up front" in this fossil. The bone thickness is comparable to that of Swanscombe. The point of maximum skull breadth is a bit more posterior than it is in Swanscombe and this involves a roughly pentagonal shape of the skull when viewed from above. This is a feature that we will see again in all Neandertal skulls of comparable and larger age. Some have chosen to argue from Fontéchevade I that heavy brow ridges were not present in the adult. Since Fontéchevade I was a young individual, this cannot be used to substantiate a modern form to the face. Again the best guess on the face would be something like Steinheim. And a face like Steinheim's on a moderately low vaulted pentagonal skull with some lambdoidal flattening, all of which Fontéchevade has, would be indistinguishable from other early Neandertals that are first recognized at this time.

Recent finds at Arago Cave near the village of Tautavel at the edge of the eastern Pyrenees by Henry and Marie-Antoinette de Lumley provide more human remains from the same time interval as Swanscombe and Steinheim. The facial skeleton shows strong resemblances to Steinheim. The posterior part of the skull is missing. The two mandibles show strong resemblances to the Mauer mandible, but the large one, Arago XIII, has much larger teeth than Mauer. These finds are consistent with a developing picture of early Europeans which are morphologically, temporally, and spatially good candidates as ancestors to European Neandertals.

REFERENCES CITED

Chang, K. 1968 *The Archaeology of Ancient China,* rev. ed. New Haven, Conn.: Yale University Press.

Day, M. H. 1965 *Guide to Fossil Man.* Cleveland: World.

Dubois, E. 1894 Über die Abhängigkeit des Hirngewichtes von der Körpergrösse. *Archiv für Anthropologie* 25:1–28, 423–441.

Hooten, E. A. 1946 *Up from the Ape,* rev. ed. New York: Macmillan.

Howell, F. C. 1960 European and Northwest African Middle Pleistocene Hominids. *Current Anthropology* 1:195–232.

Jacob, T. 1972 New Hominid Finds in Indonesia and their Affinities. *Proceedings, 28th International Congress of Orientalists, Canberra: 1971.*

Koenigswald, G. H. R. 1956 *Meeting Prehistoric Man.* Translated by Michael Bullock. London: Thames and Hudson.

Kretzoi, M. and L. Vértes 1965 Upper Biharian (Intermindel) Pebble-Industry Occupation Site in Western Hungary. *Current Anthropology* 6:74–87.

Robinson, J. T. 1953 *Meganthropus,* Australopithecines and Hominids. *American Journal of Physical Anthropology* 11:1–38.

Simons, E. L. and P. C. Ettel 1970 Gigantopithecus. *Scientific American* 222:76–85.

Tobias, P. V. and G. H. R. von Koenigswald 1964 A Comparison Between the Olduvai Hominines and Those of Java and Some Implications for Hominid Phylogeny. *Nature* 204:515–518.

Weidenreich, F. 1936 The Mandibles of *Sinanthropus pekinensis:* A Comparative Study. *Palaeontologia Sinica,* series D, vol. 7, fascicle 3, pp. 1–162. Peking: Geological Survey of China. 1937 The dentition of *Sinanthropus pekinensis.* A comparative odontography of the hominids. *Palaeontologia Sinica* n.s. D, no. 10, whole series no. 101. Peking: Geological Survey of China. 1943 The skull of *Sinanthropus pekinensis.* A comparative study on a primitive hominid skull. *Palaeontologia Sinica* n.s. D, no. 10, whole series no. 127. Peking: Geological Survey of China. 1945 *Giant Early Man from Java and South China,* vol. 40, part 1. New York: Anthropological Papers of the American Museum of Natural History.

Woo, J. 1966 The Hominid Skull of Lantian, Shensi. *Vertebrata Palasiatica* 10:14–22.

13

The Emergence of Homo Sapiens

In the preceding chapter we pursued the course of human evolutionary history through the Middle Pleistocene and into Upper Pleistocene times. The emergence of *Homo sapiens* is recognized as occurring at some point during this span of time. A numerous group of fossils dating from the Elster (Mindel) cold period can be identified as *Homo erectus,* as shown in the previous chapter. An even more numerous group of fossils, Neandertal, is now accepted by a number of authorities as early *Homo sapiens.* These date predominantly from the Würm cold period. A few come from the Riss-Würm interglacial. Between these two larger samples of fossil man there is a span of approximately 150,000 years in which known fossils are sparse. This then, in practice, is where anthropologists have preferred to divide *Homo erectus* from *Homo sapiens.* This makes the picture look very neat. But the differing interpretations of specific fossils are not so simple. Much of this chapter will be devoted to these problems of interpretation.

The interpretive problems at issue can be classified under the headings of *anagenetic* and *cladogenetic* evolution. Anagenesis refers to change that is progressive through time but that does not split the gene pool of the species into unbridgeably separate units. When one species evolves from another anagenetically, the species involved are ancestral and descendant species, chronospecies. Cladogenesis refers to branching evolution

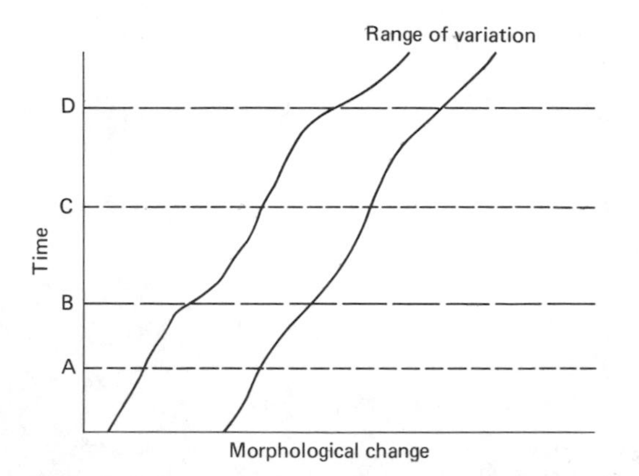

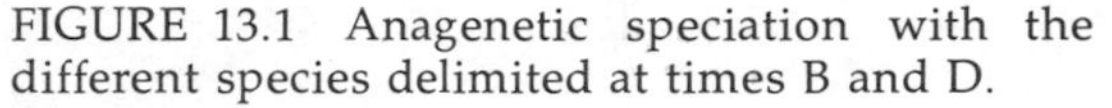

FIGURE 13.1 Anagenetic speciation with the different species delimited at times B and D.

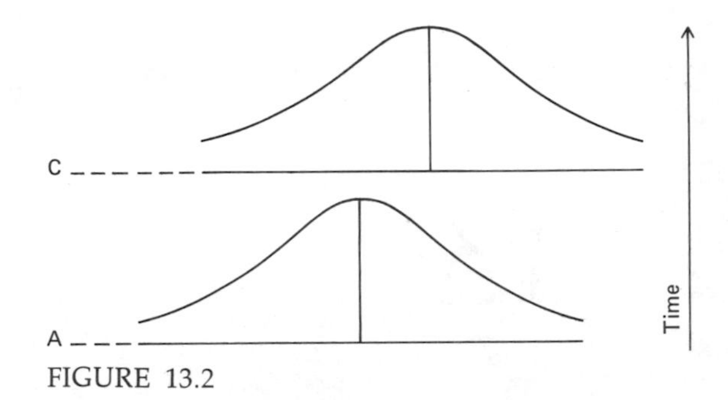

FIGURE 13.2

and occurs when one ancestral species gives rise to two or more descendant biospecies.

With respect to fossil man one set of problems obviously concerns the anagenesis of *Homo sapiens:* Where is the most appropriate place to draw the line between an ancestral *erectus* population and a descendant *sapiens* population? A different set of problems concerns the presence or absence of cladogenetic speciation in Middle or Upper Pleistocene times.

Perhaps it is best to start with hypotheses that can be rejected easily. One such hypothesis is the "cerebral rubicon." Another is an idea referred to as the parallel phyla hypothesis. A "cerebral rubicon," or threshold, of 1250 cc of endocranial capacity—give or take a bit due to differences in gross body size of individuals—has been proposed in considering fossils. If the endocranial capacity is less than 1250 cc, the fossil is considered *erectus,* if it is more, the fossil is considered *sapiens.* Unfortunately, in anagenetic evolution no one trait or no combination of several traits can be used to *identify individuals* as members of one species or the other even when these traits are used to *characterize* the species. This can be best explained by an illustration. Suppose we plot morphological change against time as is done in Figure 13.1. The morphological change plotted can be thought of as a single measure such as vault height of the skull or it may be a number representing change in several variables.

In Figure 13.1 a sample of individuals at time A would come from an earlier species, those from time C from a later species. If we represent the morphological variability within each sample in Figure 13.2, then it is obvious that some individuals in the ancestral species (at time A) are more "modern" than some individuals in the descendant species (at time C). Therefore such a typological approach to identifying individuals will fail more and more as the fossil record becomes more complete.

The solution to the above problem is simply to recognize that if a single biospecies changes over a long period of time to the point that it is useful to designate a new, anagetically produced species, then the real problem for the paleoanthropologist or paleontologist is to decide if a cladogenic split also occurred. If at any one time only one biospecies is represented, then the two chronospecies to be designated are delimited by drawing a line at one point in time. All individuals on one side of the line are members of one chronospecies. All individuals on the other side of the line are members of the other chronospecies regardless of variation among individuals. Therefore the invocation of any threshold or "rubicon," cerebral or otherwise, to assign an individual to either of temporally adjacent chronospecies is a misunderstanding of the process of evolution.

Since the conclusion we have just reached is one that has not been well recognized or observed in published statements on the origin of *Homo sapiens* it should be repeated. *If we are concerned with anagenetic speciation, the only relevant criterion for identifying individuals as members of one or the other species is chronology, not morphology.*

In branching speciation—cladogenesis—the initial point in time of speciation must be agreed upon rather arbitrarily with respect to the definition of the biospecies. That this is true in fossil man is not surprising, since it is also true among living species. In other words, the definition of the biospecies provides a principle that should regulate taxonomic practice, but it does not provide a prescription for a precise "cut-point" between species. In practice, therefore, the existence of cladogenetic speciation and its time of occurrence are inferred by interpolating back in time from a later period in which the continuing existence of two species is unequivocal.

NEANTHROPIC AND PALEANTHROPIC MAN

During the first third of this century the prevailing view of the human fossil record was that two separate lines of evolutionary development existed. These two lines had a common ancestor only at a very distant time in the past for which no fossil evidence was available. The neanthropic (new man) line included fossils that were more or less modern throughout time. The paleanthropic line included fossils that were primitive and crude throughout time. It should be noted that analogous "parallel phyla" hypotheses existed in archeology during this same period. Separate tool traditions, core tools versus flake tools, for example, were considered to be somehow mutually exclusive although existing side by side among different peoples. These separate tool traditions were imagined to produce occasional syncretisms that explained those otherwise inexplicable stone tool assemblages known to include both kinds of tools. Such syncretisms, "hybridizations," were similarly invoked to explain human fossils that did not fit well as either neanthropic or paleanthropic man.

This approach to explanation, both in archeology and in physical anthropology, rested obviously on a foundation of archetypal thinking. It also had the dubious virtue of avoiding most evolutionary problems. Real change and divergence was relegated to an indefinite, distant time (and indefinite place) that was not the object of study. It was postulated that segments of those phylogenetic lines that were the object of study existed in such a way that they displayed only very minor changes through time. This theory makes it possible to avoid facing fundamental evolutionary change and avoid facing ancestors who were less than beautiful by present-day standards.

A catalog of names of fossils that have been considered members of the neanthropic line at various times would include

Galley Hill
Piltdown
Grotte des Enfants, Grimaldi
Oldoway Man
Bury St. Edmunds
Moulin Quignon
Chichy
Mount Denise
Olmo
Ipswich
Castenedolo
Kanam
Kanjera
Swanscombe
Fontéchevade

Recent additions to this list might include Vértesszöllös, which has been discussed, and even more recent finds from the Omo River region in Ethiopia. The latter find will be discussed later. The paleanthropic line has included the finds of

Neandertal
Heidelberg
Peking
Java
Solo
Broken Hill (Rhodesia)

Recent students of fossil man will recognize all members of the latter list. But the "neanthropic" list has become remarkably decimated and many of the names can no longer be found in modern textbooks. Many were not at all "early," as proven by better dating techniques. Some (e.g., Kanam) were not modern. Some were outright frauds, the most famous being Piltdown (Weiner, 1955). Whether Kanjera was pre- or post-Neandertal is still indeterminant. The remaining claimants to "early modern man" status are Swanscombe and Fontéchevade, which have been discussed.

The "paleanthropic" list has not suffered in the same way and has indeed become further filled out. It must be concluded that the hopes and preconceptions of the early investigators had much to do with the existence of such parallel lines of man.

With the decimation of the neanthropic line the hypothesis of cladogenetic speciation in *Homo* becomes very tenuous. The only fossils the least bit enigmatic in this respect are Swanscombe and Vértesszöllös. Fontéchevade is morphologically and chronologically well within what will later be defined as the Neandertal phase of *Homo sapiens*. The Vértesszöllös individual may have had an endocranial volume approaching that of Swanscombe, which has been estimated as 1325 cc. The Vértesszöllös occipital bone has a more primitive toric structure. Bone thickness has not been given for exactly comparable points on the two fossils but is roughly equivalent and within the range of late *Homo erectus*. The teeth were more like those of *erectus* than *sapiens*. These fossils could be considered either late *erectus* or early *sapiens*. I have chosen a convenient temporal datum at which to draw a line between *erectus* and *sapiens*—the beginning of the Hoxnian warm period (the second interglacial of the Alpine sequence). This places Vértesszöllös as *Homo erectus* and Swanscombe as early *Homo sapiens*. This is in line with the conventional placement of Swanscombe and is consistent with the point made in previous paragraphs on anagenetic divisions.

NEANDERTAL

In the material that follows I will use the term Neandertal in the broad sense. The term will designate the early stages of the within-species evolution of *Homo sapiens*. In time it includes *sapiens* fossils up to approximately 40,000 years ago. Morphologically it includes populations that, with exceptions to be mentioned, are indistinguishable from those of modern man in postcranial features. The facial skeleton was heavier than in modern man, with heavier brow ridges, and not as well under the cranial skeleton. The mandible, maxilla, and teeth are heavier structures than those of modern man. In short Neandertal is the expectable form of early *Homo sapiens* out of *Homo erectus*. As a subspecies this would appear as *Homo sapiens neanderthalensis*.

Neandertal in the broad sense is specified because the term Neandertal was originally applied only to European fossils. When fossils were found comparable in age and differing a little from the European Neandertals in other parts of the Old World considerable disagreement arose over how far the label should be extended. There is still no agreement on this point. I prefer to use Neandertal in the broadest sense and to speak of the early Würm Europeans as *classic* Neandertal to be able to discuss the many questions about this particular group of Neandertals.

The first really numerous group of Neandertal fossils comes from Europe and the Middle East in Lower Würm times, the first maxima of the last major glaciation (Figure 13.3). Some Neandertal finds come from the Riss-Würm interglacial and few are known, as mentioned in the preceding chapter, from the Riss and the preceding interglacial. A note on the history of the original Neandertal find will help clarify, if not resolve, some of the controversies.

The name of this group of fossils comes from a narrow valley, the Neander valley or the Neandertal, near Düsseldorf, Germany. Prior to the find of "Neandertal Man" in this valley there was another find in Gibraltar (Forbes' Quarry), which was later identified as Neandertal. In 1856 workmen conducting limestone quarrying operations in the Neandertal were cutting back the cliff face that

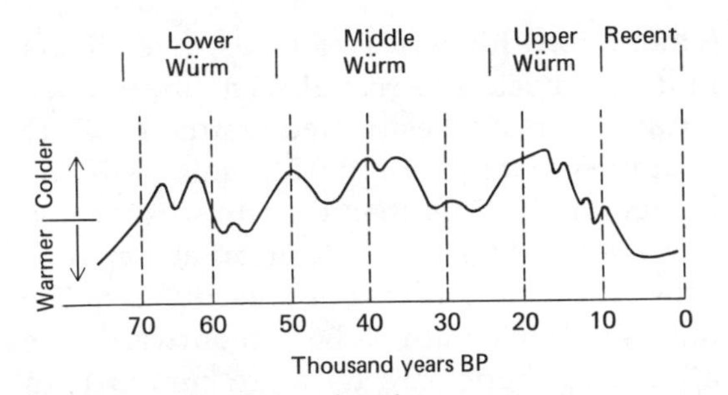

FIGURE 13.3 Generalized temperature curve, midlatitude Europe, and divisions of the Würm glaciation. Paleotemperature curve modified from Woldstedt (1962). Mean annual temperature difference between warmest and coldest periods shown is 11°–12°F.

included a cave, the Feldhofer grotto. The cave debris were simply dumped. Some of the bones from the cave were saved and given to a teacher, Johann C. Fuhlrott. This was three years before the appearance of Darwin's *Origin of Species*, but evolutionary speculations were much in the air. To Fuhlrott goes the credit for recognizing the human bones from the Neandertal as being an earlier form of man, Neandertal Man, the first to be so recognized.

Among the bones found was a calotte (skull cap) including the upper margin of the orbits (Figure 13.4), part of a pelvis, two

FIGURE 13.4 The skull from the Neandertal, front and side views of cast.

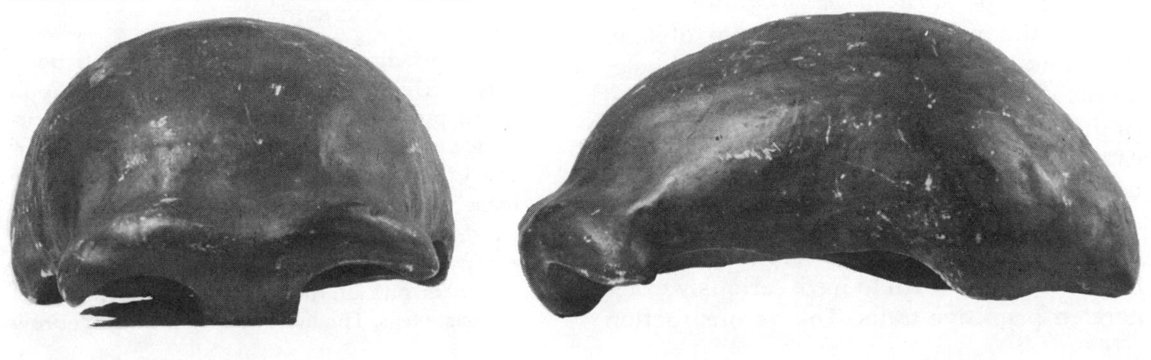

femurs, two humeri, two ulnas, five ribs, a radius, clavicle and scapula. The individual, an adult, had an estimated cranial capacity of approximately 1230 cc. The vault is slightly longer and broader than the later-discovered Steinheim skull. The postcranial remains show the marks of heavy musculature. The radius of the forearm is bowed outward. The other long bones are no more curved than they are in present-day populations.

Because of the manner of recovery, nothing is known of cultural or faunal associations with this find. A great diversity of opinion arose on the interpretation of Neandertal. Schaafhausen, who described the fossils, and Thomas Huxley held that this was a primitive *Homo sapiens*. Rudolf Virchow, the founder of pathology, considered the morphology of Neandertal to have shown the effects of rickets (Virchow, 1872). Others such as Blake (1864) suggested that Neandertal Man was not only pathological but an idiot as well. Neandertal indeed did have a crippled left arm, osteoarthritis, and some other marks that may be pathological. But when other European fossils were found, similar in general morphology, it was decided that Neandertal man could not be wholly explained by pathology.

Normally the first fossil found and used in designating a new "type" is considered the "type specimen" to which all later descriptions refer. Neandertal was established as a primitive type of *Homo sapiens*. But, perhaps due to the more complete nature of later finds and the lack of archeological and paleontological data on the original Neandertal find, a later find from La Chapelle-aux-Saints came to be used as the type specimen. This has resulted in greater difficulty in understanding early *Homo sapiens*, since the picture of La Chapelle, as reconstructed by Boule, is now known to have seriously exaggerated primitive traits. This reconstruction

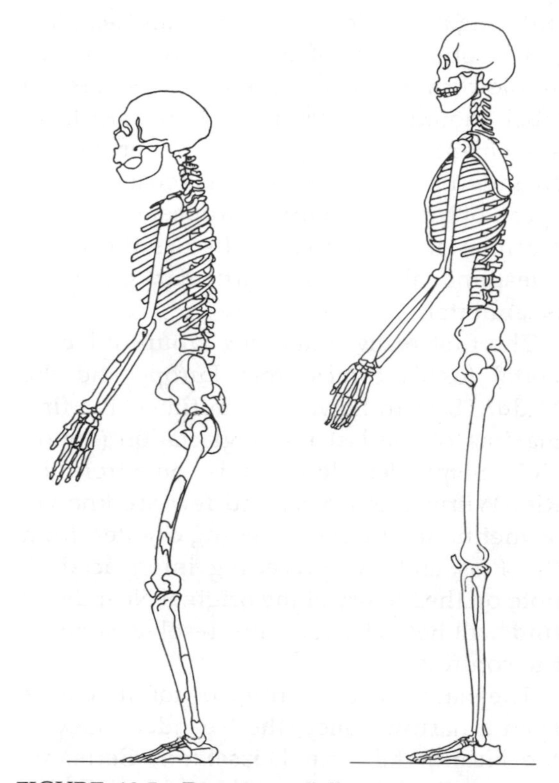

FIGURE 13.5 Drawing, following the reconstruction of M. Boule, of the man of La Chapelle-aux-Saints compared to a modern Australian. (*Source:* Boule, "L'homme fossile de la Chapelle-aux-Saints," *Annales de Paleolontologie*, 1913, 8:1–67. Courtesy of Masson et Cie, Paris.)

provided the prototype which many scholars had in mind when discussing classic Neandertal (Figure 13.5). It was dramatized by Elliot Smith, who said

> His short, thick-set, and coarsely built body was carried in a half-stooping slouch upon short, powerful, and half-flexed legs of peculiarly ungraceful form. His thick neck sloped forward from the broad shoulders to support the massive flattened head, which protruded forward, so as to form an unbroken curve of neck and back, in place of the alternation of curves which is one of the graces of the truly erect *Homo sapiens*. The heavy overhanging eyebrow

> ridges and retreating forehead, the great coarse face with its large eye-sockets, broad nose, and receding chin, combined to complete the picture of unattractiveness, which it is more probable than not was still further emphasized by a shaggy covering of hair over most of the body. The arms were relatively short, and the exceptionally large hands lacked the delicacy and nicely balanced co-operation of thumb and fingers which is regarded as one of the most distinctive of human characteristics. (Smith, 1924, 69–70.)

Studies by Patte (1955) and by Straus and Cave (1957) have shown that the postcranial skeleton of classic Neandertal was much like that of modern man. He was not stooped. He did not walk with a bent-knee gait. Nor did he walk on the outer edges of his feet in simian fashion. The differences between classic Neandertal and modern Europeans in postcranial morphology is no greater than the difference between Europeans and some of the other living populations of man.

What was classic Neandertal like? By classic Neandertal, I designate not a physical type but simply those Neandertals of the cold and wet early Würm times of Western Europe. These are the individuals who, it has been suggested by some, represent a divergent species, *Homo neanderthalensis,* which also supposedly became extinct. A list of the best known of the classic Neandertals would include finds from Neandertal (Germany); La Chapelle-aux-Saints, La Quina, La Ferrassie (France); Spy (Belgium); Gibraltar (Iberian peninsula); and Monte Cicero (Italy). These individuals were short and stoutly built, being perhaps four to five inches shorter than the average West European today but slightly heavier, quite muscular, and with short, broad digits on the hands and feet. They differed most from modern Europeans in facial structure.

The following comments on the face and skull of classic Neandertal (Figure 13.6) are relative to modern Europeans. The cranial base was large both in length and breadth, and the base was less sharply flexed (Figure 13.7). Consequent probably to these features of the base of the skull were certain differences in the skull vault. The vault height was lower and the vault more globular in outline viewed from the front or rear. Some lambdoidal flattening plus the long base produced what has been described as a "bun-shaped" occipital. Endocranial capacity in absolute terms was as great or greater than it is in modern Europeans. But relative to body size the brain size was about the same. The four larger skulls, La Chapelle, La Ferrassie, Le Moustier, and Monte Cicero, give an average value of approximately 1570 cc. Gibraltar and one of the La Quina specimens both give values of slightly over 1300 cc and have, partly because of cranial capacity, been considered female.

Correlated with less flexure of the cranial base, the facial skeleton was still a bit more anterior to the calvarium (that part of the skull other than the facial bones) than it is in individuals today. Prominent brow ridges (but smaller than in the earlier Steinheim fossil) were molded over large orbits. The most immediately apparent difference in his physiognomy was that classic Neandertal had a large, heavily structured face. The nose was very large. The external nose was broad and projected out sharply from the face. The internal nose (the area of the turbinates) was large and this combined with large maxillary sinuses gave the midfacial region a very filled-out appearance which reduced or eliminated the canine fossa.

The teeth were still larger than they are in modern man. As in many earlier fossils shovel shaped incisors are common and the molar teeth have enlarged pulp cavities (taurodontism). The mandible shows a sym-

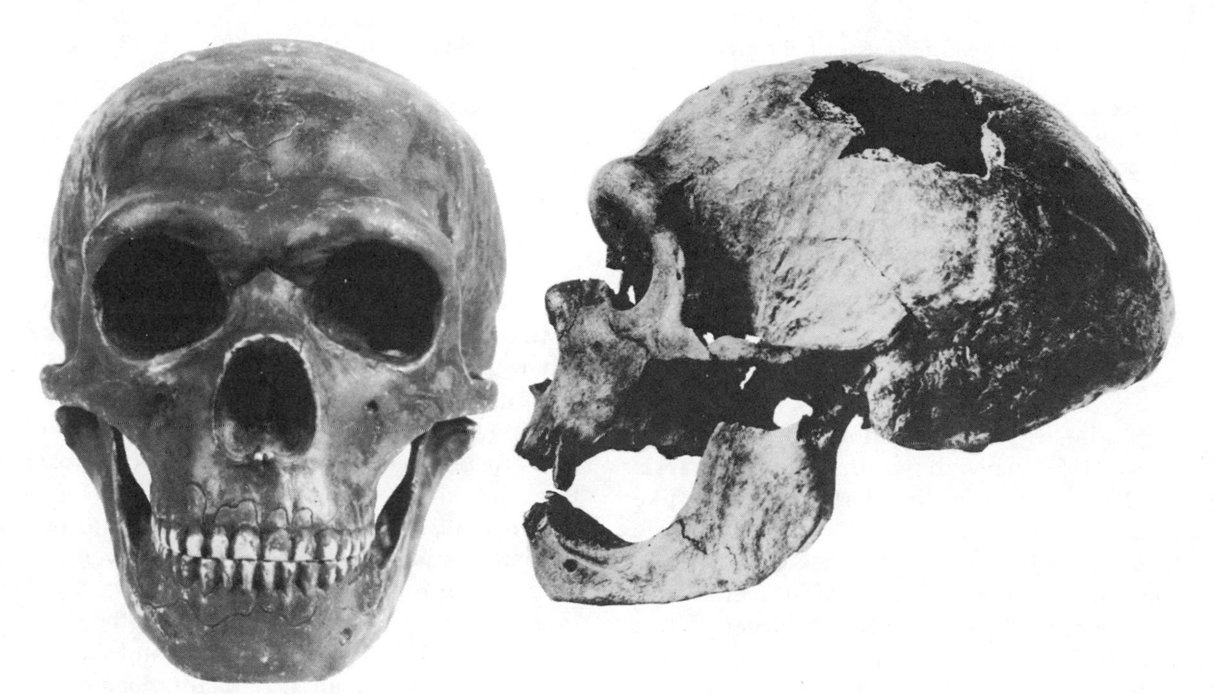

FIGURE 13.6 La Chapelle-aux-Saints skull, frontal view of cast and left lateral view of original. (*Source:* Director of the Musee de l'homme.)

FIGURE 13.7 Craniogram of La Chapelle-aux-Saints (Dotted line) superimposed over craniogram of two modern skulls: Highland New Guinea (solid line) and Amerindian (dashed line). Not to same scale.

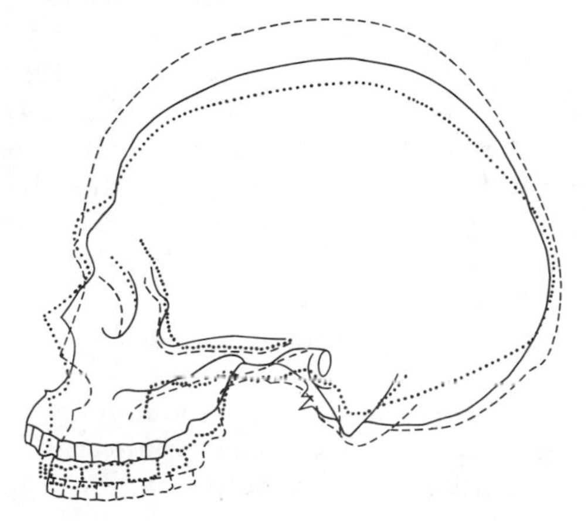

physeal region that still slopes backward, but only slightly.

There are many other minor details that anthropologists and anatomists have used to characterize classic Neandertal. The major features have been covered and a longer "trait list" of distinctions would convey, by its very detail, an erroneous sense of the magnitude of difference between Neandertal and modern man, since earlier and less similar forms have been described in less detail in earlier chapters. Classic Neandertal, as all other Neandertals, is in most features the logical transition between *Homo erectus* and present-day *Homo sapiens.* Classic Neandertal fails in only a few ways to fall into such an intermediate state and appears to be divergent. He is heavier, both absolutely and relative to his stature, than present populations in the same area and is probably heavier

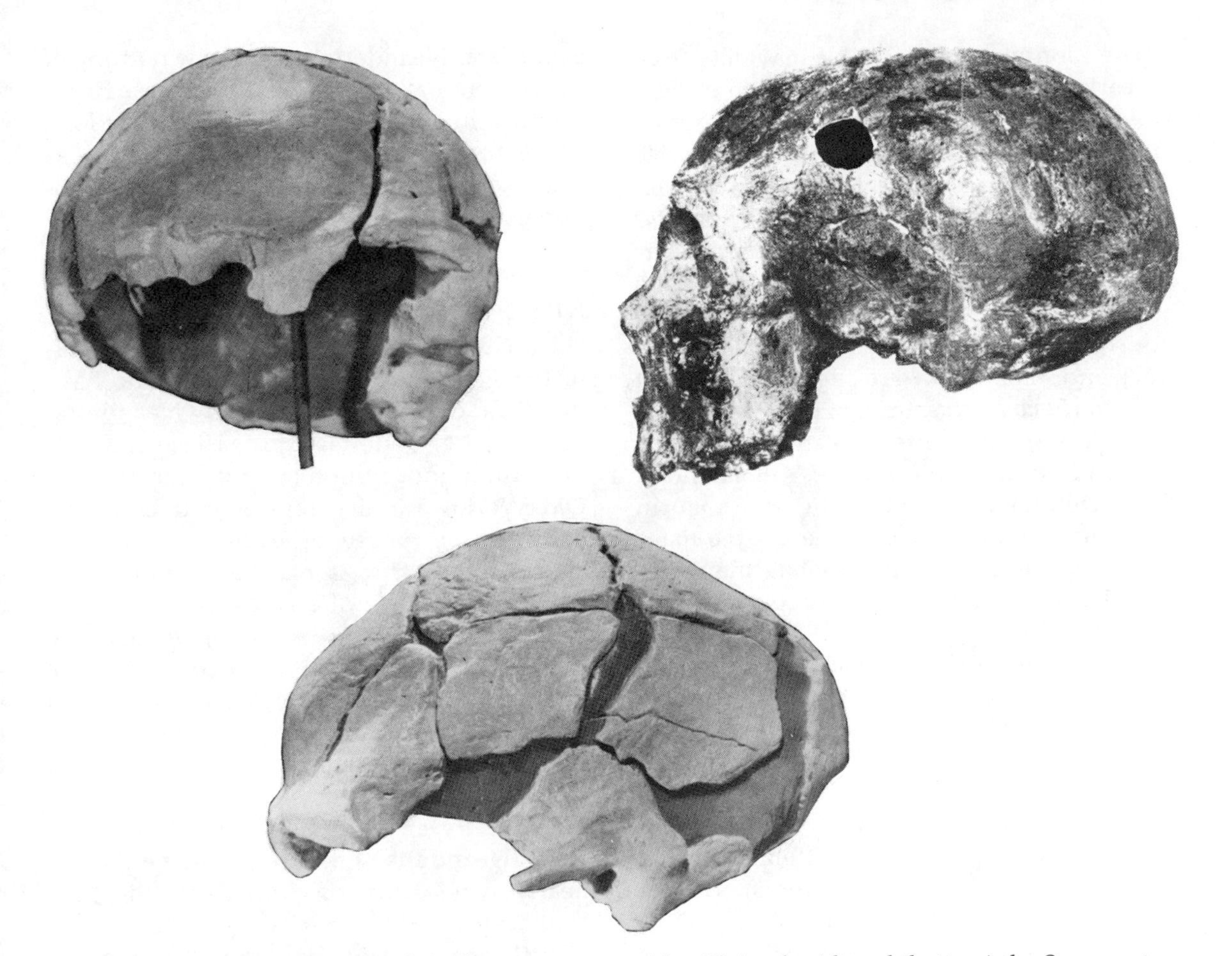

FIGURE 13.8 (Top) Ehringsdorf skull as reconstructed by Kleinschmidt and (bottom) the Saccopastore skull. (*Source:* Koenigswald, 1958; Sergi, 1931. Courtesy of Masson et Cie, Paris.)

than his earlier Neandertal antecedents. The large nose and large maxillary sinuses may also be a small divergence confined to classic Neandertal. These features have been suggested by Howell (1952) to be possibly a result of isolation of the Western European Neandertals during the Lower Würm with consequent changes due to genetic drift or to climatic selection. Carleton Coon (1962), among others, has advocated the idea that these features represent adaptations to the cold.

There are a few specimens conventionally accepted as Neandertal from this same region dating from the Riss-Würm (Eemian) interglacial. Not counting the disputed Fontéchevade specimens, the best known are the Saccopastore finds in Italy, the Ehringsdorf finds in Germany, possibly the Krapina finds in Yugoslavia, and the Montmaurin mandible found in France. The dating of the Krapina finds has been questioned, and these finds may instead derive from an interstadial of the Würm.

The Montmaurin mandible may date from an earlier period than the Eem warm period. Howell has suggested that it may come from an interstadial of the Riss or earlier (Howell, 1960). Dental dimensions are slightly smaller, all around, than they are in *Homo erectus* mandibles. The mandible resembles the earlier Mauer mandible in many ways but the ascending ramus is not so wide and the body not so thick. The symphyseal region is quite thick and slants backward slightly less than it does in *Homo erectus*. In all of these characteristics Montmaurin fits an early Neandertal phase of *sapiens* evolution. H. Vallois and F. C. Howell have noted that the Montmaurin specimen would provide a reasonable mandible for Steinheim. It also has many resemblances to the recently discovered Tautavel mandibles.

The Ehringsdorf finds included a crushed skull minus the face, an adult, and a child's mandible. The child's mandible foreshadows the features of the adult specimen, which includes a sharply sloping symphyseal region and an unusual amount of alveolar prognathism (forward protrusion of the face in the region surrounding the tooth roots). The reconstruction of the skull involves pieces whose edges do not join and may therefore involve some error. The skull can be seen to be much like the Saccopastore I skull (Figure 13.8) but with more lambdoidal flattening and a more angulated occiput. In terms of dimensions and overall form the Saccopastore skull is much like the earlier Steinheim skull. It differs in that its brow ridges are smaller and, as in the classic Neandertals, there is no canine fossa.

As pointed out by Sergi, the describer of the Saccopastore skulls, these early Neandertals form a reasonable sequence into the classic Neandertals. They had a smaller cranial base, a smaller cranial capacity (1150–1300 cc), and not as great midfacial protrusion as the later Neandertals. We know nothing of the postcranial morphology from these finds. The Krapina finds include fragmentary long bones that indicate persons who were not as thick set and muscular as the classic Neandertals.

NEANDERTALS OF AFRICA AND ASIA

I include in this category the individuals of the Eem and early last glacial who have been sometimes referred to as "Neandertaloid" or "Neandertal-like." These individuals, throughout the temperate and tropical Old World, are also representatives of the early *sapiens* phase of evolution. This includes from Africa: a mandible from Makapansgat (the Cave of Hearths as opposed to Limeworks Cave of *Australopithecus*), the Temara mandibles from Morocco, the Cave of Hercules mandible from Algeria, the Haua Fteah mandibles from Libya, the Broken Hill skull (Rhodesia), the Hopefield skull (South Africa), the Saldanha skull (South Africa), and the Eyasi skull fragments (Tanzania). Asian Neandertals include 12 fragmentary to nearly whole calvaria from Java (the Solo skulls), and a skull from Ma-pa, China.

Little need be said about the mandibles except that they are good examples of Neandertal. The Broken Hill skull, sometimes called Rhodesian Man, is the most complete. It too was originally called anomalous. The later find of the Saldanha skull provided another specimen almost identical to that of Rhodesian Man (Figure 13.9). The Rhodesian skull differs in minor features from the classic Neandertal of Europe. It has heavy brow ridges that continue to the midline of the skull, forming a very prominent glabella. As in many Neandertals there is no canine fossa. The face is long. The nose is not beaklike as in classic Neandertal. The cranial capacity is estimated as 1280 cc. The fragmen-

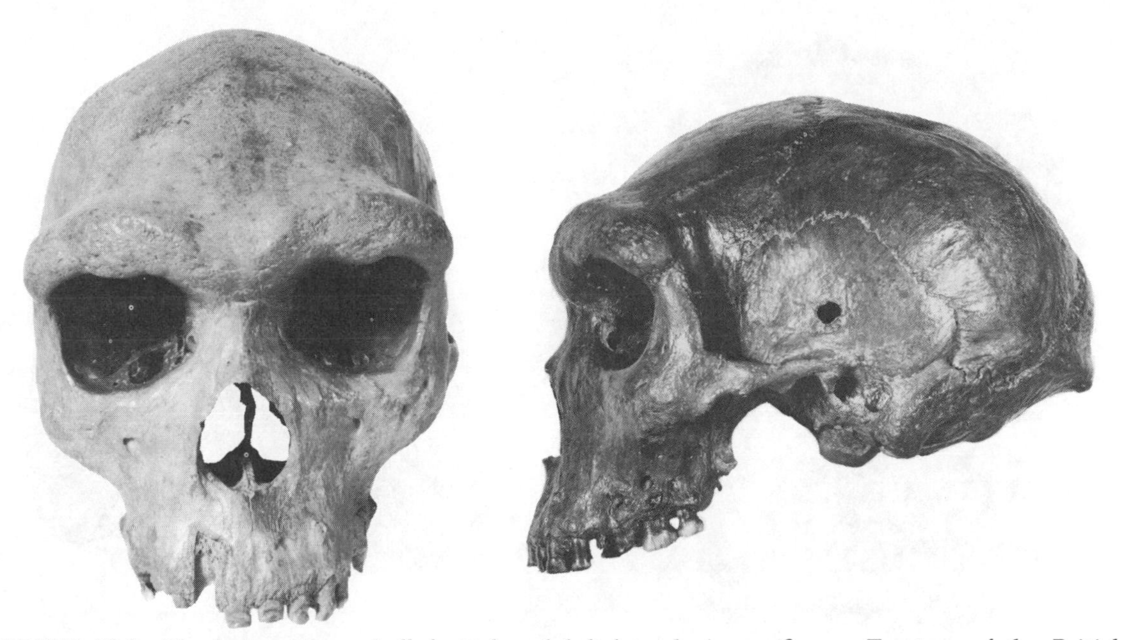

FIGURE 13.9 Rhodesian Man, skull frontal and left lateral views. (*Source:* Trustees of the British Museum, Natural History.)

tary long bones recovered with the skull indicate that the individual was taller and not so stoutly built as was the typical classic Neandertal.

Now we return to Java and the region of the early *Homo erectus* finds. The village of Ngandong is only 6 miles from the village of Trinil and is also in the Solo River valley. In 1931 Indonesian field excavators of the Geological Survey of Java began unearthing fragments of skulls. A total of 12 individuals were recovered. With the exception of skulls, the only other finds were two tibias. It has been suggested that these skulls, as well as many similarly treated Paleolithic skulls, were the victims of their own kind. No mandibles were present. The bone around the foramen magnum was broken away in most of the skulls. Cannibalism, ceremonial or otherwise, may have been involved. These fossils are not well dated. One suggestion was that they were Upper Pleistocene and contemporaneous with European Neandertals. Indeed, they have been called Asian Neandertals.

The Solo skulls (Figures 13.10 and 13.11) appear to be more primitive than Neandertal. The cranial capacity ranges from 1100 to 1300 cc. In overall outline the Solo skulls are much like those of the African Neandertals: Hopefield, Broken Hill, and Saldanha. Solo Man tends to have an occipital torus higher on the skull than did the African Neandertals. The brow ridges of the heaviest of them, Solo XI, are not as extensive as they are in Rhodesian man. The skulls show sagittal "keeling" where the parietals come together at a pronounced angle at the midline of the skull. They do not have the globular form of the European Neandertals when the skull is viewed from the rear. Weidenreich's reconstruction of Solo Man is shown in Figure 13.12. In this reconstruction Weidenreich was

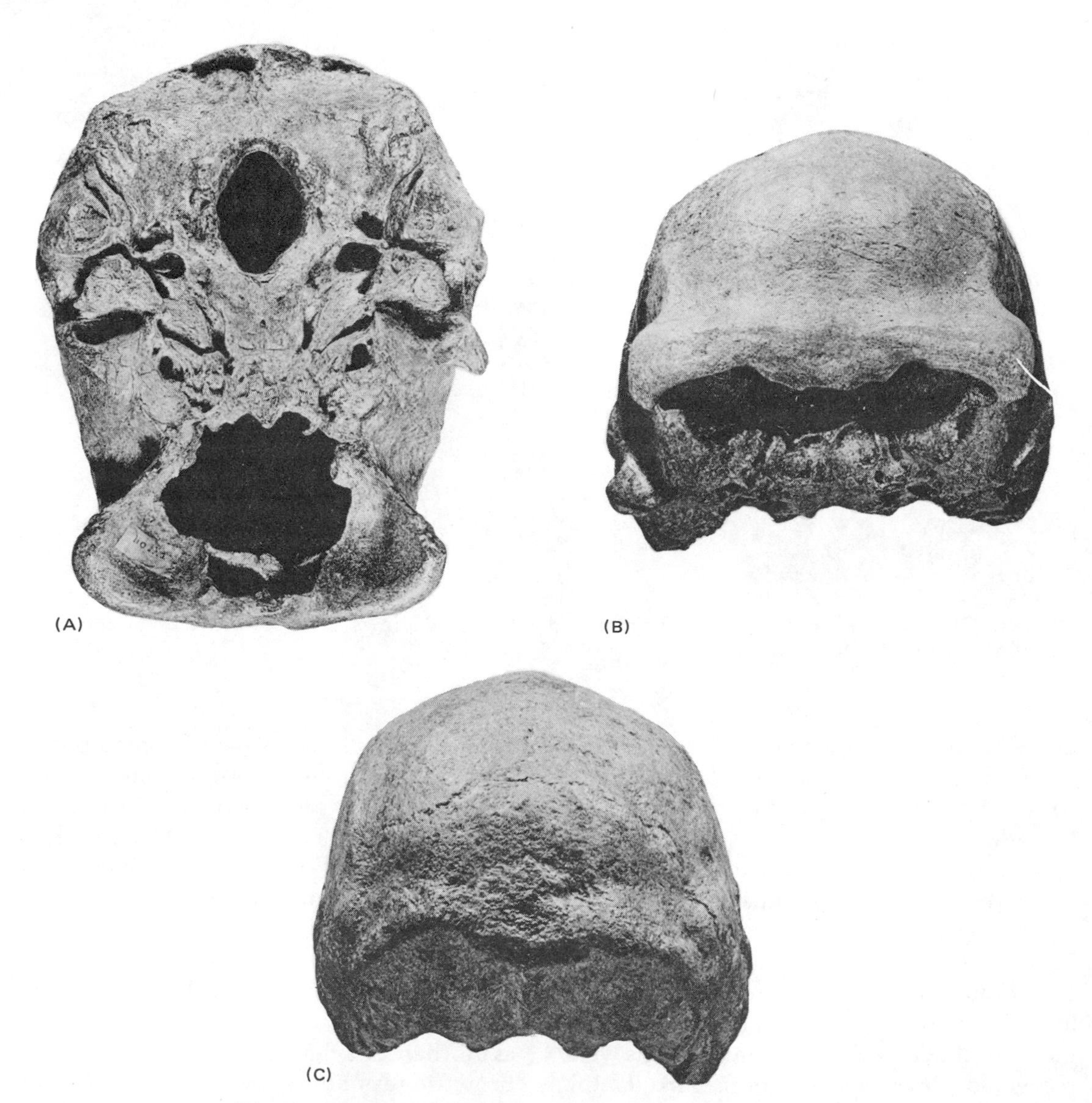

FIGURE 13.10 Solo XI skull: (A) basal, (B) frontal, and (C) occipital views. (*Source:* Weidenreich, 1951.)

strongly influenced by Sinanthropus. If the reconstruction is correct, Solo Man had a shorter face than Neandertals. It may well be that the Solo finds are intermediate in time between *Homo erectus* and the later Neandertals. As such they too would be early *sapiens*, as suggested by their morphology. This would place them with Steinheim, Swanscombe, and Tautavel rather than with Neandertal.

The generality of Neandertal as a phase of *sapiens* evolution is conformed by relatively

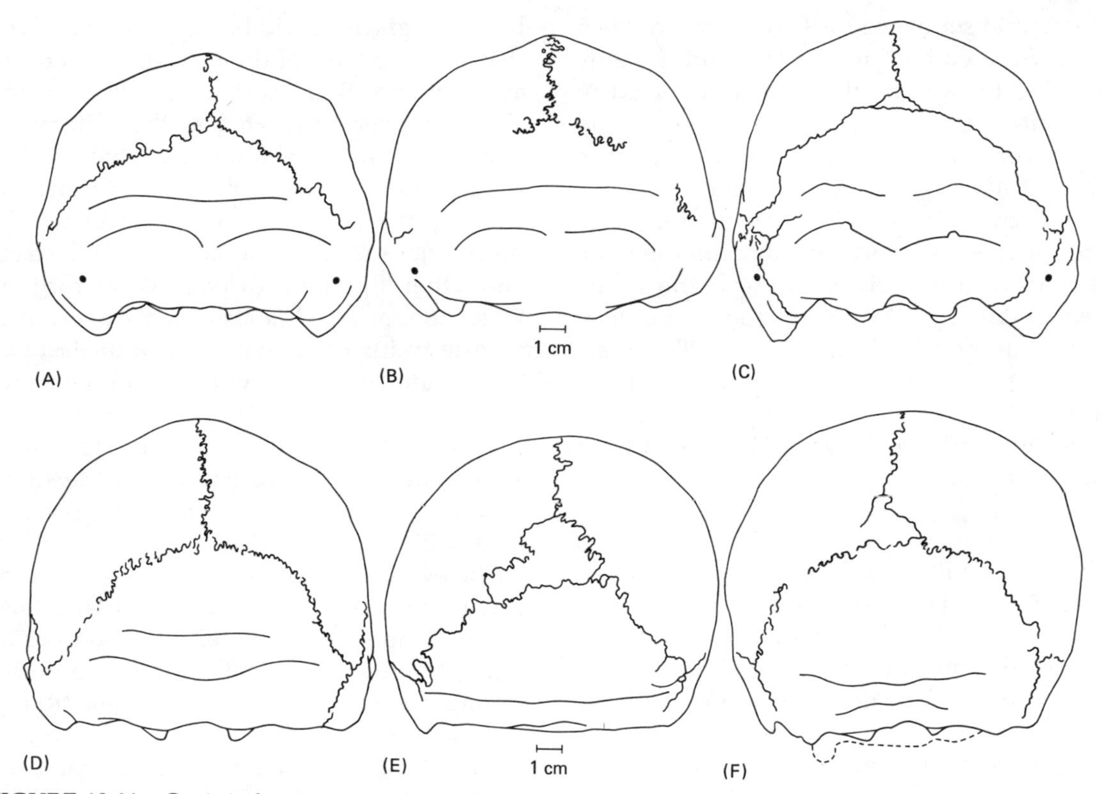

FIGURE 13.11 Occipital views comparing the (A) *Homo erectus erectus,* (B) *Homo erectus pekinensis,* (C) Solo XI (D) Rhodesian, (E) Saccopastore, (F) La Chapelle-aux-Saints skulls. (*Source:* Weidenreich.)

FIGURE 13.12 Franz Weidenreich's reconstruction of the skull of Solo Man. (*Source:* Weidenreich, 1951. Used with permission.)

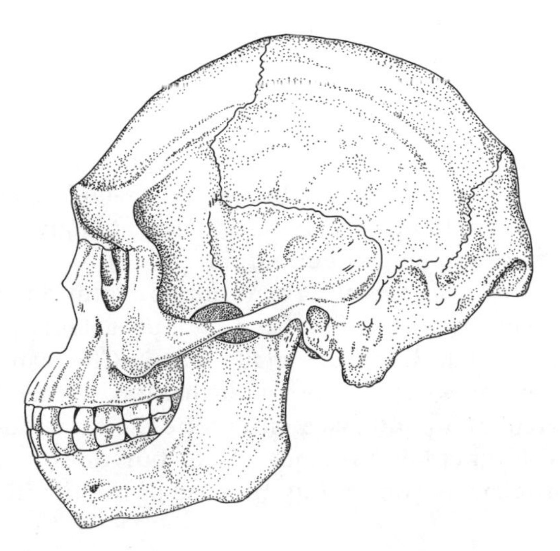

recent finds in China. In 1956 part of a femur and the right parietal of an individual were found at Sjaraosso-gol in the Ordos region where Neandertal-like teeth had been found earlier. The femoral fragment is nondiagnostic but the parietal, like the teeth, appears to be Neandertal. In 1958 a part of a Neandertal skull was recovered from a cave at Ma-pa, southern Kwangtung. The find included a fragmentary calvarium, the nasal bones, and the bony enclosure of most of the right orbit.

The faunal sequence for China has not been finely divided by time periods, and it is only possible to say that this individual lived in the Late Middle Pleistocene or early Late Pleistocene. Another cave find in 1958 was near Liu-chiang, Kwangsi province. The Liu-chiang cranium, complete except for zygomatic arches, is reportedly quite intermediate in all its features between those of the Ma-pa Neandertal and later, more modern forms of man from the Paleolithic of China. The question of transitional forms will be taken up later. At the moment it is important to note that Neandertal is now clearly known from mainland Asia.

MIDDLE PLEISTOCENE CULTURAL DEVELOPMENTS

Most of our knowledge of cultural developments continues, of course, to be based on the evidence of the evolution of tools. As has been mentioned, the control of fire is evident at about 400,000 years ago both in Asia and in Europe. The earlier unifacial choppers and bifacial chopping tools had in some areas evolved into handaxes. The occurrence of handaxes in tool assemblages came to characterize much of Europe and Africa. In much of eastern Asia rectilinear and ovate cleavers develop rather than the pointed handaxes. But even in this area assemblages containing handaxes, for example, the Padjitanian of Java, can be found. In Europe assemblages without handaxes, the Clactonian of England, for example, can be found.

In Africa there is a continuous evolution of heavy cutting tools from chopping tools to crude handaxes, originally called "Chellean" handaxes, to better made handaxes called Acheulian handaxes. In Europe the same developmental sequence exists but does not appear as continuous because of the disturbance of glacial periods. It seems, however, that it is the tools of the interglacial periods, and not the glacials, that are out of place. This has been attributed to the differential movement of stone fragments due to frost action and other erosional forces of the succeeding glacial periods that redeposited the preceding surface material in river terraces and other fluviatile (produced by running water) deposits. This also seems to occur to an extent during interstadials of the last glaciation and interferes with our understanding of the first appearance of modern man. The Acheulian handaxes in Europe, therefore, do not appear as a continuous transition from the cruder handaxes, sometimes called Abbevillian.

The evolution of more efficient tools in the Paleolithic (old stone) age was in large part the evolution of technique, aided perhaps by some evolution of greater manual abilities among the tool makers. The production of pebble tools requires very little control. A river pebble struck against a larger stone, the anvil stone, to produce a conchoidal fracture is sufficient. Such fractures and cross fractures can be carried around the entire surface of the pebble. One of man's earliest tool making discoveries must have been that some stones do not fracture as well as others for producing a sharp edge. Hard, amorphous stones such as flints and cherts work well in fashioning cutting tools and were extensively utilized from the Paleolithic through the Neolithic. In such stone, fracture stresses are transmitted along a very regular "percussion cone" and the flintknapper uses this to control the flaking process.

We have no way at present to judge the hunting effectiveness of Paleolithic man. It has been suggested that the remains of large animals occurring with *Australopithecus* may be a result of scavenging activity and that proficiency at big game hunting developed much later. Certainly the hunting of deer was important to *Homo erectus* at Chou-k'ou-tien. Acheulian tool users, undoubtedly *erectus*, at Ambrona and Torralba, Spain, were preying on *Elephas antiquus*. But it appears that they were able to take only those elephants that became mired in a swamp. It is possible to overstress the importance of hunting among "Paleolithic hunters." Among ethnographically known hunters of recent times, only those living in the far north relied on meat for the bulk of their food. Hunting groups in warm climates are known to rely heavily on gathered plant foods, and this may account for as much as three-fourths or more of their total caloric intake.

Another early Acheulian site of great interest is that of Terra Amata at Nice, France. The excavations carried out by Henry de Lumley reveal detailed information about the Acheulian residents. The site dates from around 400,000 years ago or slightly earlier. This means that the people here were near-contemporaries of those found at Chou-k'ou-tien and Vértesszöllös. Most important was the presence of huts. These were made with an oval floor plan that had branches or stakes implanted side-by-side and braced with stones. Larger posts in the center of the structure apparently supported the tops of the huts. The huts ranged from 26 to 49 feet in length and 13 to 20 feet in width.

Faunal remains at Terra Amata include small game but suggest a preference for young specimens of larger game including extinct species of elephant, ox, rhinocerous, and pig. There was a small fire hearth in each hut. Stone tools were flaked in the huts, and the impress of fur hides and one human footprint are still discernible. The huts are larger than those used by single nuclear families among hunters in recent times, and could have held from 6 to 30 individuals. This would clearly involve a social unit larger than the nuclear family. Pollen analysis showed the huts to have been occupied in spring and early summer. Huts were built successively, and probably annually, on the same floors and utilized the same fire hearths. This means that not only was there a social unit larger than the nuclear family involved but that seasonal migrations were regular and patterned within a known territory.

The development of esthetic concepts in the Paleolithic is as difficult to trace as is social structure. Even among Olduwan materials (prehandaxe) there are stone balls that have been rounded and smoothed to a far greater degree than can be explained by any possible functional requirement. If we view art, in its adjactival sense, as anything done artfully, these must be included as extremely early works of art. At Terra Amata rubbed pieces of red ochre were found. These were probably used in body decorations as red ochre has been used up to historic times. Evidence of burial and ceremonial structures ("altars") appear in Mousterian times.

A very widespread group of Middle Paleolithic tool assemblages are termed Mousterian after the rock shelter of Le Moustier in France. Mousterian is by some authors referred to as a "culture," a "complex," or "assemblage." An *assemblage* is the set of artifactual remains left by a group of people at one point in time. A *complex* is a set of related assemblages. The Mousterian appears in the Riss-Würm interglacial as a development from the Acheulian. Some Mousterian assemblages maintain handaxes throughout Lower Würm and into Middle Würm. Various types and frequencies of finely made flake tools characterize different Mousterian assemblages (Figure 13.13). These flakes were usually detached from prepared cores. The

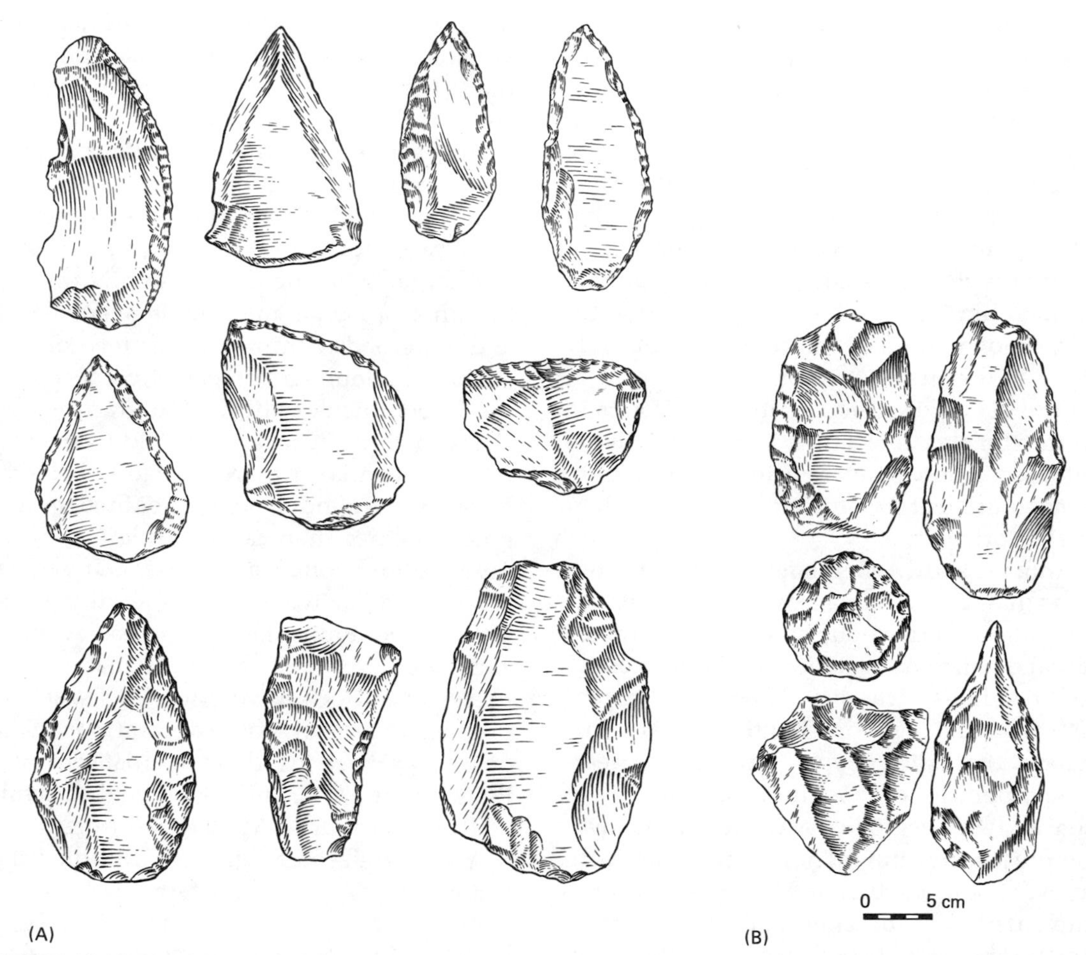

FIGURE 13.13 Middle Paleolithic tools from (A) western France (Mousterian) and (B) Ting-ts'un, China. (*Source:* Bordes, 1968 and Chang, 1968. Used with permission.)

prepared core has a flat striking platform from which flakes are struck, working around the perimeter of the core. Stages in the production of flakes or blades from prepared cores are shown in Figure 13.14. Some Mousterian assemblages included a high percentage of flake tools prepared by a Levalloisian technique. Such flakes are shaped while on the core and then detached with a single blow as shown in Figure 13.15.

The Upper Paleolithic develops out of Middle Paleolithic in Middle Würm times. The Upper Paleolithic is characterized by finely made blade tools, some of which were clearly hafted as javelins or thrusting spears; by gravers made from blades and probably used in carving and incising; by increased use of carved bone and wood; by tailored clothing in northern areas; and by specialized predation on the large herds of Upper Pleistocene grazing animals.

In northern China, at the site of Shui-tung-kou in the Ordos region, an assemblage halfway between Mousterian and Aurignacian appears. The Aurignacian is an Upper Paleolithic assemblage that becomes very widespread. It is assumed to be always associated with modern *Homo sapiens* in Upper Würm times. In Western Europe the assemblages intermediate between Mousterian and fully developed Upper Paleolithic have been termed Perigordian. Lower Perigordian is also called Châtelperronian. Upper Perigordian of East Europe is called Gravettian.

In Europe the transition from Mousterian to Lower Perigordian (Châtelperronian) takes place in the cold Middle Würm period at approximately 40,000–42,000 years ago. The Aurignacian appears to have moved, as an integrated complex, into Western Europe in the Paudorf interstadial around 31,000

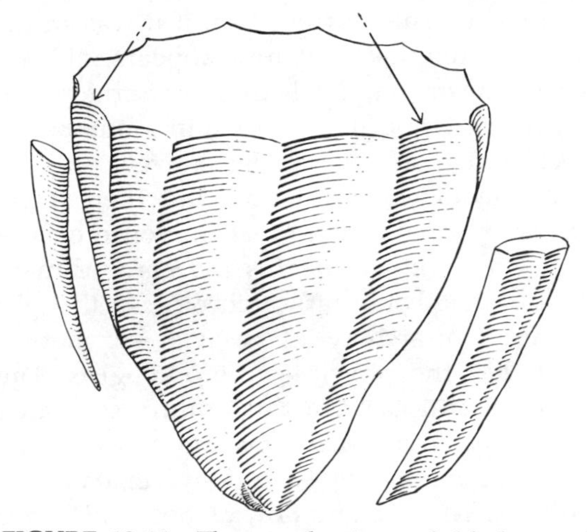

FIGURE 13.14 The production of blades and flakes from a prepared core. A striking platform is first prepared and then blades are struck off serially, leaving a fluted core from which blades continue to be removed.

FIGURE 13.15 The Levallois technique in the production of flake tools and points typical of the Mousterian.

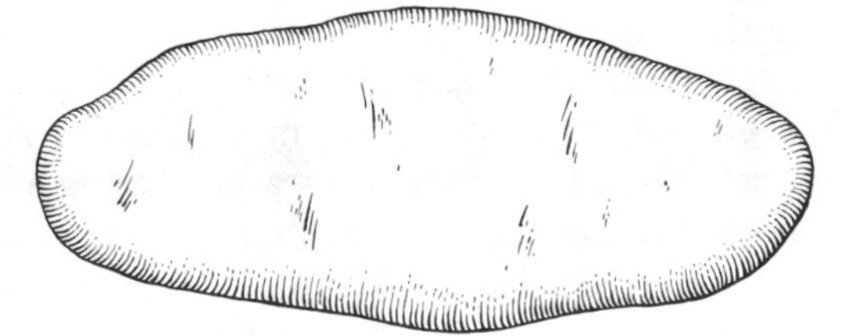

The nucleus before trimming

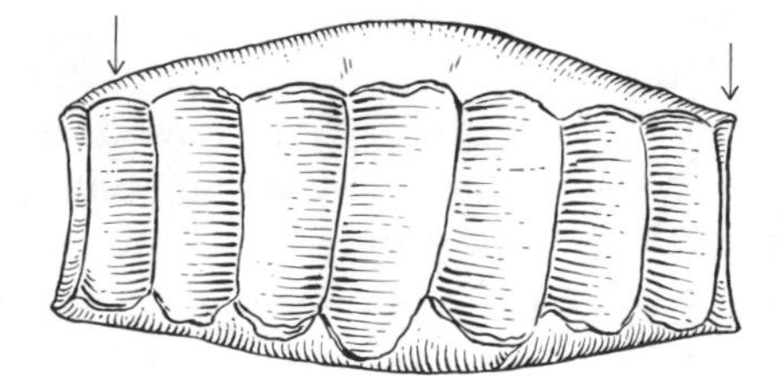

First step: Trimming the edges.

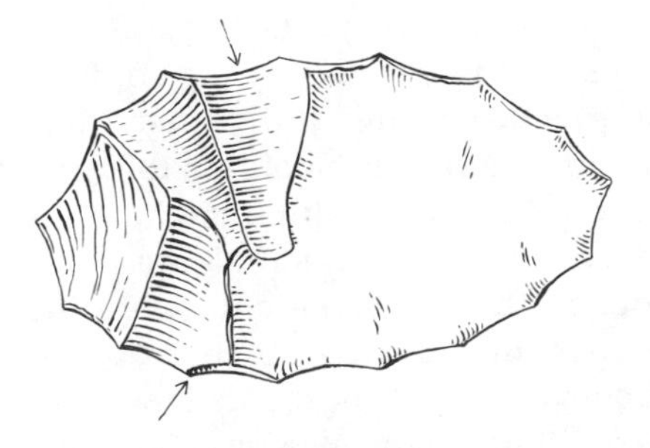

Second step: Trimming the top surface.

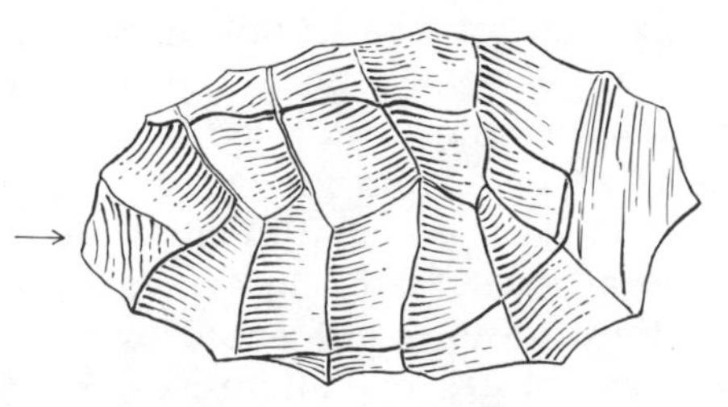

Third step: Striking the flake to be used.

years ago. The Aurignacian, if it was an intrusive complex as it now appears, did not eliminate or completely displace Perigordian, since the Upper Perigordian in a number of locales succeeds Aurignacian levels.

Archeological sequences and their interpretations have had great influence on the interpretation of the fossil record of man. Unfortunately the early archeologists thought in terms of archetypes as much—or more—than did the physical anthropologists. The Abbé Breuil early in this century separated handaxes from flake tools and conceived of them as belonging to entirely separate traditions. The ebb and flow of these industries was thought to correlated with glaciations; people with flake tools moved into Europe in glacial periods, handaxe people displaced them in interglacial times. To these "archeological phyla" were attached biological phyla because Neanthropic man was assumed to be the handaxe maker, and Paleanthropic man the flake tool maker. This archetypal distinction has now been, in the main, discarded.

Such an approach still plagues us in the Neandertal problem. The archetypal approach in archeology assumed no transition from Mousterian to Upper Paleolithic. Upper Paleolithic (Aurignacian) cultures were assumed to have moved into Europe from out of the East and to have terminated all local development. Naturally it was assumed that Neandertal man was terminated with the Mousterian and replaced by modern man having Aurignacian artifact assemblages. Although it still appears that the Aurignacian moved into Europe as a unit, it no longer appears that local developments were terminated. Nor does it appear that at any time in prehistory artifact assemblages or cultures were watertight units uninfluenced by other artifact assemblages or cultures, as is presupposed in archetypal thought.

The archeological record indicates that despite the slow pace of innovation throughout the Lower and Middle Paleolithic, new innovations did spread across continents and between continents of the Old World. In Asia *Homo erectus* continued to use cleaver-like tools, choppers, and chopping tools, rather than the pear-shaped handaxes common in Europe and Africa. But the specifics of stone tool technology did spread in such a way that a Lower, Middle, and Upper Paleolithic of Asia is correlated easily with similar phases in Africa and Europe. Fist-sized stone balls, sometimes well smoothed, were coextensive with hominid occupations throughout the Old World of the Lower and Middle Pleistocene, although we do not know the function of these balls. Finer flaking techniques, secondary flaking to dress or sharpen an edge, and use of a prepared striking platform for better control in detaching flakes developed and spread through man's range. In Middle Paleolithic times, which include the Mousterian tool assemblages of Europe, cutting tools began to show more diverse forms corresponding to more specialized functions. Blade tools appeared during this time. But again, in any major geographic area, there is ample evidence that these techniques were transmittable.

This does not mean that any archeological site of the same time period will show the same tool kit as any other site of that period. As recently as 100 years ago different small groups that illustrated a range of tools spanning practically the entire range from Lower Paleolithic through Neolithic to Early Industrial could be found. But in no inhabited continental or subcontinental area was there evidence that the residents were incapable of borrowing ideas. And there is no reason to believe that ideas could flow without the concomitant occurrence of gene flow.

The geographical "velocity" of gene flow

has been greater for hominids than for other primates throughout the Pleistocene. This is because man occupies more territory per individual and travels further than any other primate. Therefore for a given population size and amount of gene flow between populations, the genes will actually move further in human populations than in other primates. Washburn and others have shown that man as a hunter has a much less dense population than most primate species and ranges over a vastly wider area in a given season than does any other primate (Washburn and Lancaster, 1968). Since this reflects the ecological demands of a hunting existence, we can assume that such a difference has existed for as long as man has been a hunter.

The foregoing suggests that widespread and continuous gene flow has characterized the hominids throughout the Pleistocene. This would lead us to expect decreased chances of speciation. Another rather more obscure point leads in the same direction. Although it is difficult to make this a testable hypothesis it appears that, generally, speciation in mammals and birds is initially a behavioral adaptation. This follows from observations that populations that seem to represent recent speciation, or species that do not differ greatly in morphology, remain genetically isolated due to inherited behavioral differences. There have been suggestions as to why this should be the case. If different combinations of genes are highly adaptive in two different regions of a species' range, then isolating mechanisms that preserve these different genetic combinations will themselves become adaptive by reducing gene flow.

In such a situation as that envisioned in the preceding paragraph, any inherited behavioral difference that influences mating to occur at a different time, or in response to a different signal, could become quickly fixed in the different regions to create genetic isolation, that is, speciation. But what we know of man for the two million years for which we have an archeological record suggests that his major adaptation distinguishing him from other animals is his great ability to learn. From this he has developed highly successful cultural modes of dealing with his environment. In other words, man's strong point throughout the Pleistocene has been an unusual plasticity of behavior. But speciation by the evolution of inherited behavioral isolation implies a specificity of gene → behavioral phenotype inconsistent with the presumption of plasticity. This is another reason to expect the fossil record to show, as I believe it does, no speciation after the appearance of the genus *Homo.*

Other authors have taken different positions on this issue. The Paleanthropic versus Neanthropic (also called Palaeanthropi versus Phaneranthropi or the Archanthropinae versus Neoanthropinae) "school" has been mentioned. Other authorities who have not made an indefinitely long dichotomy of this sort have separated out various groups of fossils which, in their opinion, are too divergent to be members of a single biospecies and have placed them on a side branch of hominid evolution. Today no one goes to the extremes still common 30 to 40 years ago of putting all fossil hominids on side branches, thus making it possible for us to avoid admitting that any of the known fossils on the anthropologists' shelves might be our distant forbears. Those today who postulate cladogenic speciation, represented by one or more sets of hominid fossils, must also postulate extinctions. There is only one species of *Homo* today, *Homo sapiens.* Therefore if true biospecies developed in the past all but one line must have become extinct. The last possible place in our record where such speciations or extinctions could have occurred is

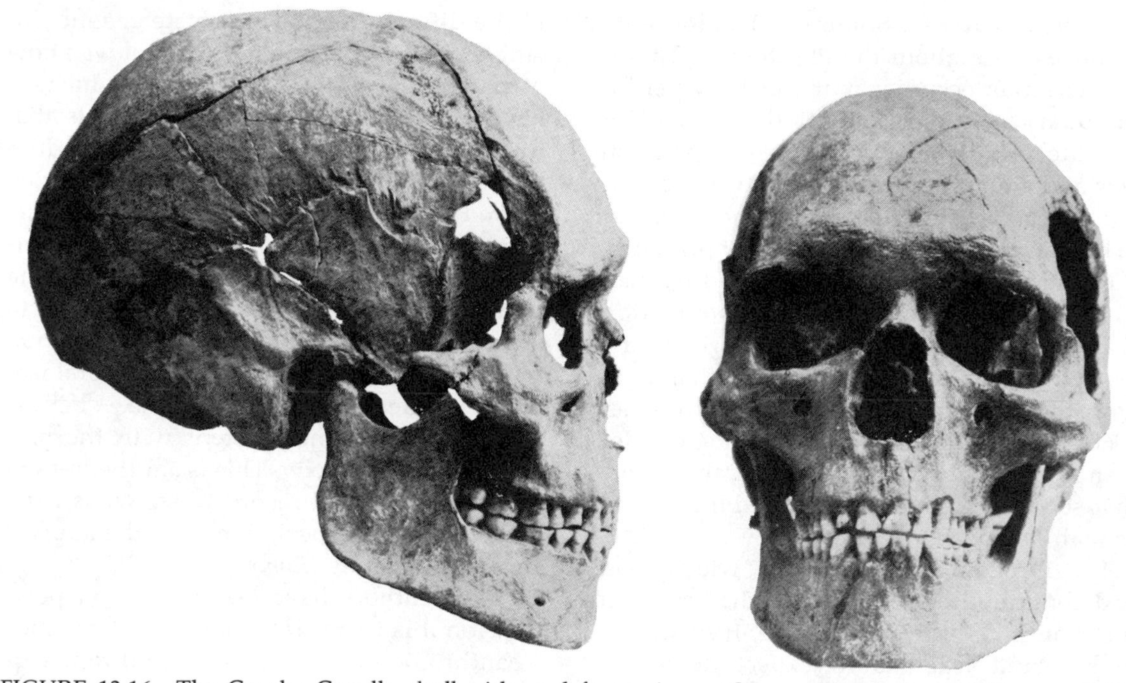

FIGURE 13.16 The Combe Capelle skull, side and front views. (*Source:* Klasatsch, and Hauser, 1910. Used with permission.)

with Neandertal. This may explain in part why the Neandertal question has attracted so much discussion among anthropologists.

If Neandertal represents an early form of our own species, *Homo sapiens,* we must expect to see a transition from this primitive *sapiens* to modern *sapiens* within the limits of the finds available. If we find no such transition we have a real problem, as there are no other fossils from which modern man could be derived. To localize a possible transition let us locate the earliest fossils that are conventionally recognized as modern *sapiens* and the latest fossils conventionally recognized as Neandertal.

The earliest dated skeleton accepted as modern *sapiens* comes from the island of Borneo. This is a very fragmentary unmineralized skull recovered in excavations in a large cave at Niah, Sarawak. The skull is probably that of a youth of 16–17 years of age. A ^{14}C date on nearby charcoal gave an estimate of approximately 40,000 years. The skull is modern in all observable features, and in comparisons to both Paleolithic and recent populations is most like the indigenous Tasmanians (Brothwell, 1960). But only the upper part of the skull can be restored and, as is the case with some Middle East finds of roughly the same time period, the face may not have been as modern in appearance as the calotte.

In Europe, where there is closer geographic juxtaposition between Neandertal and modern *sapiens,* the modern remains are somewhat younger than the Niah skull. The oldest of these and their approximate absolute dates are: Combe Capelle (ca 25,000–34,000), Předmost (ca 26,500), and Cro-Magnon (ca 20,000–25,000). The Combe Capelle finds, a burial in

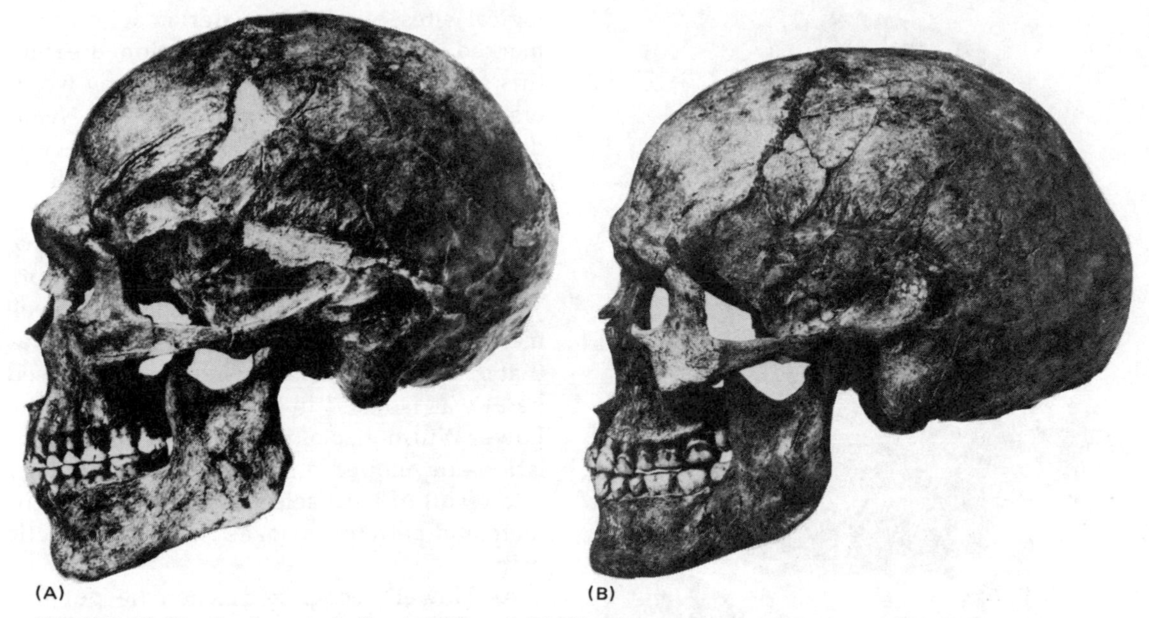

FIGURE 13.17 Predmost skulls: (A) III and (B) IX side views.

which almost all of the skeleton was preserved, came from the Combe Capelle rock shelter in the Dordogne, France. The remains were associated with either a Châtelperronian or an Aurignacian assemblage. The cave was excavated in 1909 in such a manner that it is impossible to know the exact location of the skeletal remains. In comparison to the earlier Neandertals, Combe Capelle had a greater vault height, a maximum breadth higher on the sides of the vault, a canine fossa, a rounded occiput, and a clear—although not markedly protruberant—bony chin (see Figure 13.16). The architecture of the face, with its strongly delineated brow ridges and robust malar region and mandible, was more rugged than it is in most populations today. But these features can be seen in present-day Fijian or Australian populations.

The Předmost finds of Czechoslovakia include more than 40 individuals. They are similar temporally and morphologically to finds at Brno, Czechoslovakia. The main ossuary is associated with Upper Perigordian (also called Eastern Gravettian or Pavlovian) artifacts. The occupation levels above the burials yielded the remains of mammoths numbered in various reports from 600 to "over 1000." This represents specialized and accomplished hunting at the beginning of the last intense cold period of the Würm. The individuals of the ossuary date from this time or slightly earlier. They are much like the individual from Combe Capelle with the exception that some individuals of this group had even stronger brow ridges (Figure 13.17) and approached the condition seen in Neandertal in this one respect.

Cro-Magnon is included here simply because the name is well known. He is not unusually early or unusually revealing of trends of recent hominid evolution. He did enjoy unusually good press agentry among the sci-

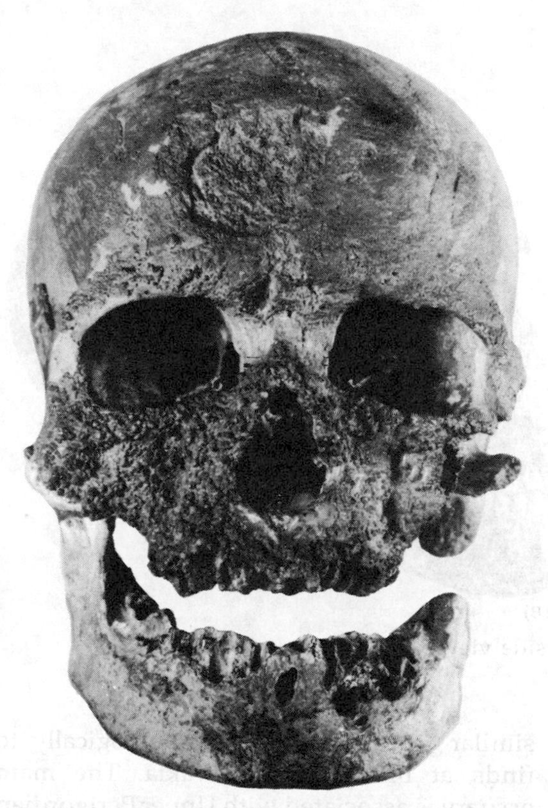

FIGURE 13.18 The "Old Man of Cro-Magnon." (*Source:* Director of the Musee de l'homme.)

entific community in the first half of this century and came to represent the "type" of a supposed Cro-Magnon race. The remains of five adults and a fetus were found in a rock shelter in the Dordogne, France in 1868. One of these, the "Old Man of Cro-Magnon" (Figure 13.18), is usually depicted. He was dolichocephalic (long-headed) but had a short, broad face and a narrow, salient nose. The associated culture was Aurignacian and a cold fauna indicates that the rock shelter was occupied during the first half of the last stadial of the Würm glaciation.

The Aurignacian blade-tool assemblages are Upper Paleolithic, are associated with modern forms of *sapiens*, and in some archeological sites overlie Mousterian levels with marked discontinuity. As mentioned earlier this set of relationships has led many workers in the past to the conclusion that Neandertal with his Mousterian culture was abruptly displaced by modern man with his Aurignacian culture coming from somewhere further east. A number of hypotheses have been put forward to explain this situation. The most reasonable was that of Howell mentioned earlier. Howell's suggestion was that Neandertal (classic Neandertal as used here) was isolated in Western Europe by the Lower Würm glacial ice spread. These populations developed a specialized morphology as a result of local selection in a difficult climate and perhaps, also, as a result of genetic drift.

As Howell recognized when he put forward this suggestion, genetic drift is unlikely to be involved. If European Neandertals had a social organization such that the effective size of the adult group within which mating occurred was only 50 or so individuals (an all-ages population of perhaps 150), then genetic drift within such a group could be an important factor. But if Europe had a population of even 45,000 individuals at the time this would provide 300 such breeding populations among which drift (chance) changes would be independent. For the region as a whole we can therefore expect no change in means as a result of drift. One effect that drift would have in such a situation would be increased variability throughout the region even though means would not change. But classic Neandertal as conventionally identified is notably homogeneous in metric features (Morant, 1927). We have to rule out genetic drift as an explanation for classic Neandertal.

Furthermore we now know that classic Neandertal was not so different from other Neandertals. Neandertals from Petralona

(Greece), Shanidar (Iraq), Amud (Israel), and Jebel Irhoud (Morocco) are morphologically like classic Neandertal but outside the area of "isolation." This latter group of Neandertals may not be as short and stocky and may not have suffered as much osteoarthritis as did classic Neandertal. Also, of course there were minor differences in morphology as there will be between any regions today.

More of the transition from *Homo sapiens neanderthalensis* to *Homo sapiens sapiens* can be seen outside of Western Europe. The best evidence of a transition comes from Mount Carmel in Israel. Two caves on Mount Carmel yielding critical fossils are Mugharet et-Tabūn and Mugharet es-Skhūl. The two Tabūn individuals were in strata that are Middle Würm in age. The ten Skhūl individuals come from strata that, on the basis of fauna, have been suggested to be up to 10,000 years later than Tabūn. A recent ^{14}C date places Tabūn at approximately 41,000 years.

Tabūn II is the mandible of an individual larger than Tabūn I. The latter individual is a female having all the characteristics of a Neandertal but without lambdoid flattening or a bun-shaped occiput. The cave of Skhūl, which is adjacent to Tabūn, contains remains that are as intermediate as is possible between Neandertal and modern *sapiens*. The teeth are like those of Tabūn. The mandibles show some development of a chin. There is quite a bit of midfacial prognathism. Skhūl V, the best preserved skull, has a high well-rounded cranial vault with an estimated cranial capacity of about 1550 cc (Figure 13.19). The Skhūl sample exhibits quite a bit of variability among individuals. This was originally interpreted as an example of rapid evolutionary change (McCown and Keith, 1939). Later it was suggested that, contrary to the view of McCown and Keith, this was an example of "hybridization" (Montagu, 1940; 1951). If, however, we accept Neandertal as only subspecifically distinct, then "hybridization" means nothing more than gene flow. Gene flow and evolutionary change are hardly to be considered mutually exclusive categories.

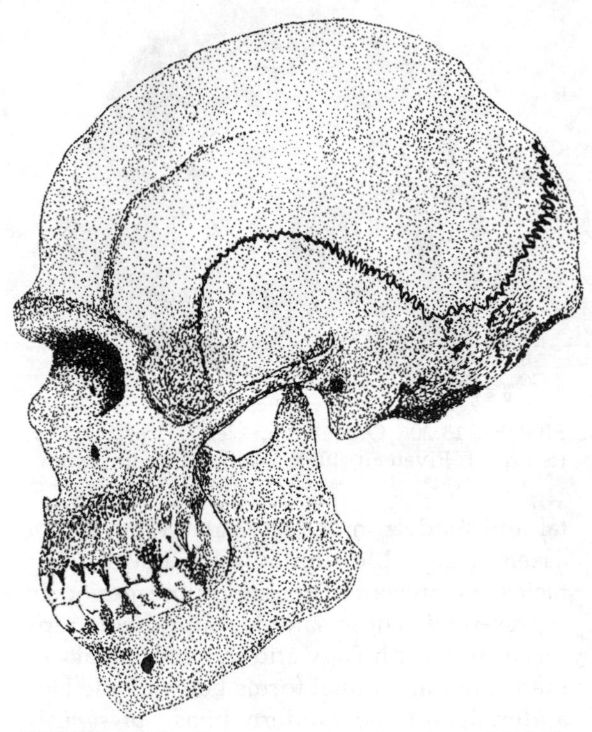

FIGURE 13.19 Reconstruction drawing of Skhūl V.

This is not to say that Skhūl does in fact represent gene flow between populations that are subspecifically distinct. Studies of present-day populations that are "hybrids" of two or more morphologically distinct ancestral populations show no increase in anthropometric variability. Therefore the implication—variability hence hybridity—is suspect to begin with.

At another cave, Djebel Qafzeh near the Sea of Galilee, we find another population considered intermediate between Neander-

FIGURE 13.20 Qafzeh 6, left lateral view. (*Source:* J. Piveteau, 1957. Used with permission.)

tal and modern man. The same kind of tool assemblage, Mousterian of Levalloisian facies, is present. The same time period is represented. These samples supposedly represent in morphology and in temporal placement the transitional forms between the Neandertal and the modern phase of *sapiens* evolution. As a matter of fact the skeletal material from Djebel Qafzeh, which is still being recovered, has never been described in print. The only skull that has been figured in publication, Qafzeh 6, does not appear "transitional" at all but, rather, like a ruggedly built fully modern man (Figure 13.20).

Recent finds from the Omo region of southwest Ethiopia may provide evidence of the transition from *Homo sapiens neanderthalensis* to *Homo sapiens sapiens*. Two calvaria and fragments of a third skill were recovered in 1967 (Leakey et al., 1969). Calvarium II is heavily built, has a sloping forehead, and a heavy occipital torus that coincides with opisthocranion. This calvarium shows sagittal keeling (that is, the midline of the skull tends to be angulated rather than a smooth curve) and has its greatest breadth very low on the skull. The endocranial volume is approximately 1435 cc. The overall impression is of an enlarged and more filled out version of Rhodesian Man. The Omo I calvarium is more modern. The nuchal region is smaller and lower on the skull and the occipital torus is correspondingly low and more modestly developed. The vault of Omo I is higher than in Omo II but it is ruggedly built and has a prominent brow ridge. The overall impression is that Omo I is much like some of the Skhūl specimens.

The two calvaria come from different sites but stratigraphically they appear to be near contemporaries. As yet only a minimum age for these important fossils can be established. Strata above the hominid bearing stratum have given ^{14}C dates of 27,000 years to greater than 37,000 years. Since the calvaria are in lower strata, they should be older. It has been hinted that they could be much older, possibly late Middle Pleistocene, which would provide evidence anew for early modern man.

Transitional forms are apparently lacking in Europe. There could be a true hiatus between the two forms, as has been the conventional assumption. But such a hiatus implies that population displacement was a much more rapid process than was gene flow. Such an occurrence can be seen in the expansion of European populations in the eighteenth and nineteenth centuries. This depended on a vastly superior European technology that was not or could not be adopted fast enough by local populations to avoid displacement. It is true that Upper Paleolithic tool technology was more economical in the production of blade tools and involved a wider variety of more efficient tools. But the archeological picture, as it is now known in Europe, does not show Mousterian Middle Paleolithic peoples being abruptly displaced throughout Europe by Upper Paleolithic peoples.

Without the movement of a foreign and superior technology entirely through Europe, the analogy to recent European expansions becomes invalid. There are other reasons for the gap in the evolutionary record in Europe. I will suggest four possible contributing factors in this situation. First, there has been a strong tendency in the past for Pleistocene archeological research in Europe to be concentrated on cave sites, the known habitation of classic Neandertal. It may well be—and is especially likely, since an interstadial time may be involved—that the transition is to be found in open air sites of the kind characteristic of the immediately succeeding Upper Paleolithic. Second, for whatever reason, the archeological picture of the Würm interstadial periods in Europe is poorly known. Third, the immediate juxtaposition of *Homo sapiens sapiens* over *Homo sapiens neanderthalensis* is more apparent than real. Most of the classic Neandertals were excavated prior to the development of some of the more precise dating techniques. Those with definitive features can mostly be summarized at most as cold fauna, Lower or Middle Würm, finds. As many as 20,000 years could have elapsed between this period and Combe Capelle.

Finally, the transitional forms may have been found but simply remained unrecognized. There has been a long and unfortunate tradition of defining Neandertal as "the man of Mousterian culture." Such a definition confounds two sets of variables, biological and cultural. If these two sets of variables are to any degree independent, then the rigid equation of physical type and cultural type can create wholly unnecessary intellectual difficulties. This may have happened. There is certainly the possibility that the transition to modern *sapiens* took place within Mousterian times. All fossil hominids from Mousterian levels have been traditionally identified as Neandertal. In 1859, only three years after the original discovery was made in the Neander valley, a mandible of Middle Würm fauna

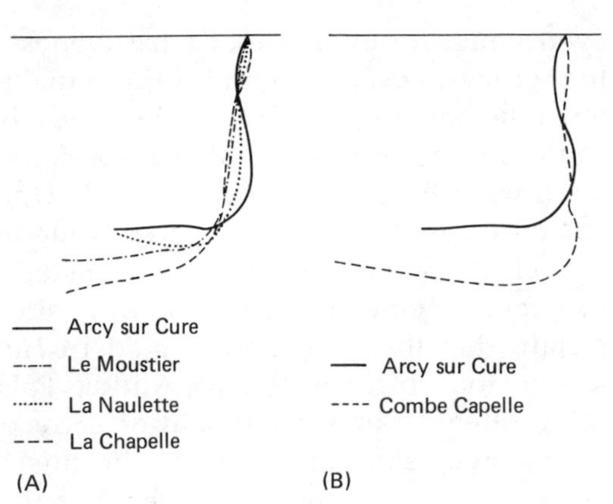

FIGURE 13.21 Arcy mandible outline compared to those of (A) various Neandertals and (B) Combe Capelle. (*Source:* Puccioni, 1912, Used with permission.)

was discovered in the Cave of the Fairies, Arcy-sur-Cure (Yonne), France. The mandible was reported to have come from the lowest (Mousterian) levels of the large cave (Figure 13.21). Morphologically it was not what the experts in the early part of this century expected and was therefore simply ignored.

The description given (Puccioni, 1912) makes it obvious that this mandible is more like modern man than it is like Neandertal. Indeed, it is, if anything, slightly more modern than the mandible of Combe Capelle. Leroi-Gourhan (1959) has confirmed the modern features of the mandible, although it has some archaic traits, and has confirmed from adhering soil particles that it did indeed come from the Arcy caves. But there is still a question as to whether it did, in fact, come from a Mousterian level.

Similarly a "Neandertal" mandibular fragment from Ochos in Moravia, Czechoslovakia appears to have had a relatively modern chin. Other fossil skulls, manifestly non-Neandertal, occurring in Mousterian levels have been assumed to be burials intrusive from higher and later levels (Bordes, 1958) even when no evidence of intrusiveness or higher levels exists (Freund, 1966). Finally it must be mentioned that neither the Niah skull nor any of the transitional forms from Southwest Asia are associated with Upper Paleolithic tools. And the apparently modern Djebel Qafzeh existed with a Mousterian-like tool assemblage. This makes it almost certain that the transition to modern *Homo sapiens* took place in the late Middle Paleolithic. Elwyn Simons has said of early primate phylogeny that some of the missing evidence is to be found on the paleontologists' shelves. Similarly it can be guessed that the final step to modern man in Europe will be discovered by excavations in the back rooms of our museums.

RESUME OF HOMO SAPIENS

There is little evidence today for cladogenetic speciation after the appearance of the genus *Homo*. The appearance of *sapiens* out of *erectus* populations can be considered an anagenetic development throughout the entirety of the Old World. If only one species was everywhere present, then the recognition of the emergent taxon *sapiens* must be made at the same time period throughout. The most convenient and generally acceptable time for the distinction is early in the Hoxnian (second) interglacial. This gives the *sapiens* species a time range of approximately 300,000 years. As this was a continuous evolution from *erectus*, the early *sapiens* retained some *erectus*-like features and are designated by the subspecific name *Homo sapiens neanderthalensis*. These Neandertal *sapiens* are well known only for Würm times. Those Neandertal populations that differed most from one another were those at the extremes of their distribution (Western Europe versus Java), as might be expected. There is now evidence, primarily from the Middle East, that *Homo sapiens sapiens* populations represent a subspecific succession from *Homo sapiens neanderthalensis*. This transition appears to have occurred in Middle Paleolithic cultures, specifically in late Mousterian. Regardless of differing opinions of the role of European populations in this sequence, there is wide agreement that *Homo sapiens sapiens* appears between 40,000–35,000 years ago.

MAN IN THE NEW WORLD

There is no evidence of human occupation of the New World until Upper Pleistocene times. Primate evolution in the Old and New Worlds has followed independent although somewhat parallel lines. A broad expanse of land has connected Siberia with North America at various times during the Pleistocene. These bridgings were the result of a drop in sea level associated with major glaciations. For instance, during the last maxima of the Würm, sea levels were as much as 400 feet below present levels. Such conditions permitted steppe-adapted grazing animals such as the horse to cross between Siberia and North America. But there is no evidence that primates, being tropical animals were able to make such a crossing until man did so. And man was able to do so only when he developed the technology needed to permit him to live the year round in a cold environment. These developments occurred in the Old World in Würm times.

Many authorities now assume that the first movement of man into the New World across a Bearing Straits land bridge occurred 20,000–25,000 years ago. There are some claims for a much earlier appearance. Since man's tools and fire hearths are more common than his bones, it is best to date his appearance in the New World on the basis of artifacts. The earliest archeological material, whose authenticity is agreed upon, dates to about 14,000 years ago. At this time there were large herds of ungulates and other Pleistocene mammals in North and South America of kinds that had become extinct much earlier in the Old World. These included now extinct species of horses, llamas, ground sloths, the mastodon and the wooley mammoth, and camels. Over much of North and South America, there was a flowering of big game hunting cultures between 11,000–12,000 years ago. Distinctive lanceolate projectile points such as the Clovis point and Folsom point are characteristic finds from the Paleo-Indian cultures. These Paleo-Indian cultures lasted in some places to 8,000 years ago. During this time the great extinctions of Pleistocene genera occurred. Some have claimed that the Paleo-Indian hunters were a more important factor in these extinctions than was climatic change.

There is some evidence for an earlier "preprojectile point" phase of technological development in the Americas. There is more evidence for preprojectile point assemblages in South America than in North America. But the earliest evidence of human occupation, accepted by a number of experts, is approximately 24,000 years ago. Sites of this age include Lake Chalco and other sites in Mexico, Old Crow in the Canadian Yukon, and Flea Cave and Pepper Cave in Peru. Some claims for greater antiquity of preprojectile point sites go back 29,000 years (Santa Rosa Island, California), 38,000 years (Lewisville, Texas), and over 40,000 years (Calico Hills, California). Most archeologists have doubted the artifactual nature of the material assigned these dates.

Animal radiations, rapid evolutionary change with diversification, often have been found to accompany the movement of species into a continental area uninhabited by that species up to the time in question. To date, however, our evidence is just the opposite for the populating of the Americas. The earliest skeletal material differs little from some Amerindian populations today. The earliest dated skeletal material is from Tepexpan, Mexico (perhaps 10,000 years or earlier), Marmes, Oregon (ca 11,000 years), Midland, Texas (perhaps 14,000–20,000 years), Laguna Beach, California (ca 17,000 years), and "Los Angeles Man" (ca 23,000 years). These particular finds do not differ from recent Amerindian populations. Some Paleo-Indian remains of slightly less antiquity do differ slightly on the average from present day Amerindians. They were more narrow-headed (dolichocephalic) and had slightly greater supraciliary (brow ridge) development.

Some have claimed that these Paleo-Indian skeletons differ not at all from recent Amer-Indians. The basis of this is to show that the skulls can be duplicated among individuals of some Amerindian populations. This ignores the fact that there is considerable physical diversity among Amerindian populations of different regions of the New World. They are not so homogeneous as is commonly imagined and, on the whole, the Paleo-Indian skulls are slightly "archaic" in comparison to recent populations. But the overall picture remains the same: man's duration in the Americas has been brief and the amount of gross morphological change small in comparison to prior developments in the Old World.

REFERENCES CITED

Blake, G. C. 1864 On the Alleged Peculiar Characteristics and Assumed Affinities of the Human Cranium from Neanderthal. *Journal of Anthropology and Sociology* 2:139–157.

Bordes, F. 1958 Le Passage du Paléolithique Moyen au Paléolithique Superieur. *In* G. H. R. von Koenigswald (ed.), *Neanderthal Centenary.*

Köln: Böhlau-Verlag. 1968 *The Old Stone Age.* New York: McGraw-Hill.

Boule, M. 1913 L'Homme fossile de la Chapelle-aux-Saints (Suite). *Annales de Paléontologie* 8: 1–67.

Brothwell, D. R. 1960 Upper Pleistocene Human Skull from Niah Caves, Sarawak. *Sarawak Museum Journal* 9:323–349.

Chang, K. 1968 *The Archaeology of Ancient China,* rev. ed. New Haven: Yale University Press.

Coon, C. S. 1962 *The Origin of Races.* New York: Knopf.

Freund, G. 1966 Comment on "Transition from Mousterian to Perigordian: Skeletal and Industrial." *Current Anthropology* 7:38–39.

Howell, F. C. 1952 Pleistocene Glacial Ecology and the Evolution of "Classic Neanderthal" Man. *Southwestern Journal of Anthropology* 8:377–410. 1960 European and Northwest African Middle Pleistocene Hominids. *Current Anthropology* 1:195–232.

Klaatsch, H. and O. Hauser 1910 Homo Aurignacensis Hauseri, Ein Paläolithischer Skeletfund Aus dem Unteren Aurignacien der Station Combe-Capelle bei Montferrand (Perigord). *Praehistorische Zeitschrift* 1:273–338.

Koenigswald, G. H. R. (ed.) 1958 *Hundert Jahre Neanderthaler. 1856–1956.* Utrecht: Kemink en Zoon N.V.

Leakey, R. E. F., K. W. Butzer, and M. H. Day 1969 Early *Homo sapiens* Remains from the Omo River Region of South-west Ethiopia. *Nature* 222:1132–1138.

Leroi-Gourhan, A. 1959 Étude des Restes Humains Fossiles Provenant des Grottes d'Arcy-sur-Cure. *Annales de Paléontologie* 44:87–148.

Matiegka, J. 1934 *Homo Předmostensis Fosilní Člověk z Předmostí na Moravě.* Prague: Nákladem Ceské Akademie Věd a Umění.

McCown, T. D. and A. Keith 1939 *The Stone Age of Mount Carmel. II. The Fossil Human Remains from the Levalloiso-Mousterian.* London: Oxford University Press.

Montagu, M. F. A. 1940 Review of T. D. McCown and A. Keith, *The Stone Age of Mount Carmel* (1939). *American Anthropologist* 42:518–522. 1951 *An Introduction to Physical Anthropology,* 2nd ed. Springfield, Ill.: C. C. Thomas.

Morant, G. M. 1927 Studies of Palaeolithic Man. II. A Biometric Study of Neanderthaloid Skulls and of Their Relationships to Modern Racial Types. *Annals of Eugenics* 2:318–381.

Patte, E. 1955 *Les Neanderthaliens.* Paris: Masson et Cie.

Piveteau, J. 1957 *Traité de Paléontologie. Tome VII. Primates. Paleontologie Humaine.* Paris: Masson et Cie.

Puccioni, N. 1912 La Mandibola di Arcy-sur-Cure Appartiene al Tipo di Neanderthal? *Archivio per L'Antropologia e la Etnologia* 42:277–281.

Sergi, S. 1931 Le Crane Neandertalien de Saccopastore, Rome. *L'Anthropologie* 41:241–247.

Smith, G. E. 1924 *The Evolution of Man.* London: Oxford University Press.

Straus, W. L. and A. J. E. Cave 1957 Pathology and the Posture of Neanderthal Man. *Quarterly Review of Biology* 32:348–363.

Virchow, R. 1872 Untersuchung des Neanderthalschädels. *Zeitschrift für Ethnologie* 4:157–165.

Washburn, S. L. and C. S. Lancaster 1968 The Evolution of Hunting. *In* Richard B. Lee and Irvin DeVore (eds.), *Man the Hunter.* pp. 293–303. Chicago: Aldine.

Weidenreich, F. 1951 Morphology of Solo Man. *Anthropological Papers of the American Museum of Natural History* vol. 43, no. 3.

Weiner, J. S. 1955 *The Piltdown Forgery.* London: Oxford University Press.

Woldstedt, P. 1962 Über die Gliederung des Quartärs und Pleistozäns. *Eiszeitalter u. Gegenwart* 13:115–124.

Glossary

Abiogenesis– the production of living organisms from nonliving components

Acclimatization– the short-term physiological adjustment of individuals to climatic change

Acetabulum– the cup-shaped socket on the pelvis into which the head of the thigh bone, femur, fits

Adaptation– the change in features or characteristics of a population in response to a specific environment. Unless otherwise specified a genetic change is implied

Adaptive radiation– the spread and diversification into many species as a result of an evolutionary innovation which permits movement of a species into new habitats

Additive gene effects– the effect of genes on quantitative (continuous) traits in which two of one kind of allele have twice the effect of one such allele

Additive genetic variance– that part of the genetic variance in a population which is ascribable to additive gene effects

Agglutination– the clumping of red blood cells

Agonistic– threatening or threatlike

Albumin– the major protein component of blood serum

Allele– the alternate form of a gene which can occupy the same locus on a chromosome

Amino acid– a class of organic compounds which are components of proteins. There are 20 common amino acids. Each amino acid has a carboxyl group (COOH) and an amino group (NH_2).

Anemia– a subnormal amount of hemoglobin in the blood

Anthropoidea– a suborder of the Primates which includes monkeys, apes and humans

Anthropometric– having to do with anthropometry; a physical measurement on a person

Anthropometry– the practice of measurement of physical traits of man

Antibody– a small protein in blood serum which is produced in response to an antigen

Antigen– a substance, usually a protein, capable of stimulating the production of antibodies

Antiserum– a blood serum containing a specific antibody

Appendicular skeleton– the bones of the appendages (arms and legs)

Arboreal– having to do with trees

Arboricolous– dwelling in trees

Arcade, dental– the arch formed by the teeth in the curve of the jaw

Archetype– a hypothetical original type or ideal type which may be used to represent a class of objects

Archetypal mentality– the tendency to assume that archetypes were once real

Artifact– in archaeology, an item manufactured or modified by human use

Assortative mating– a form of nonrandom mating in which mates tend to be more alike phenotypically, or less alike phenotypically, than would be expected by chance

Autosomal chromosome– any chromosome other than those determining sex

Axial skeleton– the skeleton of the torso and head

Bicuspid– see premolar

Bilophodont– a molar tooth surface having two ridges running transversely across the tooth

Bipedal– walking on two feet

Blood group– a classification of red blood cells determined by characteristic, inherited, antigens on the surface of the red blood cell

Breeding population– the group of individuals within which mates are found; an intermarrying group

Calvarium– the skull other than the facial bones

Canine fossa– a fossa (depression) in the maxilla above the canine tooth

Carbon-14 dating– determining the age of organic material by measurement of the amount of carbon-14 remaining

Carrier– an individual heterozygous for a recessive, usually deleterious, allele

Caudal– having to do with the tail

Ceboidea– the primate superfamily which includes all the New World monkeys

Centrum of a vertebra– the disclike central body of bone in a vertebra

Cephalic index– the ratio (100 × head breadth)/head length

Cercopithecoidea– the primate superfamily which includes all the Old World monkeys

Centromere– the point of constriction on a chromosome where spindle fibers attach

Chromatid– one strand (one-half) of a chromosome that has duplicated and divided down its length (the two chromatids are attached at the centromere)

Chromatography– a technique used to separate a mixture of proteins due to different migration rates through a porous medium

Chromosome– the threadlike structure composed of protein and nucleic acid in the nucleus of the cell

Chronology—the study of the time and sequence of events

Cingulum– an enamel ridge around the base of a tooth

Cladogenesis– the origin of new forms by the splitting (branching) of previous forms

Clinal– having the characteristics of a cline; sloping or grading from one level to another

Coccyx– the terminal bone at the base of the spinal column; it is formed of fused, rudimentary caudal vertebrae

Codominance– gene action that results in a heterozygote that can be distinguished from either homozygote

Codon– a sequence of three nucleotides which code for a specific amino acid

Concordance– an alikeness in a given trait among twins

Crossover– the exchange of DNA between members of a pair of homologous chromosomes

Culture– the traditional forms of learned behavior, usually restricted to those forms of learned behavior in *Homo sapiens* which depend upon language

Cusp– a protuberance on the occlusal surface of a tooth

Cytoplasm– the fluid of a cell consisting primarily of water, inorganic salts, carbohydrates, fats, and proteins

Demographic– having to do with numbers of people; the vital statistics, density, rate of growth, and size of a population

Dental formula– the number of teeth of each kind, from front to back, in each quadrant of the jaw

Deoxyribonucleic acid (DNA)– a long organic molecule composed of nucleotides, each containing deoxyribose sugar

Diastema– a gap between teeth; a diastema accommodating a projecting lower canine occurs between the second incisor and canine of the upper dentition

Digits– fingers or toes

Diploid– having the full number of chromosomes; twice the number of homologous pairs of chromosomes

Dizygotic twins– fraternal twins; twins resulting from two ova separately fertilized

Dominance, social– greater influence on, or control of, behavior of one animal by another

Dominant trait, genetic– a trait which appears identical in both the heterozygous and one homozygous genotype

Dorsoventral– oriented as, or extending from, the back to the stomach surface of the body

Drypothecinae– a fossil subfamily of the Pongidae; this subfamily is named for and includes the fossil genus *Dryopithecus*

Dryopithecus– a fossil genus of apes found in Africa, Europe and South Asia in the Miocene and Pliocene epochs

Ecological niche– the system of environmental features important to the survival of a particular species

Ecology– the study of the relationships between organisms and environments

Electrophoresis– a technique used to separate different molecules in a mixture of substances as a result of different rates of migration of molecules through a supporting medium when an electric current is passed through the supporting medium

Endocranial– referring to the interior of the cranial cavity

Endogamy– marriage between individuals within a defined social group

Enzyme– a protein that acts as a catalyst in biochemical reactions

Epiphysis– the end of a bone which is formed from a center of ossification separate from the body of the bone

Epistasis– the joint effect of nonallelic genes on the form of a phenotype

Epoch– a unit of geological time which subdivides a period

Equitorial plate– the plane through a cell along which cellular division occurs

Erythrocyte– red blood cell

Estrus– a period of heightened sexual excitement or receptivity in females of most mammalian species at or near the time of ovulation; heat

Ethnic group– a group of people having in common some distinguishing cultural features

Eugenics– a movement advocating change in populations by encouraging or discouraging reproduction in different phenotypes

Evolution– descent with modification; biological evolution may be defined as a change in gene frequencies of a population

Exogamy– marriage outside of a defined social unit

External occipital protuberance– a prominent point of bone on the outer surface of the occiput which is the point of attachment of the nuchal ligament

Femur– the thigh bone

Fecundity– the offspring producing capacity (potential) of a woman

Fertility– the number of offspring produced

Fetus– the individual *in utero* from the eighth week after conception to birth

Fitness– the expected number of offspring of a given genotype to be contributed to the next generation, relative to other genes or genotypes

Foramen magnum– literally "large hole"; the hole in the base of the skull accepting the spinal cord

Fossil– the impression of a formerly living organism preserved by mineral replacement of organic material, by casts of surrounding material, or otherwise

Frequency– the number of occurrences. In genetics, used as relative frequency, that is, the same as proportion

Gamete– the mature haploid germ cell. A spermatozoan or ovum

Gamma globulin– a class of proteins occurring in blood serum; gamma globulins include the immunoglobins or antibody proteins

Gene– the functional unit of inheritance; a delimited segment of the DNA molecule

Gene flow– the movement of genes between populations

Gene pool– all of the genes within a given breeding population

Genetic code– the sequence of nucleotides in DNA or RNA which determines the sequence of amino acids joined in the formation of a protein

Genetic drift– random change in gene frequency from generation to generation; a chance effect

Genetic load– a measure of the reduction in fitness of a population as a departure from optimal fitness

Genome– the genetic structure of an individual including the genes and the structure of the chromosomes

Genotype– the list of alleles at one or at all loci characterizing an individual

Genus– the taxonomic category above the species level; a genus includes one or several related species

Germinal cells– sex cells giving rise to sperm or ova

Gestation– the process of embryonic and fetal development in the womb

Gestation period– the length of time from conception to birth

Half-life– the length of time for half of the units involved in a process to be transformed; the half-life of radioactive isotopes is the time in which radioactivity decreases by half

Haploid– having half the number of chromosomes of the zygote

Haptoglobin– a protein of blood serum

Hardy-Weinberg proportions– the genotypic proportions that result from the random union of genes

Hemizygous– a description of zygosity in males for *X*-linked genes; females can be either homozygous or heterozygous, but males, having only one *X* chromosome, are spoken of as hemizygous for *X*-linked genes

Hemoglobin– the respiratory pigment of red blood cells involved in oxygen transport to the tissues and carbon dioxide transport from the tissues of the body

Hemolyze– to lyse or break open red blood cells releasing free hemoglobin

Heritability– the proportion of variation in a trait which is associated with variation among genotypes in a population

Heterodont– having groups of teeth of different form in the tooth row of the jaw

Heterosis– hybrid vigor

Heterozygous– having two functionally different alleles at a locus

Holandric– due to genes located on the *Y* chromosome

Home range– a region habitually occupied by an animal without excluding other animals

Homeothermy– warm blooded; the ability to maintain a relatively constant internal temperature in a wide range of ambient temperatures

Hominid– a member of the family Hominidae

Hominoid– a member of the superfamily Hominoidea

Homodont– having all teeth of the same general pattern down the toothrow

Homologous chromosomes– chromosomes which have similar form and pair during meiosis

Homozygous– having identical alleles at a given locus for both members of a homologous pair of chromosomes

Hybrid– the offspring of parents from genetically quite distinct lineages

Ilium– the most anterior of the three bones of the pelvis

Immunity– the existence of antibodies against foreign proteins or organisms

Inbreeding– marriage or mating of individuals having a recent common ancestor

Inion– a bony external protuberance on the occiput

Innominate– literally the "no name" bone; the hip bone or lateral half of the pelvis composed of three bones fused together: the ilium, the pubis, and the ischium

Insectivorous– feeding on insects

Interglacial– a major warm period between glacial periods

Interstadial– a minor warm period within a general time of glacial advance

Ischium– the lowest of the three pelvic bones forming the innominate bone

Karyotype– the chromosomal complement of a cell

Locus, genetic– the location of a gene on a chromosome

Loph– an enamel crest on the chewing surface of a tooth

Malar bone– the cheekbone

Mandible– the bone of the lower jaw

Mastoid process– a heavy conical protrusion of bone, pointing downward, just behind the ear

Maxilla– the bone of the upper jaw

Meiosis– the process of cell division when germinal cells begin the production of gametes

Melanin– a dark brown organic pigment; a principle component in skin and hair color

Menarche– the onset of menstruation

Mendelian population– an interbreeding population

Menstruation– the discharge of blood due to changes in the wall of the uterus occurring as a part of the estrous cycle

Metabolism– the synthesis or degradation of

substances in living organisms and the energetics involved in these changes

Metaphase– the stage of cell division where the chromosomes are aligned on the equatorial plate

Metazoa– a division of the animal kingdom which includes all multicelled animals

Mitosis– a process of cell division resulting in genetically identical daughter cells

Molar– a cheek tooth with large surface area adapted for grinding and crushing

Monogamy– having only one spouse or mate

Monozygotic twins– identical twins; twins which result from the division of a single zygote, a single fertilized ovum

Mutagen– a substance or energy source which increases the rate of mutation of a gene

Mutation– any change producing a new structure in the genetic material; often a nucleotide substitution in DNA

Niche– the part of the total environment or habitat occupied by a particular species

Nuchal– referring to the nape of the neck

Nuchal crest– the bony crest extending transversely across the nuchal area of the occiput providing attachment area for nuchal muscles and ligaments

Nucleic acid– an organic molecule formed of long sequences of nucleotides

Nucleoprotein– a compound composed of nucleic acids conjugated to protein found in living cells

Nucleotide– a unit composed of a nitrogenous base, a pentose sugar, and a phosphate

Nucleus– an enclosed spherical structure in the center of the cell and, in eukaryotic cells, containing the chromosomes

Occipital torus– a rounded ridge of bone extending transversely across the nuchal area of the occiput

Occiput– the most posterior bone of the skull

Oöcyte– an egg cell ready to begin meiotic division

Orbit– the eye socket

Organelle– a specialized structure within a cell which acts as a small organ in cellular processes

Orthogenesis– the belief that evolutionary change proceeds in a straight line to achieve some preordained form

Orthognathous– an upright alignment of the lower face; lack of forward protrusion of the chewing structures

Orthograde– erect or upright

Osteology– the study of bones

Ovum– the egg cell

Palate– in vertebrates, the roof of the mouth

Paleontology– the study of organisms of past geological periods

Panmictic– mating at random with respect to a given genetic system

Patrilocal– postmarital residence in the area of the husband's family or social group

Pedomorphic– retaining infantile features in the morphology of the adult

Pelvis– the basin-shaped bony structure of the hips

Penetrance– the frequency with which a gene is expressed in a trait when present in an individual

Peptide– two or more amino acids linked together by linking a carboxyl group to an amino group

Phenotype– a trait; the observable expression of a genotype

Phylogeny– the evolutionary descent relationships among species or higher order taxa

Placenta– an organ composed of closely juxtaposed maternal and embryonic tissue involved in the transport of food, oxygen, and wastes between fetal and maternal blood systems

Plantigrade– locomotion in which the sole of

the foot (the plantar surface) is placed flat on the ground

Plasma– the fluid portion of blood which remains after only the red blood cells have been removed by centrifugation

Pleiotropism– the expression or effect of a gene on more than one trait

Polygenic– a trait influenced by genes at multiple loci

Polygyny– the marriage of one male to two or more women

Polymorphism– the occurrence of two or more discontinuous forms of a trait

Polymorphism, genetic– the occurrence of two or more discontinuous forms of a trait in a population and in such frequencies that the rarest form cannot be maintained by mutation alone

Polypeptide– a large peptide

Postorbital constriction– a narrowed transverse diameter of the cranium visible as a constriction behind the bony orbit of the eye

Prehensile– grasping or the ability to grasp an object

Premolar– the cheek teeth just anterior to the molars; also called bicuspids

Procumbent– slanting forward; tending toward the horizontal

Prognathism– forward protrusion of the tooth region

Protein– an organic molecule composed primarily of amino acids joined by peptide bonds

Protozoan– a one-celled organism

Proximal– nearby; the part of a limb nearest to the midline of the body

Pubis– the most anterior of the three bones forming the pelvis

Purine– a nitrogen base compound having a double carbon ring structure

Pyrimidine– a nitrogen base compound having a single carbon ring structure

Quadrupedal– moving on all fours

Ramus– a branch

Random mating– mating such that the probability that two alleles unite in fertilization is proportional to their frequency in the populations

Recessive– a phenotype distinguishable only in the homozygous genotype; also applied loosely to the allele of that genotype.

Recombination– the new combinations of genes resulting from the processes of meiosis and fertilization

Ribonucleic acid– an organic molecule compounded of a long sequence of nucleotides each having a ribose sugar

Ribosome– the organelles in the cytoplasm of the cell where the genetic code is translated into the amino acid sequence of protein

Rugosity– surface roughness.

Sacrum– the bone joining the two halves of the pelvis at the rear; formed from the fusion of sacral vertebrae

Sagittal crest– a crest of bone which rises along the midline of the skull and provides area for temporal muscle attachment

Secular change– gradual change through time in some population characteristic

Sciatic notch– a deep notch on the dorsal surface of the pelvic blade

Sectorial– adapted for cutting

Segregation, genetic– Mendel's first law: the units of inheritance do not blend as a result of being in one individual but segregate out again in later generations

Sex chromosomes– an *X* or *Y* chromosome

Sex linked– due to genes located on the *X* chromosome

Sexual dimorphism– traits which differ markedly between males and females of the same species

Siblings– brothers and/or sisters

Sibling species– two or more species which are true species in terms of reproductive isolation but which are morphologically alike

Sickle cell anemia (sicklemia)– a severe anemia associated with hemoglobin S, an inherited hemoglobin variant

Simian shelf– a bony shelf or strut on the inner surface of the symphyseal region of the mandible; a pongid characteristic

Somatic cell– a body cell; any cell other than those specialized to produce spermatocytes or oocytes

Spatulate– having the shape of a spatula

Species– a population of actually or potentially interbreeding organisms that are reproductively isolated from other such organisms

Spermatozoan– the fully mature male gamete with tail

Stratigraphy– the sequence of deposits or rock formations at some point on the earth's surface, or the description of this sequence

Subspecies– a subdivision of a species; subspecies differ morphologically but are potentially capable of interbreeding

Supraorbital torus– brow ridges of the frontal bone

Suture, bone– an immobile joint seen as an irregular line where two bones of the skull meet

Sympatric species– species which occupy the same or overlapping areas

Symphysis– an immobile joint where two thick bones join as at the pubis or the center line of the mandible

Synapsis– contact between two structures; in genetics synapsis refers to the side by side conjunction of homologous chromosomes in meiosis

Taxon– a group of organisms treated as a unit for purposes of classification

Taxonomy– the classification of organisms

Territory– narrow definition: an area defended by its occupants against competing members of the same species; broad definition: an area which tends to be used by a single social group of a species

Torus– a rounded ridge of bone

Transcription– in genetics, the process of transferring the genetic code from DNA to messenger RNA

Transferrin– a blood serum protein which binds free iron

Translation– in genetics, the process of transferring the genetic code from messenger RNA to protein structure

Tritubercular– having three tubercles or cusps on the surface of a tooth

Typology– the establishment of types as representatives of groups of organisms

Variance– a measure of scatter or variation in a collection of scores; the average of the squared difference of each score from the mean score

X linked– same as sex linked; alleles on the *X* chromosome

Zygote– the fertilized ovum; also the individual who results from this cell

Index

79 80 81 82 9 8 7 6 5 4 3 2